CHEMICAL APPLICATIONS OF SYMMETRY AND GROUP THEORY

CHEMICAL APPLICATIONS OF SYMMETRY AND GROUP THEORY

Rakshit Ameta, PhD
Suresh C. Ameta, PhD

AAP | APPLE ACADEMIC PRESS

Apple Academic Press Inc.	Apple Academic Press Inc.
3333 Mistwell Crescent	9 Spinnaker Way
Oakville, ON L6L 0A2	Waretown, NJ 08758
Canada	USA

International Standard Book Number-13: 978-1-77188-398-6 (Hardcover)

International Standard Book Number-13: 978-1-77188-399-3 (eBook)

Library and Archives Canada Cataloguing in Publication

Ameta, Rakshit, author
Chemical applications of symmetry and group theory / Rakshit Ameta, PhD, Suresh C. Ameta, PhD.

Includes bibliographical references and index.
Issued in print and electronic formats.
ISBN 978-1-77188-398-6 (hardcover).--ISBN 978-1-77188-399-3 (pdf).

1. Group theory. 2. Symmetry (Physics). 3. Chemistry, Physical and theoretical. I. Ameta, Suresh C., author II. Title.

QD455.3.G75A44 2016 541'.2 C2016-905082-3 C2016-905083-1

Library of Congress Cataloging-in-Publication Data

Names: Ameta, Rakshit. | Ameta, Suresh C.
Title: Chemical applications of symmetry and group theory / Rakshit Ameta, PhD, Suresh C. Ameta, PhD.
Description: Toronto : Apple Academic Press, 2017. | Includes bibliographical references and index.
Identifiers: LCCN 2016032885 (print) | LCCN 2016035313 (ebook) | ISBN 9781771883986 (hardcover : alk. paper) | ISBN 9781771883993 ()
Subjects: LCSH: Chemistry, Physical and theoretical--Mathematics. | Group theory. | Symmetry (Physics)
Classification: LCC QD455.3.G75 A44 2017 (print) | LCC QD455.3.G75 (ebook) DDC 541.01/5122--dc23
LC record available at https://lccn.loc.gov/2016032885

Apple Academic Press also publishes its books in a variety of electronic formats. Some content that appears in print may not be available in electronic format. For information about Apple Academic Press products, visit our website at **www.appleacademicpress.com** and the CRC Press website at **www.crcpress.com**

CONTENTS

LIST OF ABBREVIATIONS

AOs	atomic orbitals
CFSE	crystal field stabilization energy
ED	delocalization energy
GMT	group multiplication table
GOT	great orthogonality theorem
IR	irreducible representation
L	ligands
LCAO	linear combination of atomic orbitals
LGO	ligand group orbitals
LOs	ligand orbitals
MOT	molecular orbital theory
NUA	number of unshifted atom
P	pairing energy
R	reducible representations
SALC	symmetry-adapted linear combination
UA	unshifted atom
VBT	valence bond theory

PREFACE

Symmetry gives us a sense of beauty, and group theory is a study of symmetry in molecules. When one is dealing with an object that appears symmetric, group theory can help us with its analysis. As the structure and behavior of molecules and crystals depend on their different symmetries, group theory becomes an essential tool in many important areas of chemistry. Group theory is a part of mathematical sciences. It is a quite powerful theoretical tool to predict many basics as well as some characteristic properties of molecules. Where quantum mechanics provide solutions of some chemical problems on the basis of complicated mathematics, group theory puts forward these solutions in a very simplified and fascinating manner.

Apart from chemical applications of group theory, it has been also applied to robotics, computers, medical image analysis, crystallography, mathematical music theory, statistics, cosmological, stellar and atomic particle abstractions, modeling of vibrational modes of virus, molecular systems biology, mathematical biology, spectroscopy, etc.

Group theory has been successfully applied to many chemical problems. Students and teachers of chemical sciences have an invisible fear from this subject due to inadvertence with the mathematical jugglery and an active sixth dimension required to understand the concept as well as to apply it to solve the problems of chemistry. The subject of group theory is difficult to understand by the readers of chemical sciences lacking strong mathematical background. The main aim of this book is to avoid mathematical complications and present it in a form that the student, teacher, as well as researcher will find friendly.

A number of chemists have helped us in finalizing the script of group theory at various stages, and worth mentioning are Dr. Jitendra Vardia, Vadodara, Dr. Dipti Vaya, Delhi, Dr. Aarti Ameta, Udaipur and would-be Dr. Meenakshi Singh Solanki, Udaipur.

Rakshit Ameta, PhD
Suresh C. Ameta, PhD

ABOUT THE AUTHORS

Rakshit Ameta, PhD
Associate Professor of Chemistry, PAHER University, Udaipur, India

Rakshit Ameta, PhD, is Associate Professor of Chemistry, PAHER University, Udaipur, India. He has several years of experience in teaching and research in chemistry as well as industrial chemistry and polymer science. He is presently guiding seven research students for their PhD theses, and several students have already obtained their PhDs under his supervision in green chemistry. Dr. Rakshit Ameta has received various awards and recognition in his career, including being awarded first position and the gold medal for his MSc and receiving the Fateh Singh Award from the Maharana Mewar Foundation, Udaipur, for his meritorious performance. He has served at M. L. Sukhadia University, Udaipur; the University of Kota, Kota; and PAHER University, Udaipur. He has over 80 research publications to his credit in journals of national and international repute. He holds one patent, and two more are under way. Dr. Rakshit has organized several national conferences as Organizing Secretary at the University of Kota and PAHER University. He has delivered invited lectures and has chaired sessions in conferences held by the Indian Chemical Society and the Indian Council of Chemists. Dr. Rakshit was elected as council member of the Indian Chemical Society, Kolkata (2011–2013) and Indian Council of Chemists, Agra (2012–2014) as well as Associate Editor, Physical Chemistry Section (2014–2016) of Indian Chemical Society. Dr. Rakshit has been elected as Scientist-in-Charge in the Industrial and Applied Chemistry Section of Indian Chemical Society for three years (2014–2016). He has written five degree-level books and has contributed chapters to books published by several international publishers. He has published two previous books with Apple Academic Press: *Green Chemistry: Fundamentals and Applications* (2014) and *Microwave-Assisted Organic Synthesis: A Green Chemical Approach* (2015), and his forthcoming books include *Solar Energy Conversion and Storage* and *Photocatalysis* with Taylor and Francis. His research areas focus on wastewater treatment, photochemistry, green chemistry, microwave-assisted reactions, environmental chemistry, nanochemistry, solar cells, and bioactive and conducting polymers.

Suresh C. Ameta, PhD
Professor of Chemistry and Dean, Faculty of Science, PAHER University, Udaipur, India

Suresh C. Ameta, PhD, is currently Dean, Faculty of Science at PAHER University, Udaipur. He has served as Professor and Head of the Department of Chemistry at North Gujarat University Patan (1994) and at M. L. Sukhadia University, Udaipur (2002–2005), and as Head of the Department of Polymer Science (2005–2008). He also served as Dean of Postgraduate Studies. Prof. Ameta has held the position of President, Indian Chemical Society, Kolkata, and is now a life-long Vice President. He was awarded a number of prestigious awards during his career, such as national prizes twice for writing chemistry books in Hindi; he received the Prof. M. N. Desai Award, the Prof. W. U. Malik Award, the National Teacher Award, the Prof. G. V. Bakore Award, a Life Time Achievement Award by Indian Chemical Society, etc. With more than 350 research publications to his credit in journals of national and international repute, he is also the author of many undergraduate- and postgraduate-level books. He has published two books with Apple Academic Press, *Microwave-Assisted Organic Synthesis: A Green Chemical Approach* and *Green Chemistry: Fundamentals and Applications*, and has two in production now with Taylor and Francis: *Solar Energy Conversion and Storage* and *Photocatalysis: Principles and Applications*. He has also written chapters in books published by several international publishers. Prof. Ameta has delivered lectures and chaired sessions at national conferences throughout India and is a reviewer of number of international journals. In addition, he has completed five major research projects from different funding agencies, such as DST, UGC, CSIR, and Ministry of Energy, Govt. of India.

CHAPTER 1

SYMMETRY

CONTENTS

1.1 HISTORY

The origin of group theory is almost two-and-a-half centuries back. Lagrange, Abel and Galois are considered the founding workers in the field of group theory. A treatise on theories des fonctions analytiques by Joseph-Louis Lagrange laid some of the foundations of the group theory; of course, it was not named as group theory at that time. It was followed by work of Niels

Henrik Abel, who showed that there is no general algebraic solution for the root of quintic equation or higher. To solve this, he invented an important branch of chemistry, which is now known as group theory. However, the credit of the foundation of group theory goes to Evariste Galois as he was the first to use the term Group (in French *Groupe*). Group theory is a branch of algebra. Basic applications of group theory were for some puzzles, like 15-puzzle and Rubik's cube.

Broadly speaking, group theory is a study of symmetry. When one is dealing with an object that appears symmetric, group theory can help us with its analysis. As the structure and behavior of molecules and crystals depend on their different symmetries, group theory becomes an essential tool in many areas of chemistry like hybridization, molecular vibration, spectroscopy, molecular orbital theory, etc. Various scientists have excellently presented this subject from time to time.

Apart from chemical applications of group theory, it has been also applied to robotics, computers, medical image analysis, crystallography, mathematical music theory, statistics, cosmological, stellar and atomic particle abstractions, modeling of vibrational modes of virus, molecular systems biology, mathematical biology, spectroscopy, etc.

1.2 SYMMETRY

Symmetry is a kind of balancing act, which generates beauty in anything, a picture, material, molecule, etc. although symmetry alone may not be enough to substantiate beauty. Its presence gives us a sense of beauty. In Koeslter's words "Artists treat facts as stimuli for the imagination, while scientists use their imagination to coordinate facts."

Symmetry becomes quite important, when it interprets the facts and delights us particularly, when it limits our study of chemistry with the world of order, pattern, beauty and satisfaction. As a matter of fact, chemistry, like any other science, resembles the art and the chemist has a potential of creativity.

1.2.1 SYMMETRY IN NATURE

There is a well-known proverb that "God always geometrizes." Symmetry is present all over the world, i.e., in plant, animals, architecture, etc.

(i) Plants

Plants have symmetric structures in their components like leaves, fruits, seeds, flowers, etc., mostly a radial symmetry. Coniferous plants show cone shape symmetry. Parts of the plants such as leaves, flowers, fruits, etc. are the best examples of bilateral, radial and multidimensional symmetries. The symmetry present in flower makes it beautiful to look at. Here are few examples,

Flowers

Leaves

(ii) Animals

All animals possess at least bilateral symmetry in their physical shapes. All most all the animals can be divided into two equal halves. If animals possess colored marks on their bodies such as tiger, zebra, etc., then the color spread bears a high degree of symmetry in length, width and angle of the marking. Besides these, birds and butterflies also represent such symmetry pattern of colors. Animals show symmetry and rhythm not only in their physique and the color, but they also exhibit a sense of order or pattern in their inhabitance and activity. For example, the reptiles sleep by folding themselves into a spiral loop and crawl in a curved path.

Snakes

The birds, when they fly in a group, also follow some rules of symmetry. While flying in the sky, the group is always lead by a single bird and then it is followed by many birds in the fashion of Pascal's triangle.

Butterfly

Birds

Zebra

Star fish

1.2.2 SYMMETRY IN ARCHITECTURE

Not only living beings such as animals and plants possess symmetry but it also exists in buildings. Normally, symmetric buildings are made because symmetry gives us a sense of beauty. For example, world's seventh wonder Taj Mahal is highly symmetric. it can be divided into two equal parts by a vertical plane. Like Taj Mahal, many other buildings of world also show a high degree of symmetry.

Architecture

The very first sight of an object might appeal to us as symmetric or unsymmetric. This is because the brain automatically does the work of different operations, i.e., rotation and/or reflection through imaginary axes and/or planes without your asking it to do so. The brain has been so educated and trained. And it becomes abstract (alike Group theory to be dealt latter), the moment one goes poetic recalling, "A thing of beauty is a joy for ever" or "Beauty is the truth and truth is the beauty…" Unknowingly, we connect "Symmetry" of the object through abstract thinking to the objects of the physical world around us. It is this concept of the application of symmetry through abstract group theory that we shall develop here to understand certain problems of chemistry in an easy way with essentials of some mathematical manipulations. Thus, the concept of symmetry will be made quantitative to simplify the problems associated with the structure (geometry) of a molecule, the bonding of its constituent atoms in it, spectral properties, etc. It is for these reasons that group theory may alternatively be considered the "Algebra of Geometry." We shall be dealing with isolated molecules.

In coordinate geometry, the ordinary Cartesian coordinates follow the left hand rule, i.e., the thumb represents X-axis, index finger represents Y-axis, and middle finger represents Z-axis. Just to have a difference, right hand rule was considered applicable in group theory. The center of a molecule (often termed as the center of gravity of the molecule) is considered to be coincident with the center of the Cartesian coordinate system, which will follow the right hand rule.

1.3 RIGHT HAND RULE

The thumb, index, and middle finger of the right hand are extended in three mutual perpendicular (⊥) directions. The directions, in which the thumb, index and middle fingers point, are positive (+) X, Y and Z directions (Axes), respectively.

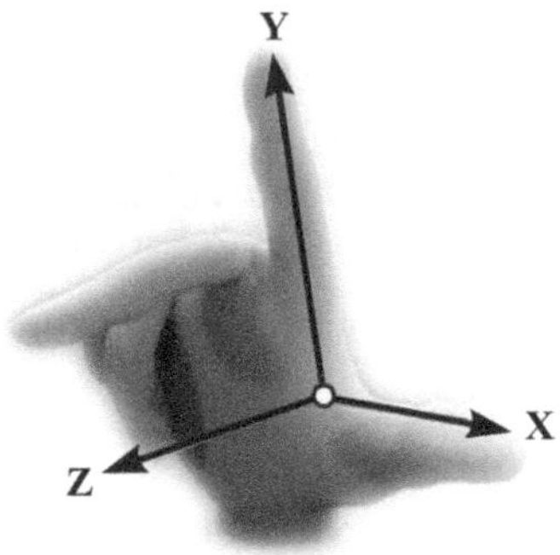

Right hand rule

Z-axis is always kept aligned with the principal axis of the molecule. If Z-axis is in the plane of the molecule, then X-axis will be perpendicular to the planc while if the Z-axis is perpendicular to the plane of the molecule, then X-axis is the axis, which passes through maximum number of atoms. Y-axis is then accordingly placed in the molecule.

It will be used throughout this book. No physical importance should be attached to any coordinate system because in calculating observable quantities of a molecule, it turns out to be immaterial; how the original coordinate system was chosen?

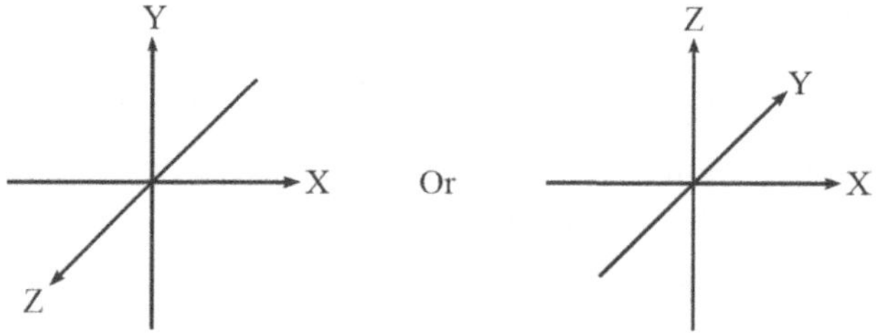

To make the idea of molecular symmetry as useful as possible, some rigid criteria of symmetry should be developed. To do this, we first consider the kind of symmetry elements, a molecule can have. These symmetry elements are geometrical entities, which are intricately related to the effect of symmetry operations. These include a line, a plane or a point with respect to which one or more symmetry operations can be carried out. The symmetry elements, of their own, do not ask one to perform the operation, although the same is implied. The symmetry elements are static entities, while the operations are dynamic in nature. It is the final products (effects) of the operations that indicate the existence of any of the symmetry elements.

So to generate this kind of effect, we have a more intuitive approach of symmetry operators. An operator is a symbol for a rule for transforming a given mathematical function into another function.

1.4 MATHEMATICAL OPERATORS

Mathematical operator is a symbol, which alone does not have any value. But when an operator is attached to a function, then it gives another function.

For example,

$$\frac{\mathrm{d}}{\mathrm{d}x}(\sin x) = \cos x$$

Here, $\frac{\mathrm{d}}{\mathrm{d}x}$ is an operator, which transform sin x function into its derivative cos x with respect to x. Some such other operators are + (carry out addition), – (subtraction), × (multiplication), ÷ (division), etc. Operators are designated by putting a circumflex (^) over the symbol of a symmetry element. When more than one operators are there, then rules of operator algebra is followed, according to which, operations are to be carried out from right to left. Thus, operator asks for a particular operation to be performed on the function to produce the resultant function, a new function.

When a symmetry element does not have circumflex on it, then it is simply a geometrical entity, but when circumflex is written on it, that symmetry element becomes symmetry operation and acts as an operator. This operator operates on molecule to result into an another equivalent or identical figure or structure.

A symmetry operation is a movement of the body (an object, a figure or a function) such that after the movement has been carried out, every point in the body is coincident with an equivalent point (or the same point) of the body in its original form. This direction to carry out the movement is provided by an operator, which acts on the body. When the original points of the body coincide with the equivalent points, the resulting configuration is called equivalent configuration. When these are the same points, the resultant is termed as identical configuration.

1.5 EQUIVALENT SYMMETRY ELEMENTS AND ATOMS

If a symmetry element A is carried into the element B by an operation generated by a third element X, then B can also be carried back into A by the application of X^{-1}. Then two elements A and B are said to be equivalent.

If A can be carried into still a third element C, then there will also be a way of carrying B into C, and the three elements A, B and C form an equivalent set of elements. In general, any set of symmetry elements, in which any member can be transformed into each and every other member of the set by the application of some symmetry operation, is said to be a set of equivalent symmetry elements.

In a planar triangular molecule (BF_3), each of the two fold symmetry axis (C_2) lying in the plane can be carried into coincidence with each of the others by rotation of $2\pi/3$ $\left(\text{or}\,\frac{360}{3}\right)$ or $2 \times 2\pi/3$, which are symmetry operations. Thus, all these three two fold axes are said to be equivalent to one another.

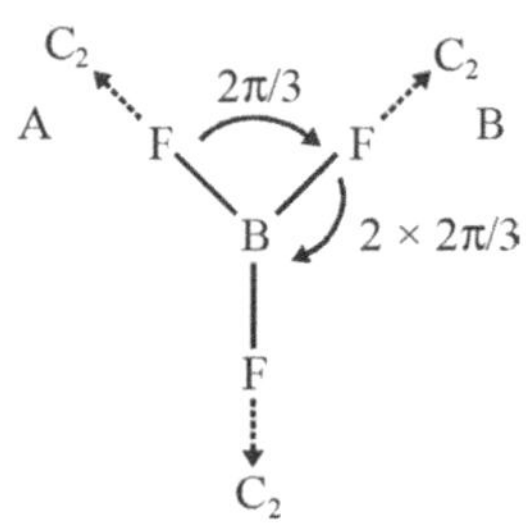

In a square planar AB_4 molecule, there are four two-fold axes (C_2) in the molecular plane and one four-fold axis (C_4) perpendicular to all the two-fold axes. Two of two-fold axes are along BAB and other two are bisecting the angles BAB. Let us place set of two-fold axes along B-A-B bonds in C_2 group while other set bisecting BAB angles in C_2' group. Then C_2 group axes are equivalent to each other, while C_2' group axes are also equivalent to each other. But C_2 and C_2' groups axes are not equivalent to each other.

Such a molecule also contains four symmetry planes, each of which is perpendicular to the molecular plane and intersects it along one of two-fold axis. Two planes, (σ) which are along C_2 group of axes are equivalent to each other, while other two (σ′) are placed along C_2' group of axes are also equivalent. But σ and σ′ planes are not equivalent to each other.

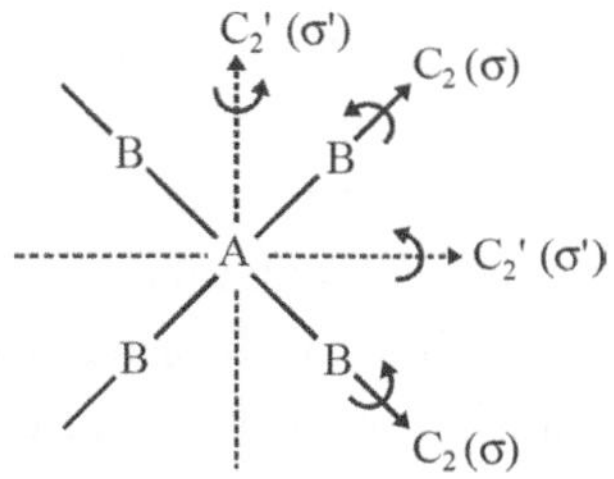

All the three symmetry planes in BF_3, which are perpendicular to the molecular plane are equivalent to each other. However, two reflection planes (molecular and vertical planes) in water are not equivalent to each other, i.e., there is no operation, which can carry σ_v to σ_v' and vice-versa.

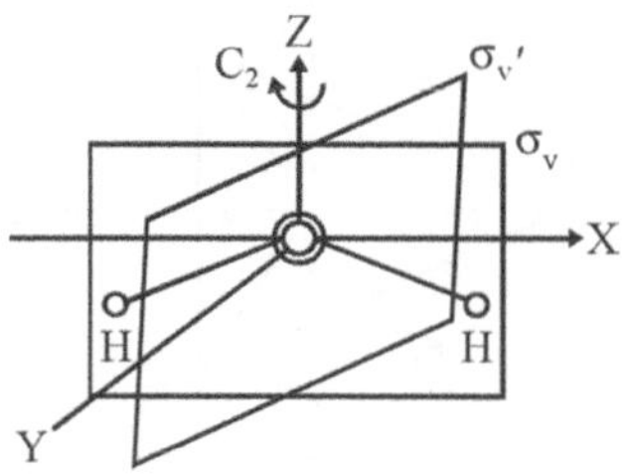

In benzene molecule, there are two sets of three two-fold axes. 3 C_2 axes passing through two opposite carbon atoms form a set and 3 C_2' axes, which bisect two opposite C-C bonds of the benzene form another set of equivalent axes. Similarly, there are two sets of 3-reflection planes, along C_2 and C_2' axes, which are equivalent to each other.

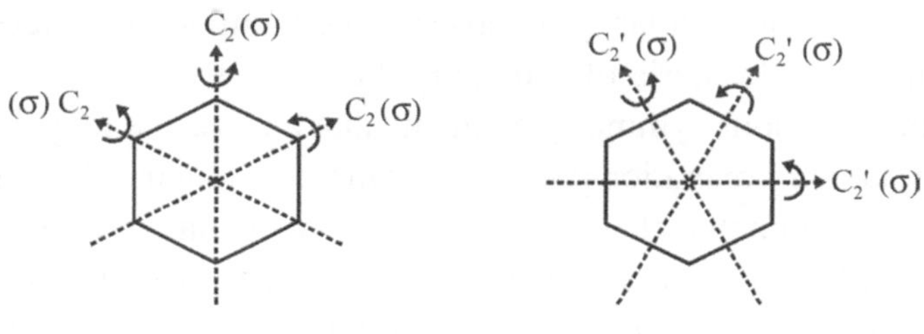

If an atom of a molecule can be interchanged with another atom of the molecule, then these atoms are said to be equivalent atoms. Of course, equivalent atoms must be of same chemical species. Thus, all the hydrogen atoms in methane, ethane, benzene or cyclopropane are equivalent to each other. Similarly, all the fluorine atoms in SF_6 (regular octahedron) are equivalent to each other. All the carbon and oxygen atoms in $Cr(CO)_6$ are also equivalent.

However, all five atoms in the PF_5 (trigonal bipyramid) are not equivalent because axial P-F bonds are longer than the equatorial P-F bonds. Axial F atoms (two atoms) are equivalent to each other while equatorial F atoms (three atoms) are also equivalent to each other. No operation can interchange an equatorial fluorine atom with axial fluorine atom, but it can interchange with another equatorial fluorine atom by rotation around C_3 axis. Hence, three equatorial fluorine atoms are equivalent to each other.

Similarly, an axial fluorine atom can interchange with another axial fluorine atom by rotation around C_2 axis. Hence, these two axial fluorine atoms are equivalent to each other.

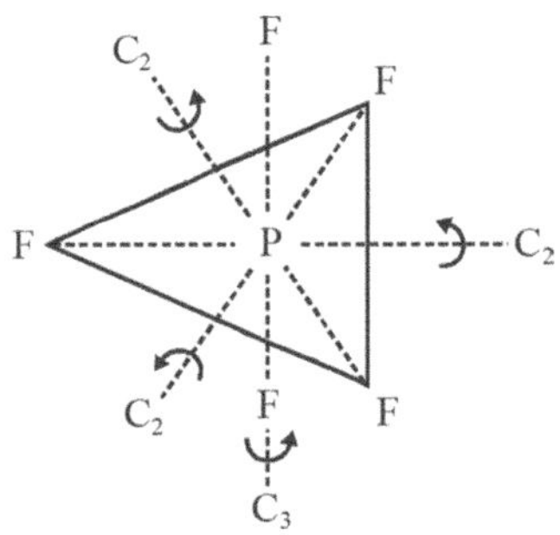

1.6 SYMMETRY OPERATIONS AND SYMMETRY ELEMENTS

By more inspection of a molecule, one can say that a particular molecule have high symmetry, low symmetry or no symmetry. In order to have better knowledge about molecular symmetry, it is very important to develop some rigid mathematical criteria of symmetry. For this purpose, one should first discuss about various symmetry elements that a molecule may possess and then the symmetry operation produced by those symmetry elements.

The term symmetry element and symmetry operation are often creates a state of confusion and are interchangeably used as these terms are inextricably related. Therefore, it becomes essential to have a clear understanding of the difference between them.

1.6.1 SYMMETRY OPERATION

Symmetry operation is a movement of a molecule from its original orientation to an equivalent orientation (sometimes identical orientation also). It can be rotation around an axis, reflection through a plane and inversion at a point or any combination of these operations. After movement, every point of molecule is coincident with equivalent point of the molecule in its original orientation. It can be said that the orientation and position of a molecule before and after operation are indistinguishable, i.e., one cannot determine the difference between molecules, before and after operation. In other words, symmetry operation results into equivalent structure (configuration). A symmetry operation, which brings the molecule to its starting original position, is termed as identity operation.

1.6.2 SYMMETRY ELEMENT

A symmetry element is a geometrical entity such as an axis, a plane, or a point (a center) with respect to which symmetry operations are carried out. One or more than one symmetry operations can be operated at a time in a molecule. Thus, symmetry element is associated with one or more symmetry operations and these two terms are interrelated. One of the simplest way to distinguish a symmetry operation and symmetry element is the presence of a circumflex (cap, ^) written over the symbol of that symmetry element, but in common practice, it is seldom used.

Five kinds of operations and the symmetry elements are normally used (Table 1.1).

TABLE 1.1 Symmetry Elements and Operations

Element	Operation	Symbol
Identity	To leave molecule as it is (unchanged)/No operation	E
Proper axis	Rotation about an axis through an angle θ	C_n
Plane	Reflection in a plane	σ
Centre	Inversion through center	i
Improper axis	Rotoreflection, i.e., rotation followed by reflection in a plane ⊥ to the rotation axis	S_n

(i) Identity (E)

This element is obtained by an operation called 'identity operation.' It is a 'doing nothing' operation. After this operation, the molecule remains as such. This situation can be visualized in two ways. Either (i) we do not do anything to the molecule, or (ii) we rotate the molecule by 360° ($\theta = 360°$).

So we can write,

$$C_1 = E, \sigma^2 = E, i^2 = E, C_n^{\,n} = E$$

The product of the element and its inverse also gives identity. It can be represented as,

$$(C_n)\,(C_n)^{-1} = E$$

Here, C_n is an axis of symmetry and $(C_n)^{-1}$ is its inverse symmetry element.

In case of H_2O molecule,

$$C_2.C_2^{-1} = E$$

Therefore, it can be concluded that if the product of any two operations is identity, then it means that the two operations are inverse to each other.

Plane of symmetry (σ) is inverse of itself. Therefore, product of plane of symmetry give identity, i.e., $\sigma.\sigma = E$. The product or combination of identity operation (or element) with any operator (or element) always gives the same operator (or element). For example,

$$C_n.E = E.C_n = C_n$$

$$E.\sigma = E.\sigma = \sigma$$

Relationship between identity operation and other operations.

$C_n^n = E$ If n = Even or odd

$S_n^n = E$ If n = Even

$S_n^n = \sigma_h$ If n = Odd

$S_n^{2n} = E$ If n = Odd

$\sigma_n^n = E$ If n = Even

$\sigma^n = \sigma$ If n = Odd

$i^n = E$ If n = Even

$i^n = i$ If n = Odd

An element generated by this 'leave-it-alone' operation is trivial and is as important as the other symmetry elements. Every molecule has this element of symmetry and it coexists with the identity of the molecule and hence, it has been named as 'identity element.' This is denoted by a special symbol E (the first letter of the word Einbeit from German) or I (Identity).

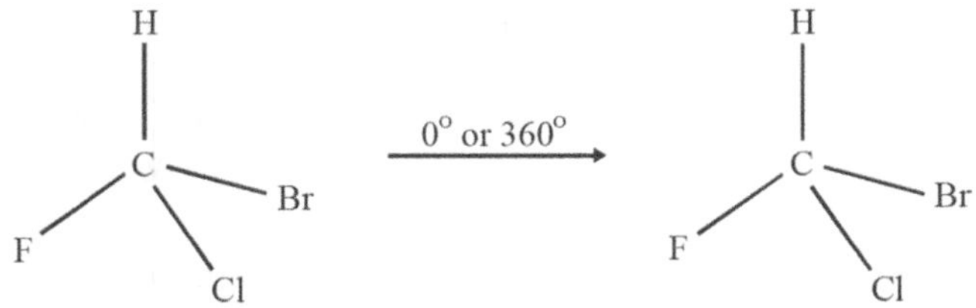

(ii) Axis of Symmetry (C_n)

When a molecule is rotated along an imaginary line or axis and it gives an equivalent or identical configuration, then such an operation is called proper rotation and the imaginary line, with respect to which molecule is rotated, is called proper axis of symmetry. The symbol C_n is used for designating both; the proper axis of rotation and the proper rotation operation. Here, subscript n denotes the order of axis. Order is highest value of n, when molecule is rotated through 2π/n (360°/n) to give an equivalent configuration. Hence, n can be represented as,

$$n = 2\pi/\theta \text{ or } (\theta = \text{Minimum angle of rotation})$$

when n is equal to 2, 3 or 4, then it is called two-fold, three-fold or four-fold axis of symmetry, respectively. C_2 is a symmetry element. When H_2O molecule is rotated through 180 °C around this axis, it is known as C_2 symmetry operation. When H_2O molecule (a) is rotated through 180° around this C_2 axis, which bisects the ∠ HOH angle, gives structure (*b*) and it is indistinguishable from the starting one. If this rotation is done once again, then orientation (*c*) is obtained. Here, (a) and (*b*), as well as (*b*) and (*c*) are equivalent structures, but (a) and (*c*) are identical structures.

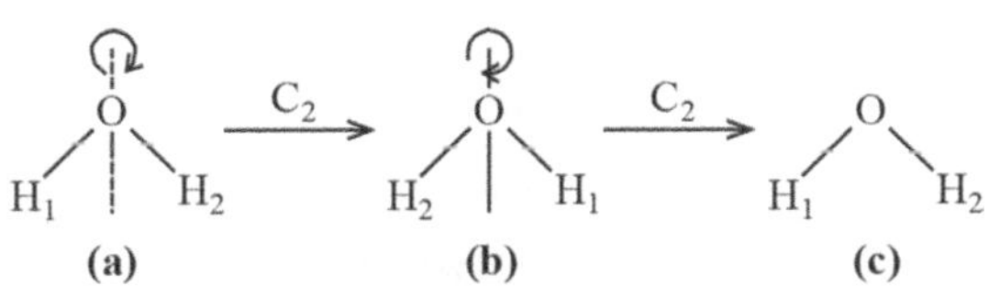

BF_3 molecule has the following structure.

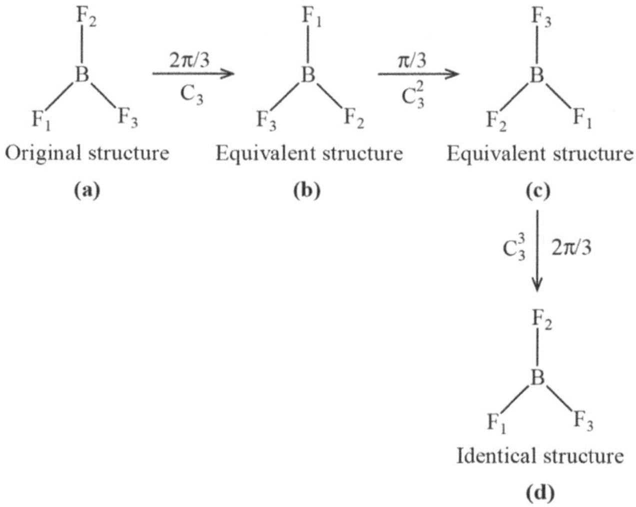

Here, a line perpendicular to the plane and passing through the center of BF_3 molecule (B atom) is a proper axis of rotation with n = 3, because complete rotation of molecule through an angle 360° (in 3 steps of 120°) around this axis gives two equivalent and one identical configuration.

In this example, minimum angle of rotation is 120°, to have an equivalent configuration and therefore, the order of the proper axis is 360°/120° = 3. Thus, three symmetry operations can be carried out in succession as shown above, which are written as C_3^1, C_3^2, and C_3^3. The last operation gives original configuration back, and hence, it is equal to identity operation, i.e., $C_3^3 = E$. Thus, a C_3 axis of symmetry is present in BF_3 molecule and it is called a three-fold axis of symmetry.

It can be concluded that axis of symmetry of order n, (C_n), has n – 1 symmetry operations to give equivalent configurations while C_n^n operation gives identical or same configuration.

Besides this axis of three-fold symmetry, there are three more axes of two-fold symmetry (3 C_2). These are passing through the central boron atom as well as one of the fluorine atoms.

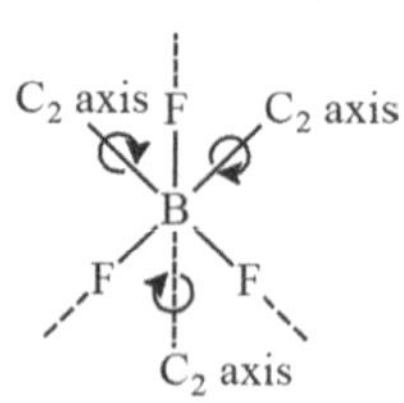

These three C_2 axes are perpendicular to C_3 axis.

Similarly, in case of benzene, the n may be 6, 3, 2 (all these axes, C_6, C_3 and C_2, coincide with each other), i.e., on rotation by 60°, 120° and 180°, we get equivalent configurations.

A molecule can possess more than one axis of rotation. The axis with highest order is called principal axis. Therefore, in benzene, C_6 axis is considered as principal axis of symmetry. Let us consider the operation C_6^2, which is one of operations generated by C_6 axis. This is a rotation by $2 \times 2\pi/6 = 2\pi/3$ and it may be written as C_3. Similarly, operations C_6^3 and C_6^4 can be also written as C_2 and C_3^2, respectively.

Thus, C_6, C_6^2, C_6^3, C_6^4, C_6^5, C_6^6 can be written as C_6, C_3, C_2, C_3^2, C_6^5 and E.

Examples of Symmetry Axis

(i) Many molecules do not have any axis of proper rotation, i.e., FClSO, but they contain identity element, which is equivalent to C_1.

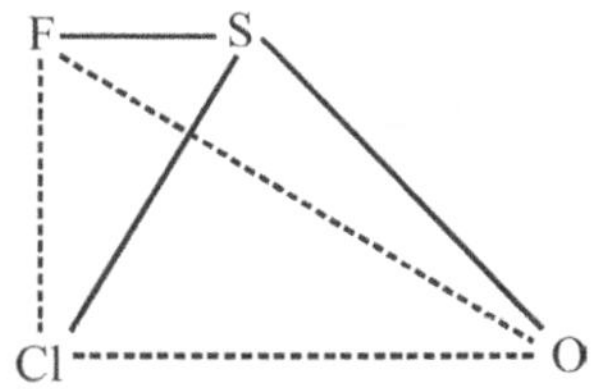

(ii) Linear molecules have infinite fold axis of proper rotation, which is collinear with molecular axis. Since all the atoms are in a line, then any minimum angle of rotation (almost approaching zero, i.e., less than a degree, minute or second) through this linear molecular axis will give its equivalent configuration, the n = 360°/0 = ∞ and hence, this axis of symmetry be designated as C_∞ ($\cdots O = C = O \cdots$).
It means, when θ tends to zero ($\theta \rightarrow 0°$), then n tends to infinity ($n \rightarrow \infty$). Examples for C_∞ axis of symmetry are H_2, CO_2, HCl, OCS, HCN, etc.

(iii) Water molecule has one two-fold axis of symmetry, which is bisecting the angle HOH and passing through the oxygen atom. Rotation around this axis through an angle of 180° will give equivalent configuration or indistinguishable configuration. This axis of rotation is called C_2 (i.e., n = 360°/180° = 2).

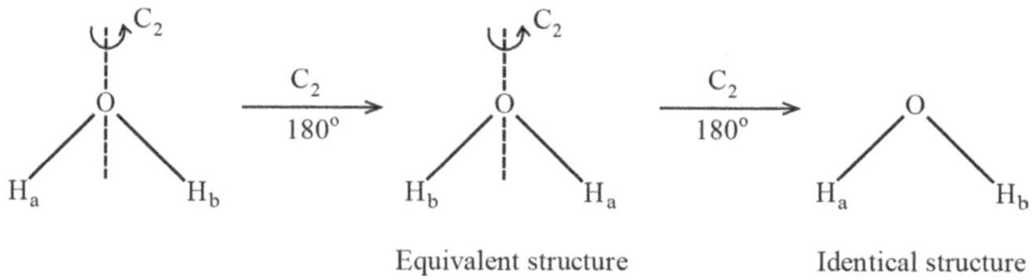

(iv) If more than one axes of same order are present, then the axis passing through maximum number of atoms is called Principal axis of symmetry. Ethylene molecule has three two-fold axes of symmetry (3 C_2). One C_2 axis of symmetry, which is collinear with C = C axis, of the molecule is designated principal axis of symmetry. The second axis is perpendicular to plane of molecule and bisecting the C = C bond. The third one is in the plane of the molecule but perpendicular to first two and intersecting both the C_2 axes at the center of the C = C bond.

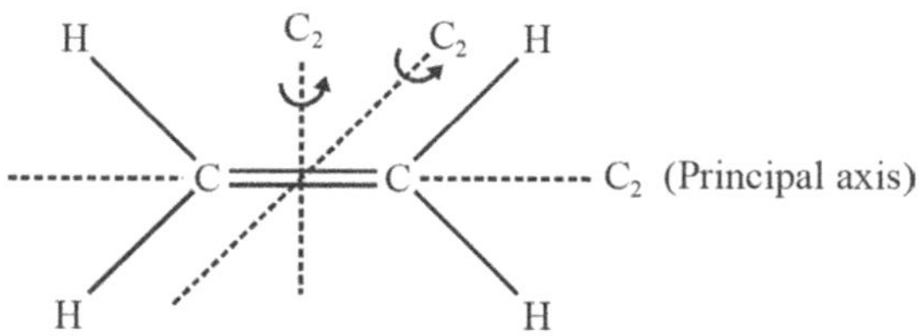

The principal axis of symmetry will be that C_2 axis, which is passing through C = C bond and cutting two carbon atoms, because other two C_2 axes are not cutting any atom.

(v) CH_2Cl_2 molecule also has one two-fold axis of symmetry (C_2) bisecting the opposite angles H-C-H and Cl-C-Cl.

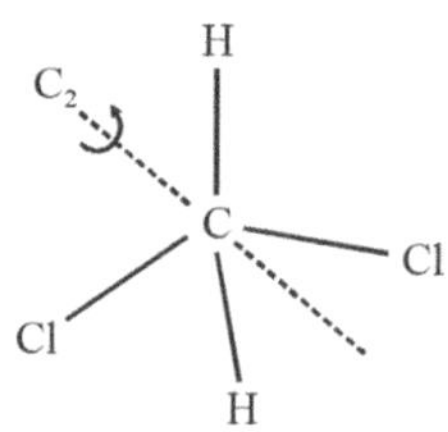

(vi) Allene (CH_2 = C = CH_2) molecule possess three two-fold axes of symmetry (3 C_2). One of them is passing through three carbon atoms along the molecular axis. Other two axes (C_2) are perpendicular to the molecular axis and passing through the central carbon atom. C_2 axis, which is passing through three carbon atoms, is called principal axis of symmetry, as the other two C_2 axes pass through only one carbon atom (central C atom).

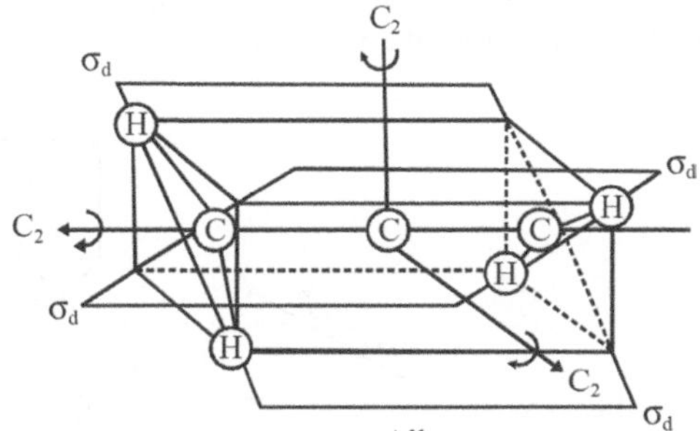

Allene

(vii) Pyramidal AB_3 molecule like NH_3 has only one three-fold axis of symmetry passing through N atom and perpendicular to the plane of three H atoms. Although the angle H-N-H is around 107° in case of ammonia, but rotation around this axis through an angle of 120° will give equivalent configurations and hence, this axis of symmetry is called C_3.

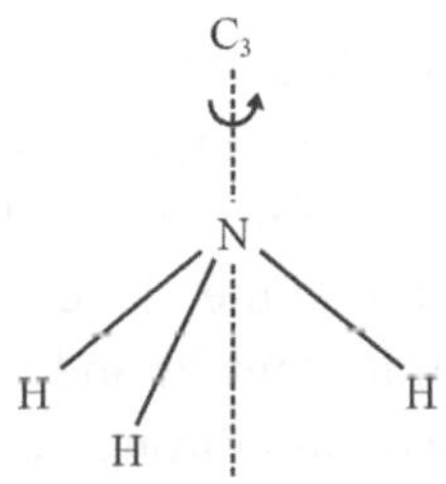

(viii) Planar AB3 molecule like BF3 has one three-fold axis of symmetry (C3) passing through the B atom and perpendicular to the plane of three F atoms. BF3 molecule also possesses three two-fold axes of symmetry (3 C2), which are perpendicular to the three-fold axis and passing through each of the B-F bond.

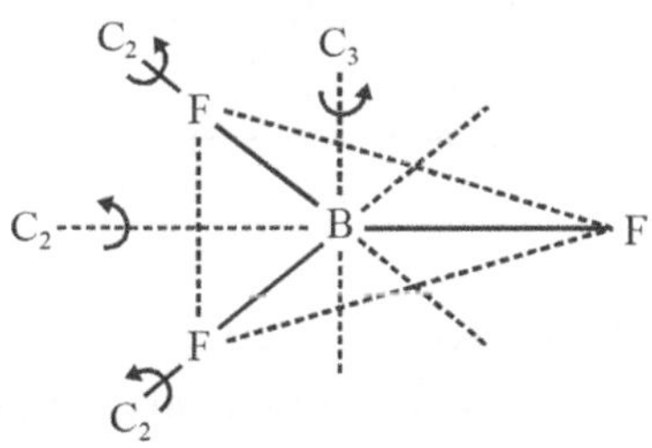

(ix) Tetrahedral AB_4 molecule like CH_4 molecule has four three-fold axes (4 C_3), each containing central atom C and one of the H atom, i.e., collinear with C-H bond. Besides this, CH_4 has three two-fold axes of symmetry (3 C_2) passing through central carbon atom and bisecting the two opposite H-C-H angles.

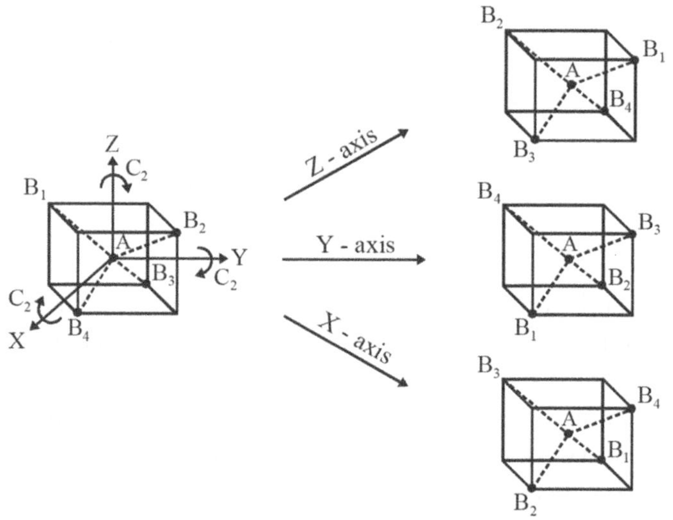

(x) Planar AB_4 molecule like $PtCl_4^{2-}$ ion has one four-fold axis (C_4) passing through the central atom Pt and perpendicular to the plane of molecule. It also has four two-fold axes (4 C_2) perpendicular to this C_4, all of them are in the plane of the ion. Two of them (2 C_2) are containing central atom Pt and pass through two opposite Cl atoms. The other two pass through the central atom Pt and bisecting the opposite Cl – Pt – Cl angles.

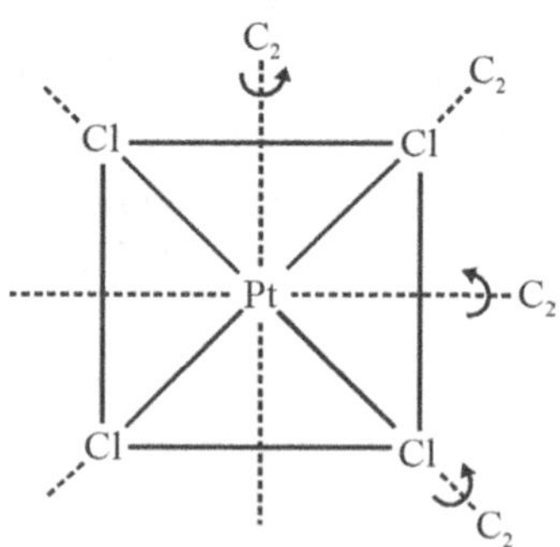

(xi) Pentagonal AB_5 molecule like $C_5H_5^-$ ion has one five-fold axis (C_5) passing through the center and perpendicular to the plane of molecule (or ion). Besides this, pentagonal molecule has five two-fold axes (5 C_2) in the molecular plane. These axes are perpendicular to the principal axis C_5.

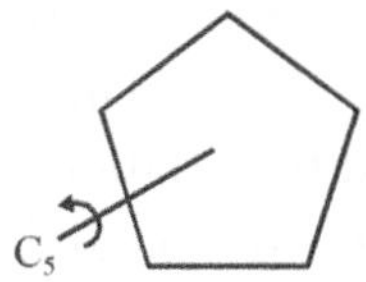

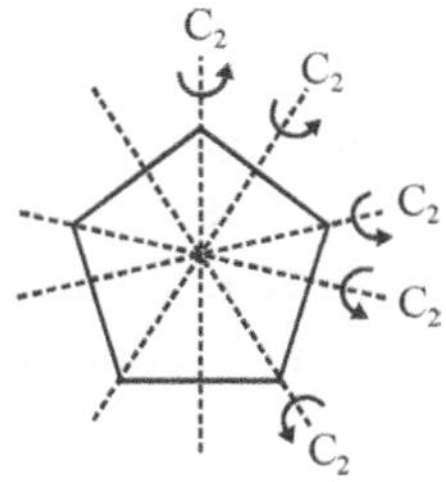

(xii) Hexagonal AB_6 molecule like C_6H_6 has one six-fold axis (C_6) passing through the center of molecule and perpendicular to molecular plane. It also has six two-fold axes (6 C_2) perpendicular to principal axis C_6, all of these are in plane of molecule. Three of them (3 C_2) are containing center and passing through two opposite C-atoms. The other three are passing through center and bisect two opposite C-C bonds.

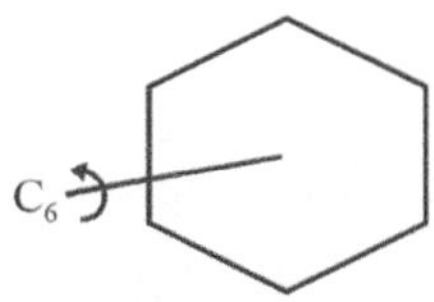

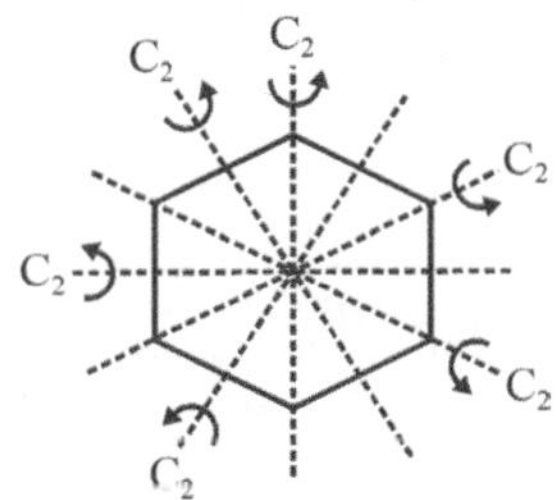

(xiii) Octahedral AB_6 molecule like SF_6 has three four-fold axes (3 C_4) passing through the central atom S and 2 F atoms located at trans-positions. It coincides with C_2 axes also. A regular octahedral molecule also has four three-fold axes (perpendicular to each other) (4 C_3) passing through the center of opposite triangles of three F atoms.

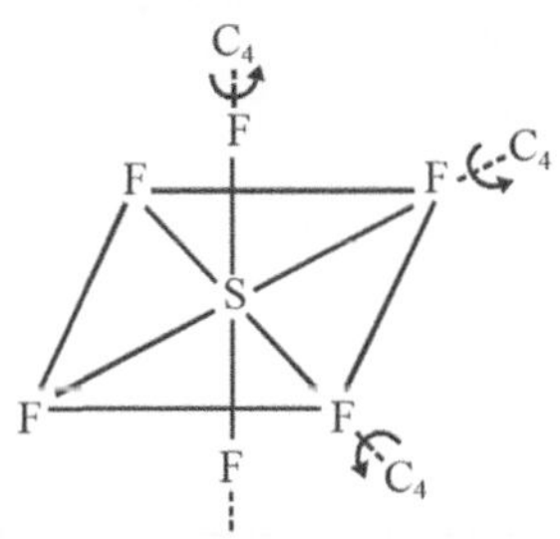

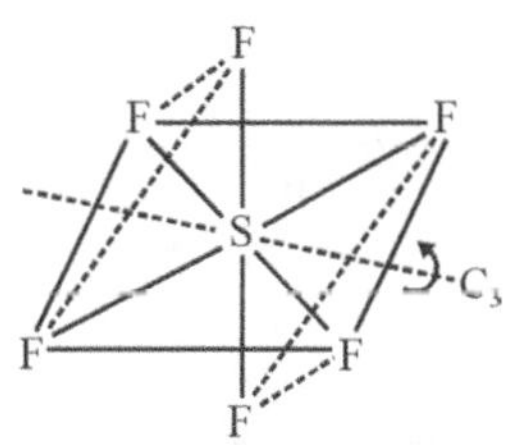

(xiv) Planar (C_7H_7) (Tropylium) ion possesses seven-fold axes of symmetry (C_7).

(xv) $(C_8H_8)_2V$ (Vanadocene) like molecule possesses eight-fold axis of symmetry (C_8) (Table 1.2).

The rotation of 3D body by 360°/n results into equivalent configuration, which comprises a C_2 symmetry operation. If some operation is performed twice in successive steps ($C_{n.}\ C_n$), then it is equal to 2, which may be written as C_n^2. In case n is even, the n/2 is integer and the rotation reduces to $C_n/2$. In essence, a C_n^m can be reduced by their least common division, such as

$$C_6^3 = C_2^1$$

Here, n/m = 6/3 = 2

TABLE 1.2 Axis of Symmetry

Molecule	Order (n) $= \frac{360^\circ}{\theta}$	Representation	Proper axis
H_2O	$\frac{360}{180} = 2$	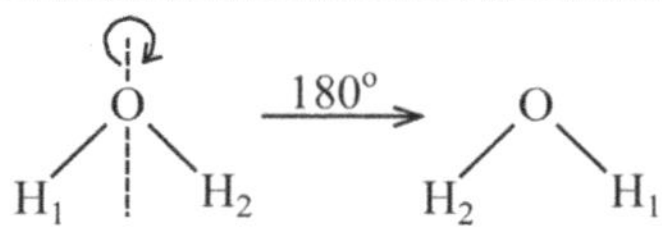	C_2
NH_3	$\frac{360}{120} = 3$	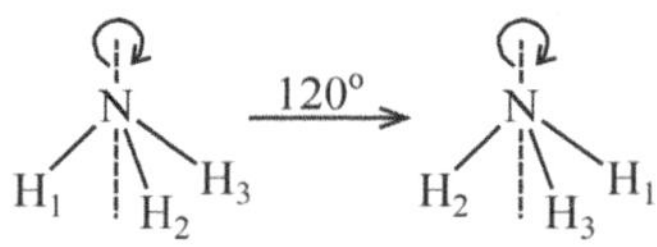	C_3
$[Ni(CN)_4]^{2-}$	$\frac{360}{90} = 4$	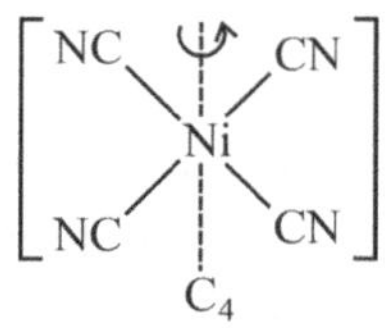	C_4
$C_5H_5^-$	$\frac{360}{75} = 5$	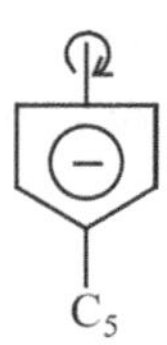	C_5
C_6H_6	$\frac{360}{60} = 6$		C_6

The set of operations generated from continued rotation in successive steps by 360°/n can be given as,

$$C_n, C_n^2, C_n^3, C_n^4, \ldots C_n^n \ (= E)$$

As $C_n^{\ n}$ means 360° rotation and therefore, it gives identical configuration (E). Thus,

$$C_n^{n+m} = C_n^n \cdot C_n^m$$
$$= E.\, C_n^m$$
$$= C_n^m$$

(iii) Plane of Symmetry (Mirror plane) (σ)

It is defined as an imaginary plane that bisects the molecule in such a way that the two parts (two halves) are mirror images of each other. This element of symmetry is represented by the symbol σ. The corresponding operation to a mirror plane is reflection. It should be noted that the operation of reflection gives a configuration equivalent to original one. If the operation is carried out twice on the molecule, then we get the original configuration. Hence, a mirror plane generates only one distinct operation $\sigma^n = \sigma$, if n is odd and $\sigma^n = E$, if n is even.

Plane of symmetry and the corresponding reflection operation both; are denoted by σ. It has certain properties such as

- The symmetry plane should always be present within the molecule, i.e., the plane can't exist completely outside the body of molecule.
- The atoms lying in the plane of the molecule makes a special case because reflection of these atoms in the plane does not move any one of them from their original positions. Therefore, any planar molecule is bound to have at least one plane of symmetry; namely its molecular plane.
- All atoms of a given molecule, which do not lie in the plane must occur in even numbers. Since each one must have a twin on the other side of the plane.

When symmetry operation σ is carried out once, we get configuration equivalent to the original one, but the application of the same σ twice produces a configuration identical with the original.

$$\sigma^n = E \text{ (If n is even)} \quad \text{e.g., } \sigma^2 \text{ or } \sigma^{2n} = E$$

$$\sigma^n = \sigma \text{ (If n is odd)} \quad \text{e.g., } \sigma^3 \text{ or } \sigma^{2n+1} = \sigma$$

Plane of symmetry can be classified into three types,

(a) Vertical plane of symmetry (σ_v)

The plane passing through or containing the principal axis is called vertical plane of symmetry. It is represented as σ_v.

(b) Horizontal plane of symmetry (σ_h)

The plane perpendicular to the principal axis is called horizontal plane of symmetry. It is represented as σ_h.

(c) Dihedral plane of symmetry (σ_d)

The plane passing through the principal axis and bisecting the angle between two C_2 axes is called dihedral plane of symmetry or if the angle between two planes of symmetry is bisected by a C_2 axis, then that set of planes is also called dihedral plane of symmetry. It is represented as σ_d.

Let us consider benzene molecule to explain clearly all the three types of planes. Benzene is a planar molecule having principal axis of symmetry C_6. The plane passing through all the six carbon atoms is perpendicular to principal axis (C_6), and hence, it is called horizontal plane of symmetry (σ_h). Besides this, it has six planes, each passing through C_6 and one of the C_2 axis. Hence, there are two sets of these six planes. Three of them, passing through two opposite C atoms are represented as σ_d planes. Other three planes passing through the center of opposite edges or bisecting C-C bond are represented as σ_v planes.

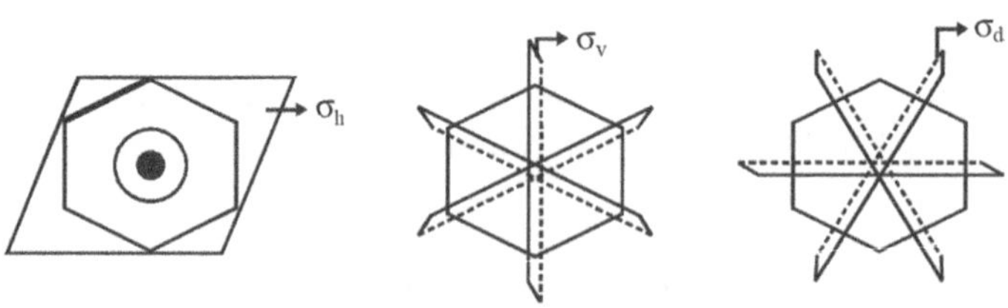

Examples of Planes of Symmetry

(i) Molecules, which are not planar and have odd number of all the atoms, do not contain any symmetry plane, for example, FClSO.

(ii) Linear molecule possesses an infinite number of symmetry planes. These planes are passing through molecular axis of the molecule and hence, all these will be σ_v ($\infty\,\sigma_v$) planes.

(iii) Molecule like F_2SO has only one symmetry plane, which passes through S and O atoms and is perpendicular to the F-F-O plane and bisecting the angle F-S-F.

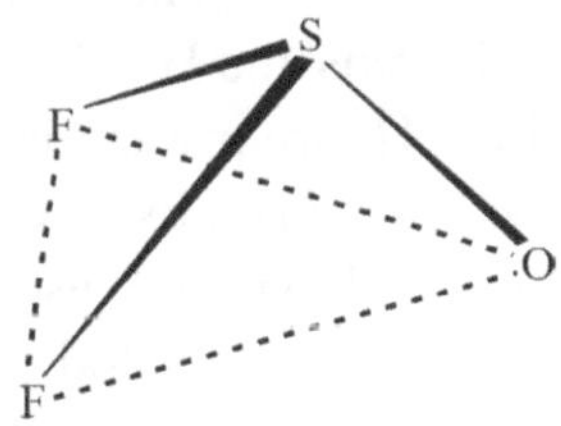

(iv) The V-shaped molecule like water has two symmetry planes. One is the molecular plane, which does not move any of the atoms. Another plane is passing through O atom and is perpendicular to the molecular plane and bisects angle H-O-H. Both these planes are along the principal axis C_2, and hence, these are vertical planes (σ_v).

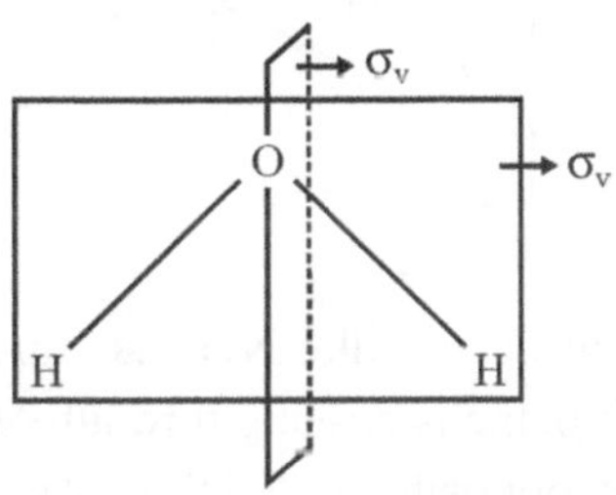

(v) Tetrahedral AB_2C_2 type molecule like CH_2Cl_2 has two mutually perpendicular planes of symmetry. One plane is passing through H-C-H atoms and reflection through it will leave these three atoms unchanged while reflection interchanges the two Cl-atoms. The same is true for other plane, which passes through Cl-C-Cl atoms and reflection through it interchanges H atoms and three atoms Cl, C and Cl remain unchanged.

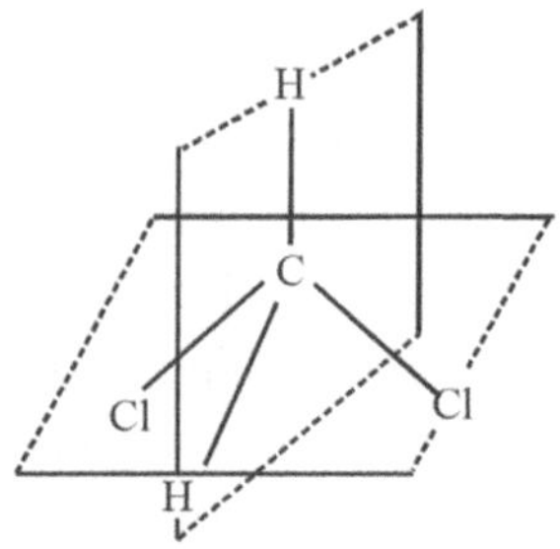

(vi) Allene molecule has three C_2 axes, one passing through molecular axis and other two are perpendicular to it. The planes passing through the molecular axis and H_1, H_2 or H_3, H_4 are planes of symmetry. H_1, H_2 and H_3, H_4 are present in different planes and these planes lie in between subsidiary axes (C_2). Thus, the two planes are passing through principal axis and bisecting the angle between C_2 axes. Hence, these are called dihedral planes of symmetry (σ_d).

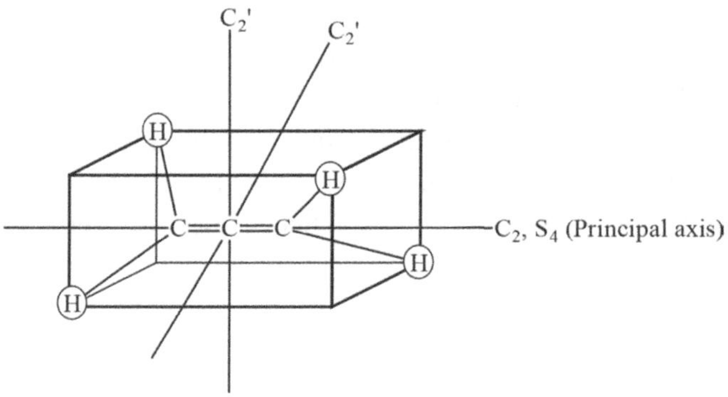

(vii) Pyramidal AB_3 molecule like NH_3 has three vertical planes of symmetry (σ_v). Each plane is passing through N atom and one of H atoms and bisecting the opposite angle HNH. As all these planes contain the principal axis (C_3), these are called vertical planes of symmetry (3 σ_v).

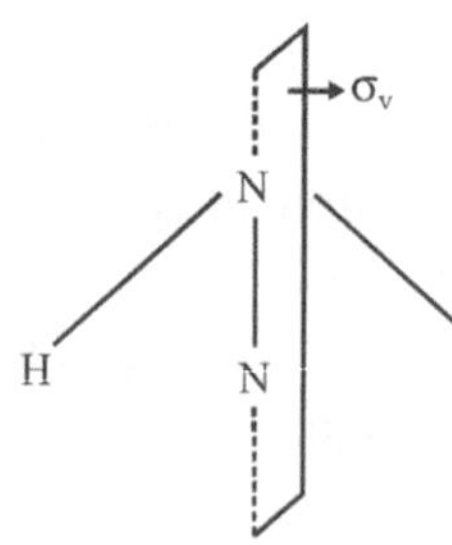

(viii) Planar AB_3 molecule like BF_3 possesses four symmetry planes. One plane is perpendicular to principal axis (C_3). It is in molecular plane and is termed as horizontal plane (σ_h). Other three planes are perpendicular to the molecular plane and passes through central B atom and one of F atom and bisect opposite angle FBF. All these planes are vertical planes of symmetry ($3\sigma_v$).

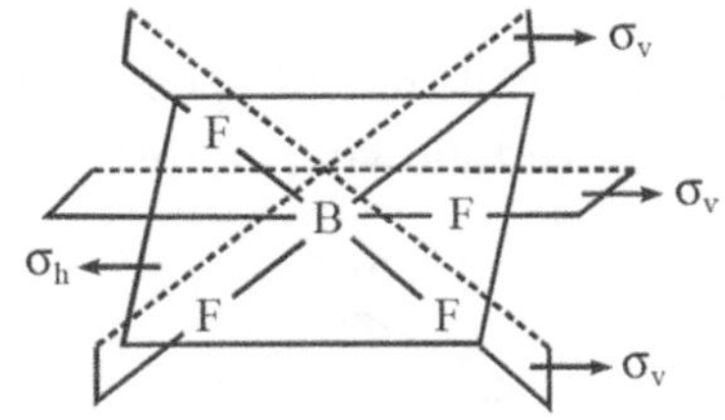

(ix) A planar molecule AB_4 like $PtCl_4^{2-}$ ion has five planes of symmetry. One plane is perpendicular to principal axis (C_4). It is the molecular plane of $PtCl_4^{2-}$ and termed as σ_h. Other four planes are perpendicular to the molecular plane and bisect the opposite ClPtCl angles (two of them) or passing through diagonal (the other two planes). All these there planes are called vertical planes ($4\ \sigma_v$).

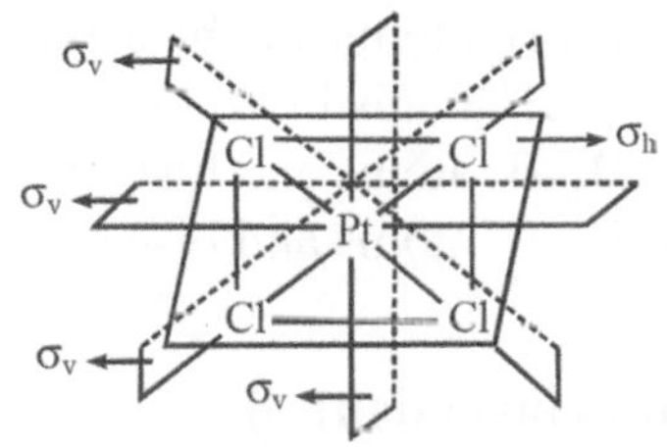

(x) Tetrahedral AB_4 molecule like CH_4 possesses six planes of symmetry. Symmetry planes containing the following atoms (i) CH_1H_2 (ii) CH_1H_3 (iii) CH_1H_4 (iv) CH_2H_3 (v) CH_2H_4 and (vi) CH_3H_4. These are six planes of symmetry. All these planes bisect the angle between the remaining two H-C-H angles, for example, plane passing through C, H_1 and H_2 atoms bisects the angle H_3CH_4 and the plane containing C, H_2 and H_3 atoms bisects the H_1CH_4 angle and so on.

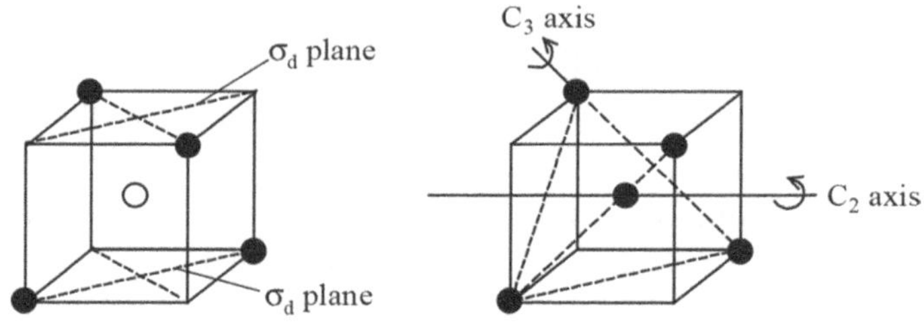

(xi) Octahedral AB_6 molecule like SF_6 possesses nine symmetry planes. These planes involve following atoms:

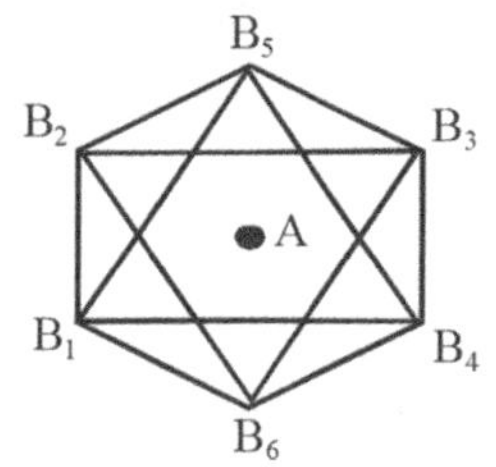

- Plane including A, B_1, B_2, B_3, B_4 atoms.
- Plane including A, B_2 B_4, B_5, B_6 atoms.
- Plane including A, B_1, B_3, B_5, B_6 atoms.
- Plane including A, B_5, B_6 atoms and bisecting B_1–B_2 and B_3–B_4 lines
- Plane including A, B_1, B_3 atoms and bisecting B_2–B_5 and B_4–B_6 lines.
- Plane including A, B_2, B_4 atoms and bisecting B_3–B_5 and B_1–B_6 lines.
- Plane including A, B_5 B_6 atoms bisecting the molecule.
- Plane including A, B_2, B_4 atoms and bisecting the molecule.
- Plane including A, B_1, B_3 atoms and bisecting the molecule.

(iv) Centre of Symmetry (Inversion) (i)

Centre of symmetry or inversion center can be explained with the help of Cartesian coordinate system. Suppose the value of Cartesian coordinates of atom are x_i, y_i, and z_i and let this atom be taken to a point, where value of its coordinate become $-x_i$, $-y_i$, and $-z_i$. If by doing such an operation with all atoms of the molecule, an equivalent configuration is obtained, then it is called the inversion center of molecule. In other words, inversion center is a point from which if a straight line is drawn from every atom of a molecule on one side and extended to an equal distance on the other side, it must come across another identical atom. A molecule with such a point is said to possess a center of symmetry or an inversion center.

When a center of inversion exists in a molecule, then certain restrictions are placed on the number of all the atoms or all but one atom. Since the center is a point, only one atom may be present at the center of the molecule. If there is an atom at the center of the molecule, it is unique from the point of view that it is the only atom in the molecule, which does not shift on performing inversion operation. All other atoms must occur in pairs. Each other atom must have a twin, with which it is exchanged, when the inversion is performed. From this, it is clear that one need not bother to look for a center of symmetry in molecule, which contains an odd number of more than one species of atoms.

The symbol for the inversion is i. Like a plane of symmetry in the molecule, a center is also an element, which generates only one operation. The effect of carrying out the inversion operation n times may be expressed as,

$i^n = E$, when n is even.

$i^n = i$, when n is odd.

Another way to express inversion center is a rotation through an angle 180° followed by a reflection in the plane perpendicular to the axis of rotation. A molecule symmetrical with respect to this transformation is also said to have a center of symmetry.

So that $i = S_2 = C_2$

Examples of Center of Symmetry

(i) A linear molecule of ABA type has an inversion center at center of atom B, while in ABB type molecule, there is no inversion center.

(ii) Planar AB_4 molecule has inversion center at center of atom A. Similarly, trans – AB_2C_2 and regular octahedral molecule AB_6 also have inversion center of the molecule at center of atom A.

(iii) Benzene molecule has an inversion center at the center of the molecule.

(iv) Regular tetrahedral AB_4 molecule does not contain an inversion center though the number of B atoms are even and an atom A is at the center of the molecule.

(v) Planar pentagonal ($C_5H_5^-$) ion has no inversion center.

(vi) Planar AB_3 type molecule has no inversion center as these two structures are not equivalent or identical.

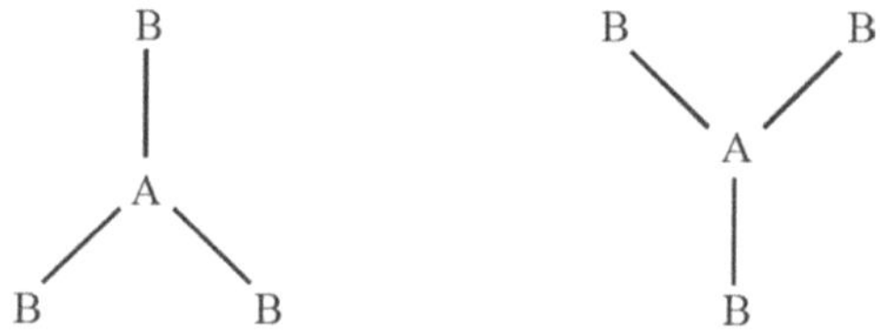

(vii) Trans-dichloroethylene has a center of symmetry while cis-dichloroethylene does not. By doing inversion on the trans-form, the atom Cl_a shifts to the position of Cl_b; H_a to the position of H_b and C_1 to the position of C_2 and vice versa.

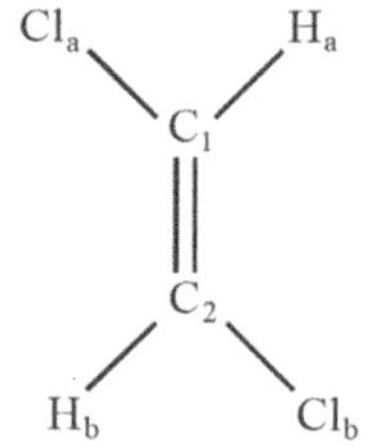

Trans-form

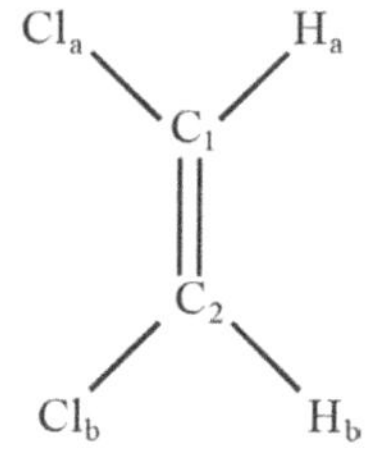

Cis-form

Thus, it gives an equivalent structure of the original molecule, which is indistinguishable. On the other hand, a similar operation on the cis-form causes the atom Cl_a to shift to the position of H_b; H_a to Cl_b and C_1 to the position of C_2 vice versa.
This structure can be easily distinguished from the original structure and therefore, this molecule in not symmetric with respect to inversion or it does not contain the center of symmetry.

(v) Improper Axis of Symmetry (Alternate axis) (S_n)

An improper rotation may be thought of taking place in two steps. In this case, the molecule is rotated along C_n axis by an angle $2\pi/n$ and then all the atoms are reflected through a plane perpendicular to the C_n axis. In some cases, these two operations can be carried out in reverse order also, but not in all. If by doing so, an equivalent configuration is obtained, then this axis is called an improper axis of symmetry and it is symbolized as S_n.

In other words, a molecule is said to possess an improper axis of rotation of the order n, if rotation by $2\pi/n$ about this axis is followed by a reflection in a plane perpendicular to this axis, then it leaves the molecule in an indistinguishable orientation. The axis, about which the rotation is carried out, is called an axis of improper rotation. This operation is also known as rotation-reflection symmetry operation and the axis is called a rotation-reflection axis. The operation of improper rotation by $2\pi/n$ is denoted by symbol S_n.

Mathematically, it is represented as,

$$S_n = \sigma_h.C_n \text{ or sometimes } S_n = C_n.\sigma_h$$

where n is the order of improper axis. Operation is carried out from right to left.

An inversion operation is the simplest rotoreflection (rotation followed by reflection) operation and is given the name S_2 or i.

If in any molecule, an axis C_n and a perpendicular plane exist independently, then S_n must also exist in that molecule. However, S_n can exist in a molecule even if the C_n or the perpendicular plane does not exist separately. In such cases, S_n becomes important because C_n axis and a perpendicular plane to it (σ_h); both are absent in the molecule.

The element S_n, in general, generates a set of operations S_n^1, S_n^2, $S_n^3 \ldots S_n^{n-1}$, S_n^n. There are differences in the sets generated for even and odd numbers. So these two cases should be considered separately. Let us consider that S_n axis (n is 4) is collinear with Z-axis of coordinate system and that the plane, through which the reflection operation is carried out, is the xy plane. An improper axis S_n of even order generates sets of operations S_n, $S_n^3 \ldots S_n^{n-1}$. S_n^n means that C_n and σ_v operations are carried out in sequence (n = 1, 2, 3…) until in all, C_n and σ each have been carried out n times.

$$S_n^n = C_n^n \cdot \sigma^n$$

where n is even and hence,

$$\sigma^n = E$$

$$S_n^n = C_n^n \cdot E = C_n^n$$

but $C_n^n = E$.

Hence, $S_n^n = C_n^n. E = E.E = E$

Beyond S_n^n, if S_n^{n+1} and S_n^{n+2} are also taken, representation of operations will take form of S_n^1 and S_n^2, respectively.

$$S_n^{n+1} = S_n^n.S_n^1 = E.S_n^1 = S_n^1$$

$$S_n^{n+2} = S_n^n.S_n^2 = E.S_n^2 = S_n^2$$

Thus, in any set of operations generated by an even order S_n, certain S_n^n may be written in some other ways, for example, if we take the set of S_6, this will generate – S_6, S_6^2, S_6^3, S_6^4, S_6^5, S_6^6.

(i) S_6^1 can be written in no other way than S_6.

(ii) $S_6^2 = C_6^2.\sigma_h^2 = C_6^2.E = C_6^2 = C_3$

(iii) $S_6^3 = C_6^3.\sigma_h^3 = C_2.\sigma = S_2$ or i

(iv) $S_6^4 = C_6^4.\sigma_h^4 = C_3^2.E = C_3^2$

(v) S_6^5 can be written in no other way than S_6^5

(vi) $S_6^6 = C_6^6.\sigma_h^6 = E.E = E$

So S_6 generates a set of S_6, C_3, S_2 or i, C_3^2, S_6^5 and E. We can make a useful observation here. This set contains C_3, C_3^2 and E, which are just the operations generated by a C_3 axis also. Hence, the existence of the S_6 axis automatically requires that the C_3 axis exists in the molecule. In general, the existence of S_n axis of even order always requires the existence of a $C_{n/2}$ axis and a center of inversion.

Now let us consider an improper axis of rotation S_n, (n is odd), which is collinear with Z-axis of a coordinate system and that the plane is the xy plane through which, the reflection operation is carried out. The most important property here is that an odd order S_n requires that C_n and σ perpendicular to this axis must exist independently. The elements S_n generates the operations S_n, S_n^2, S_n^3..., S_n^{2n}.

Let us consider S_n^n, when n = odd,

$S_n^n = C_n^n.\sigma^n = E.\sigma_h = \sigma_h$; since n is odd.

$$S_n^{2n} = C_n^{2n}.\sigma_n^{2n} = E.E = E.$$

In other words, the element S_n^n generates a symmetry operation σ, but if the symmetry operation σ exists, the plane, to which it is referred, must be a symmetry element, in its own sight.

An improper axis of rotation S_n with odd order generates certain distinct operations. Consider an example of S_5 axis.

S_5^1 can be written in no other way then S_5

$$S_5^2 = C_5^2.\sigma_h^2 = C_5^2.E = C_5^2$$
$$S_5^3 = C_5^3.\sigma_h^3 = C_5^3.\sigma_h = S_5^3$$

$$S_5^4 = C_5^4.\sigma_h^4 = C_5^4.E = C_5^4$$
$$S_5^5 = C_5^5.\sigma_h^5 = E.\sigma_h = \sigma_h$$
$$S_5^6 = C_5^6.\sigma_h^6 = C_5.E = C_5 \quad (C_n^{n+1} = C_n)$$
$$S_5^7 = C_5^7.\sigma_h^7 = C_5^2.\sigma_h \; S_5^2$$
$$S_5^8 = C_5^8.\sigma_h^8 = C_5^3.E = C_5^3$$
$$S_5^9 = C_5^9.\sigma_h^9 = C_5^4.\sigma_h = S_5^4$$
$$S_5^{10} = C_5^{10}.\sigma_h^{10} = E.E = E$$

From the above example, it is clear that if n is odd, then total operations to reach identity is not S_n^n (unlike, if n is even) and it requires 2n operations. In other words, we can say that $S_n^n \neq E$, if n is odd. In this case, $S_n^{2n} = E$. Out of ten operations, four can be represented by proper axis C_n, four by S_5 and one each by σ and E operations.

Examples of Improper Axis of Symmetry

(i) A regular tetrahedral molecule does not contain C_4 axis and perpendicular plane, separately but an improper axis S_4 exists in such molecules.

Let us consider the example of the CCl_4. A rotation of 90° about the Z-axis followed by a reflection in a plane perpendicular to Z-axis (σ_{xy}) results into a configuration, which is indistinguishable from the original one. Hence, CCl_4 molecule possesses an improper axis of order 4, which is represented as S_4.

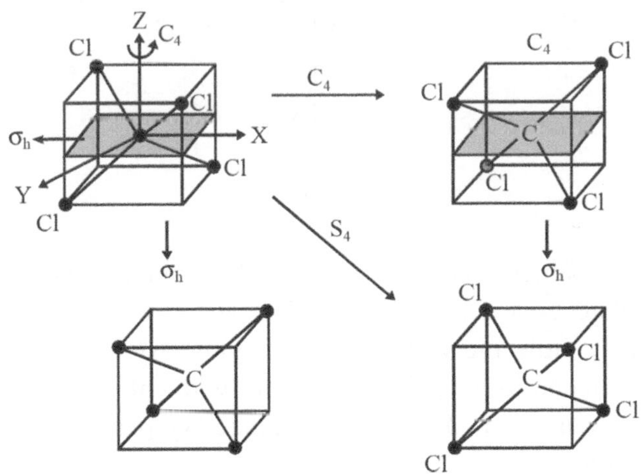

From the figure, it is also clear that a rotation of 90° about the Z-axis or a reflection in xy plane alone may not give the indistinguishable configuration, but the combination of these two gives indistinguishable orientation. CCl_4 molecule has three equivalent S_4 axis at right angles to one another viz S_4 (*x*), S_4 (*y*), and S_4 (*z*).

(ii) The planar benzene molecule possesses S_6 axis, which is coincident with the C_6 axis.

(iii) Consider ethane molecule in its staggered conformation, which has S_6 axis, where as in it eclipsed conformation, it has a S_3 axis. As the diagram shows that (*b*) and (*c*) are equivalent to each other but neither of them is equivalent to first, for example, neither σ nor C_6 by itself is a symmetry operation in this case, but the combination of both in either order, $C_6.\sigma_h = \sigma_h.C_6 = S_6$, is a symmetry operation. It produces structure (*d*), which is equivalent to structure (a). Also S_6 implies the existence of (i) C_3 axis coincident with the S_6 and (ii) center of symmetry (i).

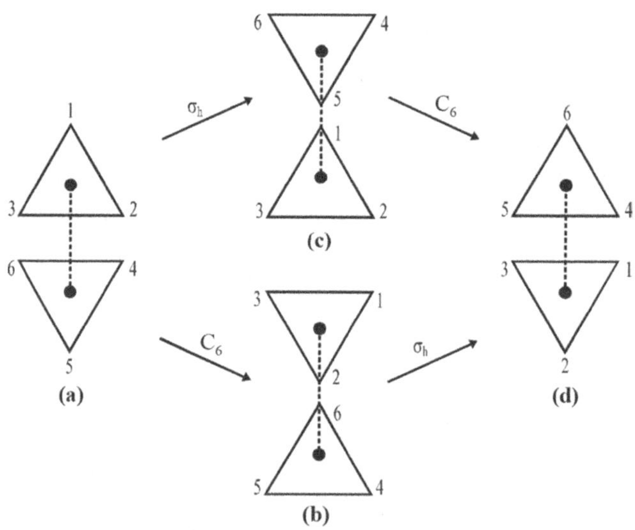

The distinct operations are S_6^1 and S_6^5, since weknow that S_6 axis will generate S_6, C_3, S_2 or i, C_3^2, S_6^5 and E

(iv) Every molecule with a plane of symmetry only, has S_1 axis perpendicular to the plane of symmetry. In chloroethylene, the plane of symmetry is the molecular plane (xy plane) and the Z-axis is then S_1 axis.

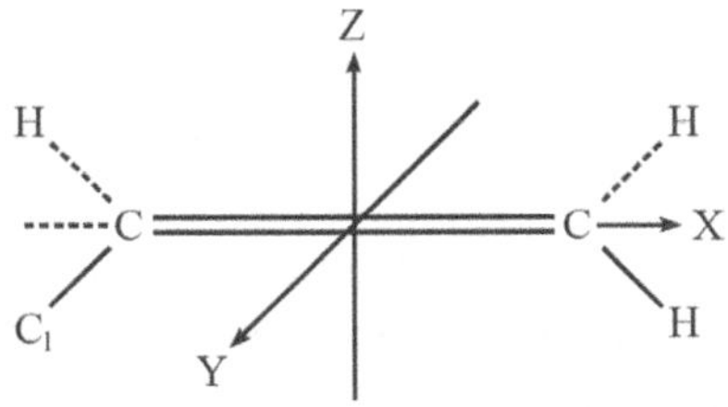

In this way, symmetry elements can be determined in any given molecule.

1.7.SYMMETRY IN ENGLISH ALPHABETS

It was very interesting to notice that plants, flowers, birds, molecules, etc. have various symmetry elements, in the same way capital letters of English alphabets also possess different elements of symmetry. English alphabets can be classified into various point group, which may be said as "Alphabetical Point Groups." Alphabetical point groups depend on their symmetry.

According to point groups or combination of symmetry elements, 26 English alphabets can be categorized into four sets:

First set consists of 11 alphabets (A, B, C, D, E, M, T, U, V, W, Y). These alphabets possess identity (E), two-fold axis of symmetry (C_2) and two vertical planes of symmetry (2 σ_v) elements and thus, belongs to C_{2v} point group. C_2 and 2 σ_v symmetry elements in these alphabets are,

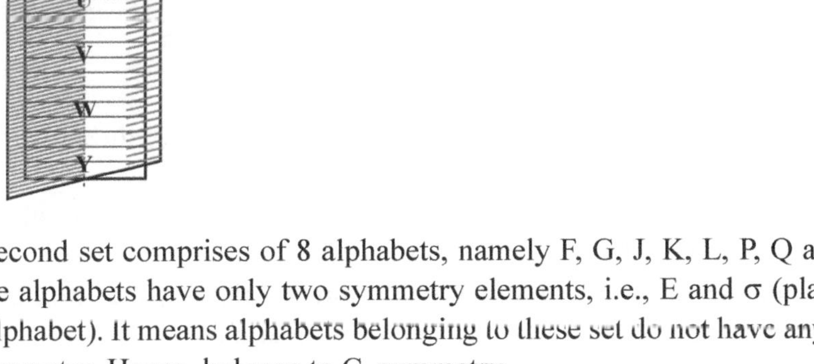

Second set comprises of 8 alphabets, namely F, G, J, K, L, P, Q and R. These alphabets have only two symmetry elements, i.e., E and σ (plane of the alphabet). It means alphabets belonging to these set do not have any axis of symmetry. Hence, belongs to C_s symmetry.

σ

F	G	J	K	L	P	Q	R

Third set belong to C_{2h} point group and includes alphabets N, S and Z with identity (E), two-fold axis of symmetry (C_2) as principal axis passing through inversion center (i) and a horizontal plane, i.e., ⊥ to the plane of the letters.

N S Z

Last four remaining alphabets, i.e., H, O, I and X belong to last set with D_{2h} point group. It consist of $E + C_2 + 3\ C_2 + 2\ \sigma_v + \sigma_h + i$. Here, it is interesting to notice that O is not a perfect circle and lines of X are not intersecting at exactly right angle (90°). Therefore, these are placed in D_{2h} point group.

1.8 SYMMETRY AND OPTICAL ACTIVITY

A compound is said to be optically active, if its mirror image is non-superimposable upon structure of the original compound. For example, ethane consists of infinite numbers of conformations. Let us consider the conformation of almost eclipsed form of ethane. This conformation is in between staggered and eclipsed conformation and it is optically active because it is not superimposable to its mirror image. Now, ethane remains optically active only, if ethane is frozen in this state.

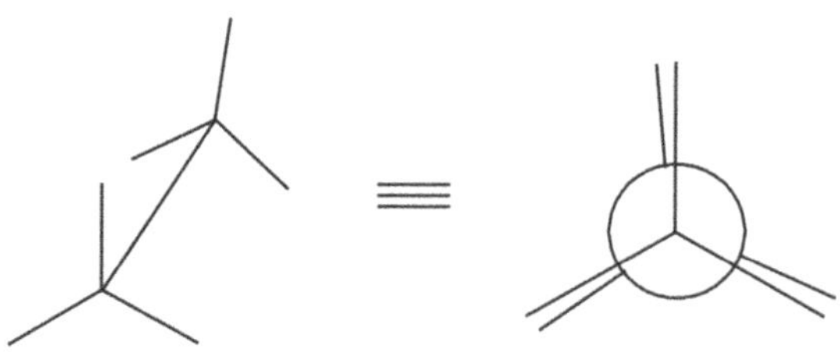

Almost eclipsed conformation of ethane (Optically active)

But as ethane has a C – C single bond, on which it can easily rotate even in presence of small activation energy. Rotation around C – C bond gives number of other conformations (more precisely six conformations), i.e., three staggered and three eclipsed conformations.

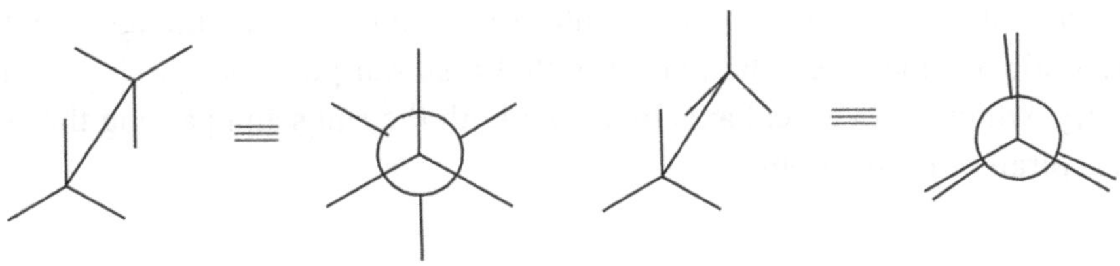

Staggered Eclipsed

Staggered form of ethane has plane of symmetry as well as center of symmetry whereas eclipsed form has only plane of symmetry, which shows that mirror image of staggered form is superimposable on original staggered conformation and similarly, mirror image of eclipsed form is superimposable upon its original. It means ethane is optically inactive because mirror image of every conformation is superimposable upon its original.

One should know, whether a compound has superimposable mirror image or not, i.e., it is optically active or not? The presence of asymmetric center in a molecule (without symmetry) is essential, that is with an atom with all different valences or molecule should be least symmetric case of dissymmetric. It is considered the criterion for a compound to be optically active. A molecule with carbon with four different valences, C_{abcd}, is non-superimposable to its mirror image; here, the central carbon atom is called "asymmetric carbon atom."

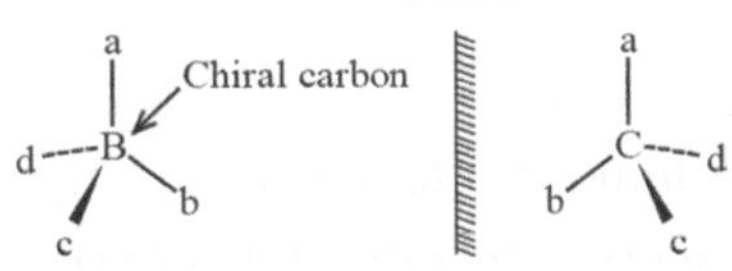

Asymmetric (Chiral) carbon

In the case, when molecule is complicated, sometimes it becomes quite difficult to identify, whether the four groups attached to a particular carbon are different. Apart from this, many optically active compounds do not have an asymmetric carbon atom.

A single criterion is sufficient to prove any compound to be optical active, i.e., absence of S_n axis of symmetry. A molecule with S_n axis of order higher then two-fold such as in case of spiran, which do not possess either a plane of symmetry or a center of symmetry. Even though, spirans are optically inactive. The reason is that a spiran possess S_n axis of symmetry, which is a vertical axis bisecting both the rings and passing through the spiran nitrogen atom.

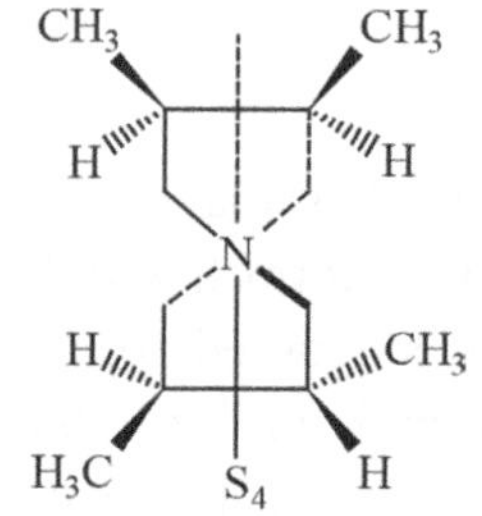

Spiran (Optically inactive)

Spirans have S_4 axis of symmetry but not C_2 axis and therefore, it is optically inactive. Some other examples having S_n axis are allenes, cummulenes and cyclohexane (chair and boat forms).

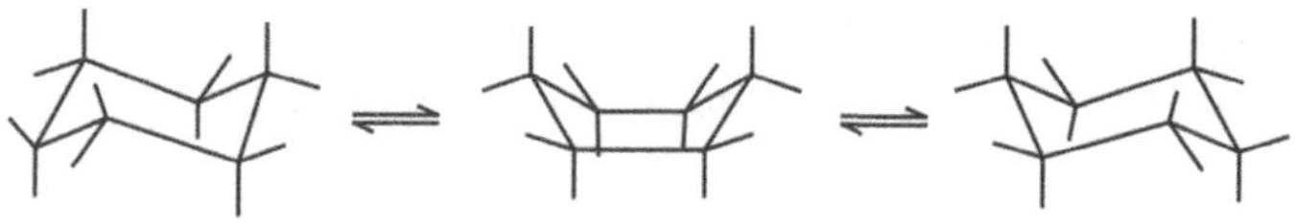

S_1, S_2 and S_6 axes are present.

On the contrary, it is important to note that compounds, which possess an ordinary axis of symmetry (element of symmetry) but no S_n axis of symmetry are optically active, i.e., trans-1,2-dichlorocyclopropane.

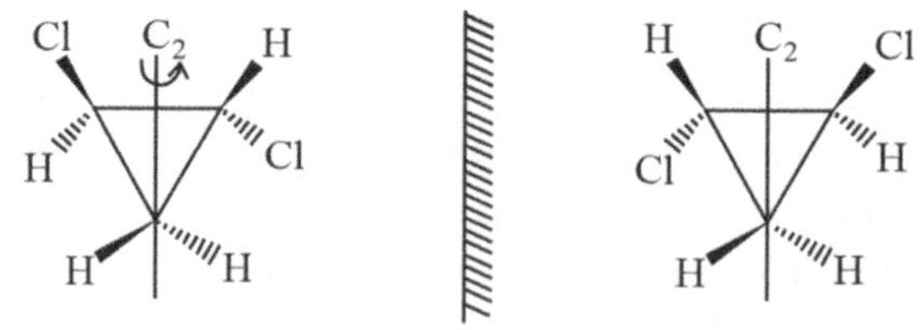

Non-superimposable mirror image (Optically active)

Cis-dimethylketopiperazine has C_2 axis of symmetry perpendicular to the plane of ring, but no S_n axis. Hence, it is also optically active. Whereas, its trans-form is optically inactive due to presence of center of symmetry ($S_2 \equiv i$).

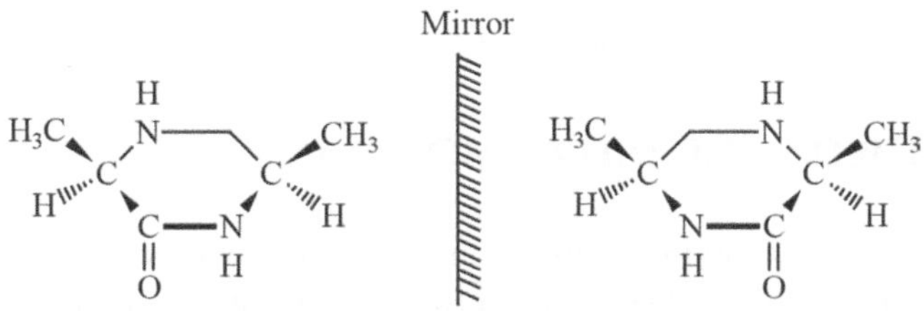

Non-superimposable mirror image (Optically active)

So, it may be concluded that optical active compounds need not be asymmetric (without symmetry), but must be dissymmetric (without S_n axis of symmetry of any order). All asymmetric compounds are dissymmetric but reverse may not always be true.

Most of the compounds having point groups like C_1, C_n and D_n show optical activity, i.e., $[Co(en)_3]^{3+}$.

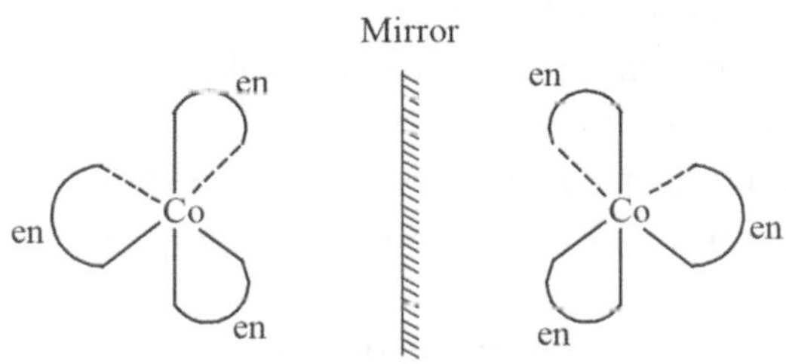

H_2O_2 belongs to C_2 point symmetry and it is optically inactive because of free rotation around O-O bond, which permits its mirror image to be superimposable on original one. A special case is biphenyl, when S_n axis is absent, still this compound is optically inactive, because when it is rotated around 4, 4′-C-C single bond of mirror image, then new orientation of molecule is superimposable on its original structure and thus, creates pseudo-S_n axis.

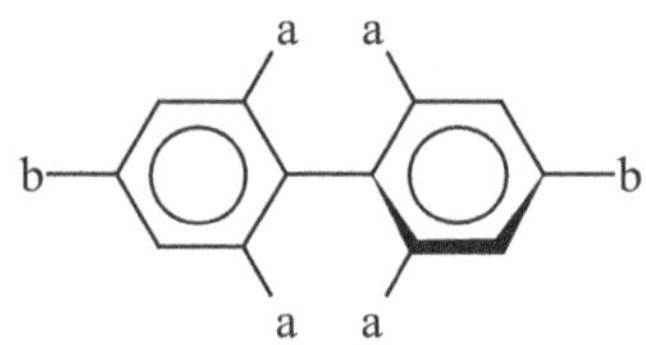

Biphenyls (Optically inactive)

1.9 SYMMETRY AND DIPOLE MOMENTS

When sum of all of the individual bond moment vectors is not equal to zero (non-zero), then that molecule will possess a dipole moment. If a molecule contains a center of symmetry, then charge on one side of the molecule gets canceled by the equal and opposite charge on the other side of the molecule. Hence, overall dipole moment becomes zero.

If a molecule has more than one element of symmetry, then also dipole becomes zero. When more than two C_n axes are present in a molecule, then dipole cannot exist because at a time, dipole vector cannot lie along more than one axis. The presence of a horizontal plane also prevents from having dipole moment or such molecule has no dipole.

Therefore, it can be concluded that if a molecule has an inversion center, and symmetry axis in a plane not parallel to the principal axis, then it does not have dipole. It means that a molecule having a symmetry plane and more than one axis of symmetry does not have dipole.

So, dipole moment is present only in such molecules, which have dipole along symmetry axis and a plane of symmetry, i.e., however, presence of one or more vertical mirror plane (σ_v) do not prevent a molecule from having dipole moment.

Group with symmetry allowed dipole moment $\Rightarrow C_1, C_2, C_n, C_{nv}, C_s$

Group with symmetry forbidden dipole moment $\Rightarrow C_i, S_n, D_n, D_{nh}, D_{nd}, T_d, O_h, I_h$

Example of molecules with zero dipole moment is trans-N_2F_2, staggered ferrocene, etc.

Molecule with dipole moment are cis-N_2F_2, O = N-Cl, etc.

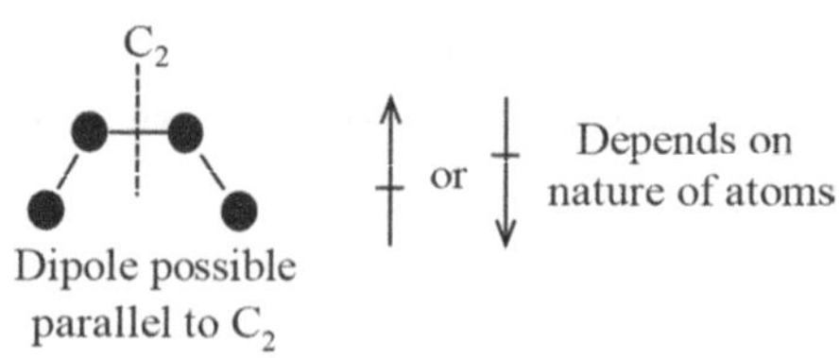

Molecules like S = C = Te, cis-FN = NF, CO, SbH_3, $FClO_3$, etc., have small, but finite dipole moment.

1.10 SOME MORE EXAMPLES

Examples of some molecules are given here, which have slightly complicated structures.

(i) Cyclohexane (Chair form)

The chair form of cyclohexane (C_6H_{12}) consists of 6 axial bonds (C-H_a) and six equatorial bonds (6 C-H_e).

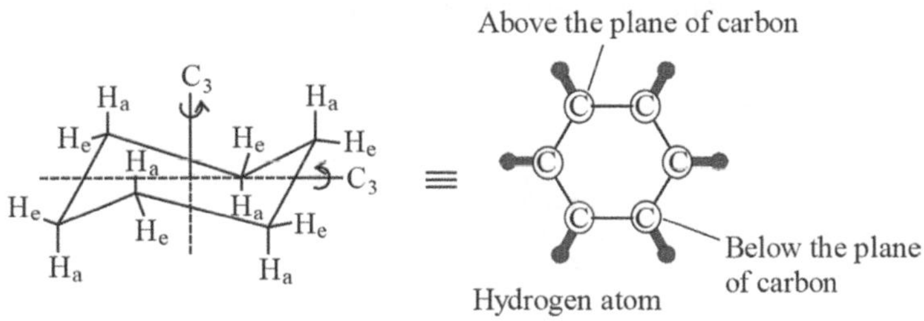

- Cyclohexane has a C_3 axis, which passes through the center of the molecule in such a way that 120° rotation along this axis will give equivalent (indistinguishable) configuration. 6 C-H_a bonds are parallel to C_3 axis.
- Three C_2 axes are present, which are perpendicular to C_3 axis in such a manner that each C_2 passes through the middle of the two opposite C-C bond. The remaining 6 C-H_e bonds are somewhat perpendicular to C_3 axis.
- Cyclohexane consists of three σ_d reflection planes passing through two opposite carbon atoms.
- An inversion center is also present in chair form, which is absent in boat form of cyclohexane.
- S_6 axis of symmetry is also present, which is coincident with C_3 axis.
- Therefore, the chair form of cyclohexane contain E + C_3 + 3 $C_2 \perp C_3$ + S_6 + 3 σ_d + i.

(ii) Spiropentane (C_5H_8)

In spiropentane, each carbon atom has sp^3 hybridized form.

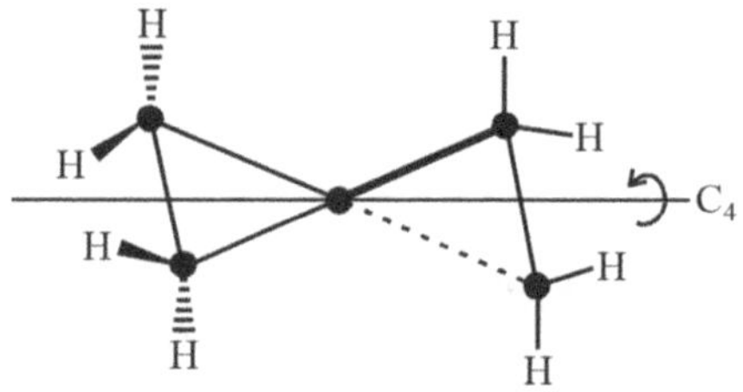

- It has a C_2 axis as principal axis, which passes through central carbon atom and middle of the two opposite C-C bonds.
- Two σ_v are present, which pass through the C_2 axis.
- Thus, C_5H_8 have E + C_2 + 2 σ_v symmetry elements.

(iii) Cyclooctatetraene (C_8H_8)

- Cyclooctatetraene have tub like structure.

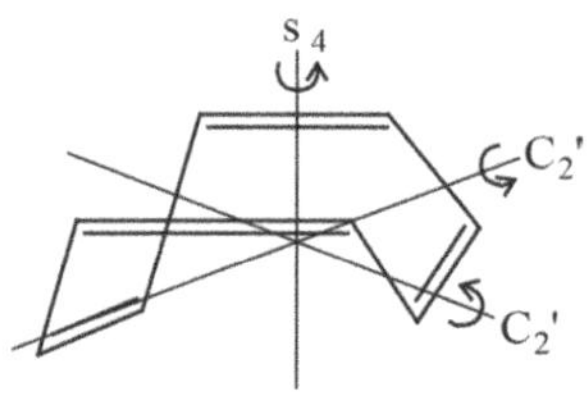

- This molecule consists of a S_4 axis.
- Coincident with this S_4 axis, there also exists a C_2 axis.
- Two more C_2' axis perpendicular to C_2 (collinear with S_4) are present in a plane perpendicular to S_4-C_2 axis are present.
- This molecule has two σ_d planes of symmetry, each bisecting two opposite double bonds and passing between the $C_{2'}$ axis.
- Thus, symmetry elements of cyclooctatetraene are E + C_2 + 2 $C_{2'}\perp$ C_2 + 2 σ_d.

(iv) 1, 3, 5, 7-Tetramethylcyclooctatetraene

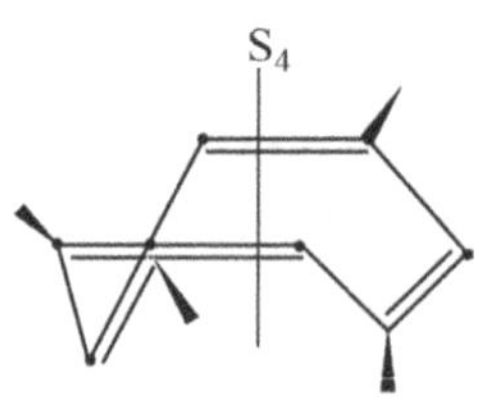

- It has a S_4 axis of symmetry and horizontal C_2 axis as described in C_8H_8. Due to presence of methyl groups, all the vertical planes are absent.
- It is important to note that this molecule does not have an inversion center, or plane of symmetry, even though it is not dissymmetric.
- Therefore, the element present in the molecule are $E + C_2 + S_4$.

(v) S_8 Molecule

This molecule have cyclic crown structure.

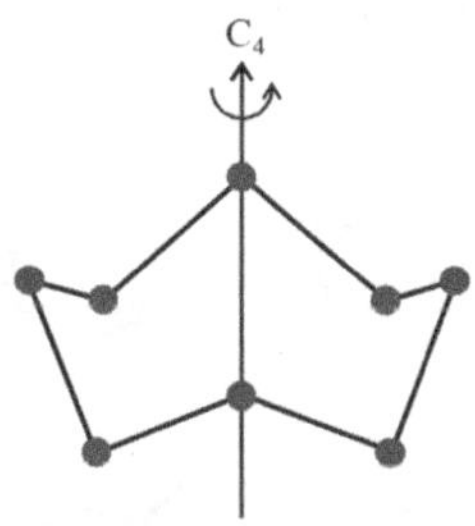

From top view, it appears staggered pair of square of sulfur atoms.

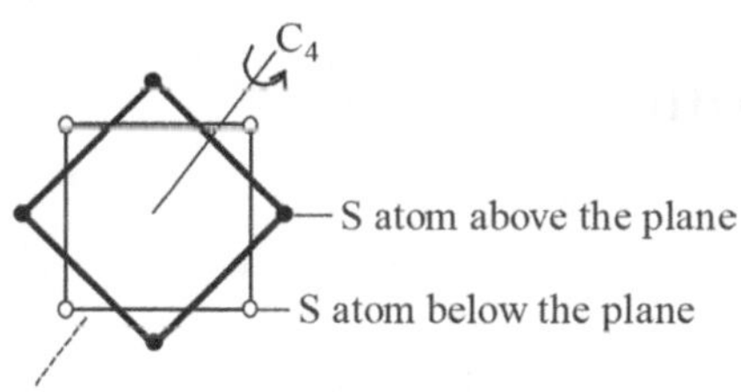

- There is a C_4 axis of symmetry, which passes through center of the crown conformation.
- There are 4 $C_2 \perp C_4$, passing through two opposite sulfur atoms. The molecule on rotation by 180° gives indistinguishable configuration.
- Four σ_d planes passing through a pair of sulfur atoms present diagonally on the either of the square planes.
- S_8 axis is present collinear with C_4 axis

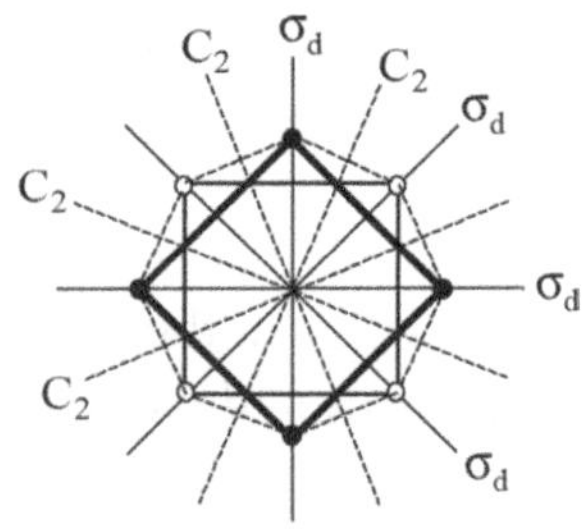

Top-view of S_8

(vi) Tetraphenylmethane, $C(C_6H_5)_4$

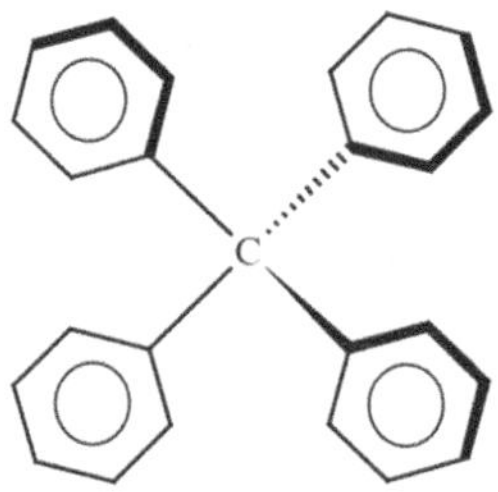

This molecule has only one S_4 axis, and therefore, possesses only E + S_4.

(vii) Boric acid, $B(OH)_3$

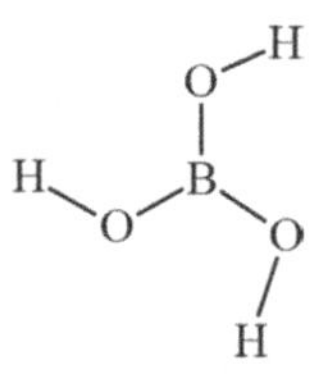

- The boron atom in boric acid is sp^2 hybridized with planar geometry. This molecule has only three elements of symmetry, for example, E + C_3 + σ_h.
- C_3 axis of symmetry passes through center of molecule perpendicular to the plane of the molecule. σ_h plane is a molecular plane, which is perpendicular to the principal axis of symmetry, C_3.

(viii) Ru (1, 10-phenanthroline)$_2$ Cl$_2$

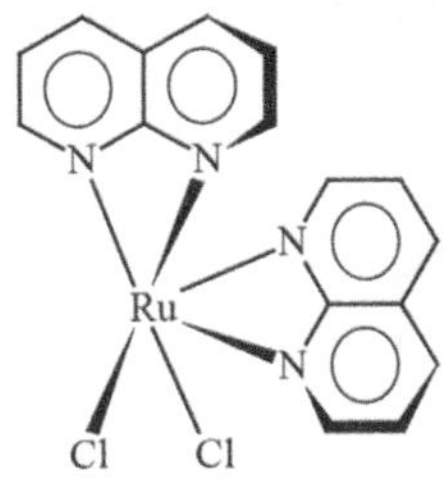

(a) Cis-[Ru(1,10-phenanthroline]$_2$ Cl$_2$

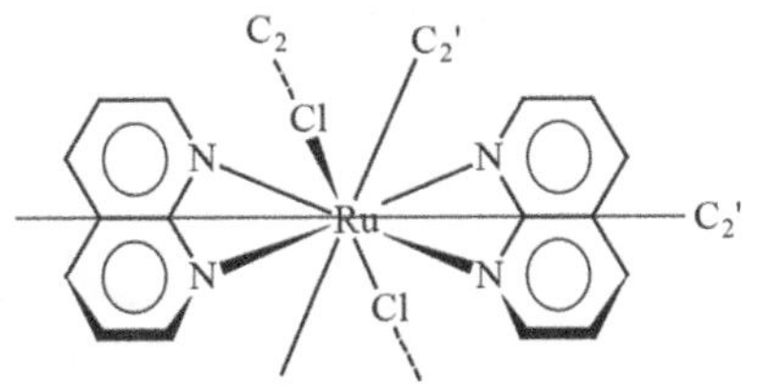

(b) Trans- [Ru(1,10-phenanthroline]$_2$ Cl$_2$

- Cis-isomer of the complex has only one C_2 axis, which bisects the Cl-Ru-Cl angle. Therefore, this low symmetry molecule has E + C_2 symmetry elements.
- In trans-isomer of the complex, two phenanthroline rings are coplanar. In this molecule, C_2 axis is present as a principal axis and passes through the Cl-Ru-Cl bonds.
- 2 $C_{2'} \perp C_2$ are also present in trans-form.
- σ_h plane and inversion center are also present in trans-form of this complex. Therefore, trans-isomer has E + C_2 + σ_h + i.

(ix) fac- and mer-[RhCl$_3$(pyridine)$_3$]

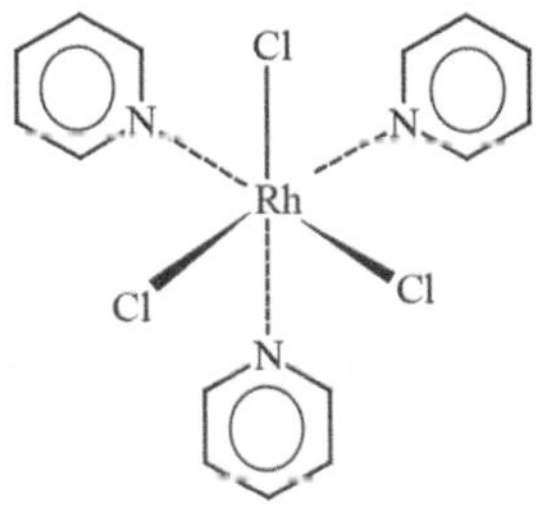

fac-[RhCl$_3$(pyridine)$_3$]

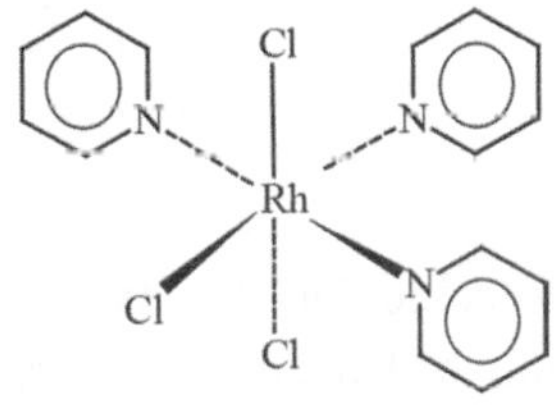

mer-[RhCl$_3$(pyridine)$_3$]

- Fac-isomer, possesses C_3 axis and three σ_v planes passing through Cl atom and pyridine ring and this molecule has E + C_3 + 3 σ_v.
- In mer-isomer, three N atoms are coplanar, and in the same way, 3 Cl atoms are also coplanar. Therefore, mer-isomer has C_2 axis passing through N-Rh-Cl and have two vertical planes (2 σ_v). One σ_v includes the 3 Cl, central N donor atoms and other includes 3 N and the central Cl donor atom. So this molecule has E + C_2 + 2σ_v.

(x) Triphenylphosphine (PPh_3)

Triphenylphosphine has three phenyl groups arranged in propeller like structure about the trigonal pyramidal P atom.

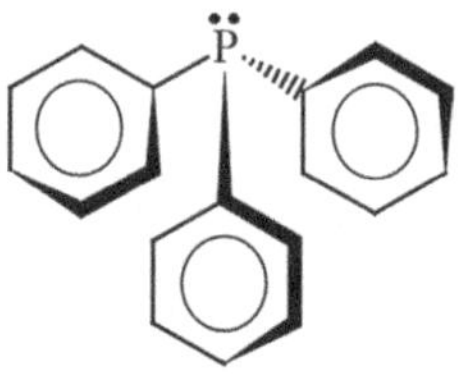

- This molecule has only C_3 axis. Thus, it has only two symmetry elements, i.e., $E + C_3$.

(xi) [Ru(1,10-phenanthroline)$_3$]$^{2+}$

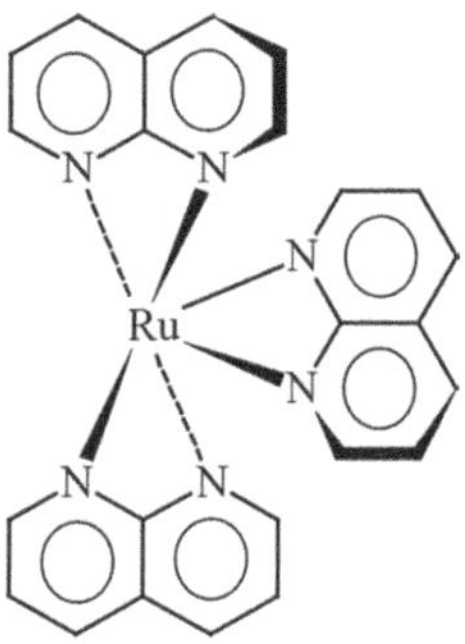

This molecule has C_3 axis (principal axis of symmetry), which passes through center of Ru atom and perpendicular to the plane of paper and 3 $C_2 \perp C_3$. Thus, it has E, C_3, 3 $C_2 \perp C_2$.

KEYWORDS

- **Axis**
- **Center**
- **Identity**
- **Improper axis**

- **Inversion**
- **Plane**
- **Reflection**
- **Rotation**
- **Symmetry element**
- **Symmetry operation**

CHAPTER 2

POINT GROUPS

CONTENTS

2.1 MOLECULAR POINT GROUPS

A molecule consists of an assembly of symmetry operations. All the symmetry elements possessed by any molecule, pass through a fixed point in the molecule and this point do not change its position during any operations. Therefore, all the operations generated by symmetry elements is said to form a symmetry or point group. It means point group is a symbol, which represents symmetry elements present in that molecule. The fixed-point group is also a group in its mathematical sense. Hence, it must also satisfy the necessary characteristics of a group in general.

Basically, there are two sets of nomenclature to represent symmetry operations. Schoenflies nomenclature, which is useful aid in probing the properties of the molecule and is often used in spectroscopy and Hermann-Mauguin nomenclature, which is used in describing structure or crystallography.

Most of the molecules can be classified into 32-point group symmetry (with few exceptions) by short hand notations using Schoenflies symbols. The assignment of molecule to an appropriate point group can be purely formal, i.e., it satisfies certain conditions of mathematical basis, such as there must be an identity element (E) in group, the existence of an inversion operation ($A.A^{-1} \equiv 1$), the product of two operations is also another operation is the group, and associative multiplication of operations.

2.2 CLASSIFICATION

Molecules can be classified into point groups based on various possible combinations of symmetry elements possessed by them. In general, there are four major types of point groups and these are given as follows:

(i) Groups with very high symmetry (cubic point groups);

(ii) Groups with low symmetry;
(iii) Groups with n-fold rotational axis (C_n);
(iv) Dihedral groups (D_n).

2.2.1 GROUPS WITH VERY HIGH SYMMETRY (CUBIC POINT GROUPS)

Cubic point group is a group with large number of characteristic symmetry elements. In general, these point groups are related to regular geometries. They possess more than two proper axes of order greater than two. It includes three types of point groups with cubic symmetry, namely,

- Tetrahedral (T_d, T_i and T_h);
- Octahedral (O_h and O);
- Icosahedral (I_h) (Dodecahedral).

2.2.1.1 Tetrahedral Group

A regular tetrahedral molecule (AB_4) has E + 8 C_3 + 3 C_2 + 6 S_4 + 6 σ_d symmetry operations. The tetrahedral molecule has four C_3 axes, three C_2 axes, six mirror planes, and three S_4 improper rotational axes.

Therefore, molecules with regular tetrahedral geometry belong to T_d point group, for example, methane.

If a center of symmetry (i) is also present in tetrahedron, then T_i group is formed. The element of T_i are same as those of tetrahedron group, i.e.,

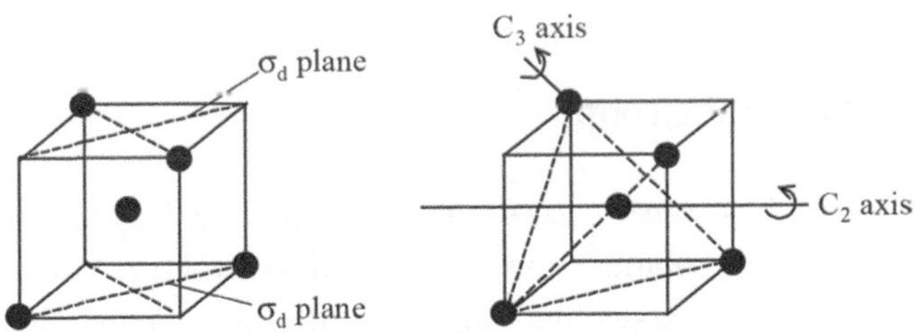

Point group T is further divided into 2 point groups.
(a) T_d group;
(b) T_i or T_h group.

(i) T_d Group

The addition of σ_d plane in T group generates the T_d group, for example, methane, CCl_4. T_d group has following elements:

- Three S_4 axes coinciding with X−, Y− and Z− axes, each of which generates S_4, $S_4^2 = C_2$, S_4^3 and $S_4^4 = E$.
- Three C_2 axes coinciding with X−, Y− and Z− axes and each of which generates an operation C_2. However, these have already been generated by S_4^1.
- Four C_3 axes, each of which passes through one apex and the center of the opposite face. Each of them generates C_3 and C_3^2 operations, i.e., eight operations in all.
- Six plane of symmetry (6 σ_d), each of which generates a symmetry operation. These planes lie on six faces of the cube.

This entire set of the operations will also include E. Six improper rotations (S_4^1's and S_4^3's), three two-fold proper rotations, eight three-fold proper rotation (C_3's and C_3^2's) and six reflection planes having 24 operations in all.

(ii) T_i or T_h Group

If a center of symmetry is added to T group, then T_i group is formed. The elements of T_i are same as those of T group and also each of those operations multiplied by i.

Six planes of symmetry (6 σ_d) are present. Each generates a symmetry operation. These planes lie on six faces of the cube.

Finally, there is one more group in T, which has additional set of σ_h plane, which contain pair of C_2 axes. It is designated by T_h.

2.2.1.2 Octahedral Group

Octahedral group has four C_3 axes, three C_4 axes, six C_2 axes, four S_6 axes, three σ_h planes, six σ_d planes, and a center of symmetry. In addition, there are three C_2 and three S_4 axes that coincide with the C_4 axes. Therefore, total symmetry operations for a regular octahedral molecule are 48.

Example of molecules with O_h point group are SF_6, $[Co(NH_3)_6]^{3+}$, $[AlF_3]^{6-}$, $Mo(CO)_6$, etc.

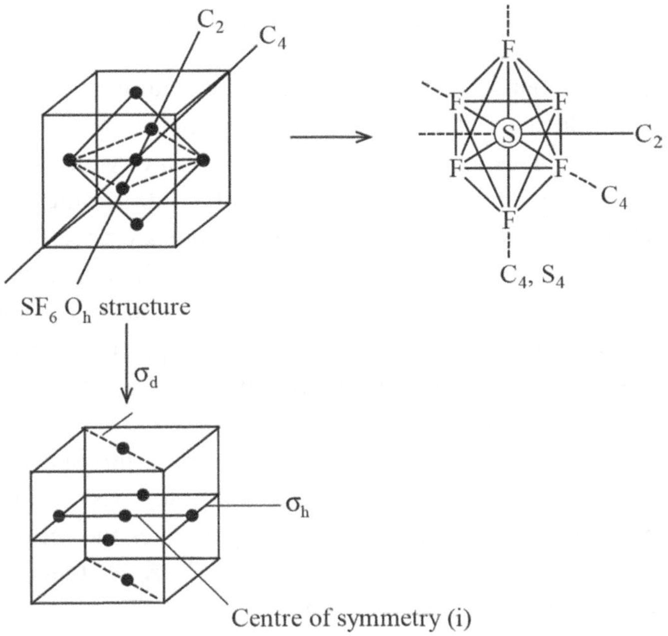

(i) 4 C_3 axes pass through opposite apices of octahedron, i.e., Axes passing through center of the two opposite faces of the octahedron. As there are 8 faces in an octahedron, there are 4 C_3 axes present.

(ii) 3 C_2 and 3 S_4 axes, which are collinear with the C_4 axis of symmetry.

(iii) 6 C_2 axes bisecting the opposite edges. There are 12 edges and each pair of them generates an operation C_2.

(iv) 4 S_6 axes coincident with 4 C_3. Each of them is passing through the center of pair of opposite triangular faces and generates a set of operations S_6, C_3, i, C_3^2 and S_6^5.

(v) 3 σ_h – There are 3 C_4 proper axes of symmetry and each horizontal plane is perpendicular to a particular C_4 axis.

(vi) 6 σ_d – Dihedral planes passing through two apices and bisecting opposite edges.

(vii) 2 S_4 axes, each of them is passing through a pair of opposite apices generating the operations S_4, C_2 and S_4^3.

(viii) 3 C_4 – Axes passing through the center of opposite faces of the cube.

(ix) i is inversion center. This operation is generated by each of the S_6 axis.

2.2.1.3 Icosahedral (I_h) and Dodecahedral Group

(i) These groups have five-fold axes in addition to three-fold and two-fold axes. The addition of a σ_h plane perpendicular to the two-fold axes leads to the center of symmetry and then the group is called I_h group. This group has total 120 symmetry operations. These symmetry elements and operations are as follows:

(ii) 6 S_{10} – A set of six S_{10} axes is present. In the dodecahedron, these axes pass through the opposite pairs of pentagonal faces, while in the icosahedron, they pass through the opposite vertices. Each of the S_{10} axis generates these operations S_{10}, $S_{10}^2 = S_5$, S_{10}^3, $S_{10}^4 = S_5^2$, $S_{10}^2 = i$, $S_{10}^6 = C_5^3$, S_{10}^7, $S_{10}^8 = C_5^4$, S_{10}^9, E.

(iii) 10 S_6 – In dodecahedron, these axes pass through opposite apices of vertices, while in icosahedron, they pass through pairs of opposite faces. Each of these axis generates the following operation S_6^1; $S_6^2 = S_3$, $S_6^3 = S_2 = i$, $S_6^4 = S_3^2 = C_3^2$, S_6^5, E. Out of these, i and E have already been observed.

(iv) 6 C_5 – There are six S_5 axes collinear with S_{10} axis. They generates C_5, C_5^2, C_5^3 and C_5^4 operations, which have already been counted under S_{10}.

(v) 10 C_3 – There are ten C_3 axis collinear with S_6 axis. These generate C_3 and C_3^2 operations, which have been already counted with S_6.

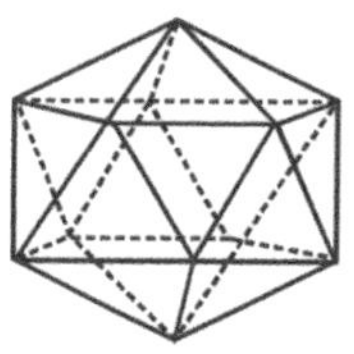

Icosahedral

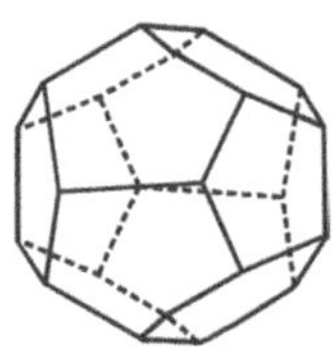

Dodecahedral

(vi) 15 C_2 – Each of these C_2 axes bisects opposite edges. These axes generate 15 C_2 operations.

(vii) There are 15 mirror planes, each one of them contains two C_2 axes and two C_5 axes. They generate 15 reflection operations.

(viii) In all, there are 120 elements and these are given by E, 12 C_5, 12 C_5^2, 20 C_3, 15 C_2, i, 12 S_{10}, 20 S_6 and 15 σ. These elements constitute a group that is called I_h group.

- The group T_d and a pure rotational subgroup T is of order 12. It consists of the following classes – E, 4 C_2, 4 C_3^2, 3 C_2.
- The group O_h has a pure rotational subgroup O of order 24. It consists of the following classes E, C_3, 6 C_4, 3 C_2, 8 C_3, 6 C_2.
- The group I_h has a pure rotational subgroup I of order 60. It consists of the following classes – E, 12 C_5, 12 C_5^2, 20 C_3 and 15 C_2.

All taken together, there are following seven groups, which contain multiple high order axes.

- T, T_h and T_d
- O and O_h
- I and I_h

2.2.2 GROUP WITH LOW SYMMETRY

Low symmetry groups possess only one or two symmetry elements. There are three group of low symmetry.

2.2.2.1 C_1 Group

One-fold rotational axis. The molecule has only one symmetry element, i.e., E. All the irregular molecules or chiral molecules with an asymmetric center belong to this point group. Examples of this point group are:

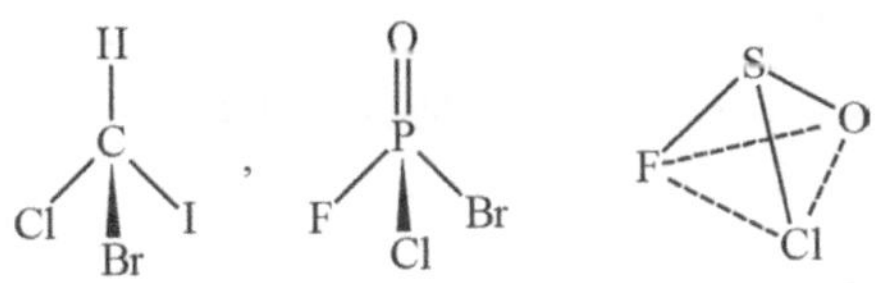

2.2.2.2 C_s Group

Only symmetry element plane is present in the molecule. This molecule has two operations E and σ. Thus, although they have very low symmetry, they are not chiral. Order (h) of this group is equal to two. Examples of this point group are:

The combination of E + σ = C_s (or S_1).

2.2.2.3 C_i Group

Molecules with this point group has element i (inversion) along with E, for example, $(E + i = C_i)$. This is also a group with order two and it equals to S_2. Example of this point group is:

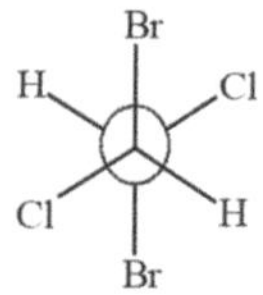

1, 2-Dibromodichloroethane

2.2.3 *GROUPS WITH N-FOLD ROTATIONAL AXIS (C_n)*

2.2.3.1 C_2 Group

C_n group has n fold axis of symmetry (operation) besides the identity operation (E). Here n varies from 2–6. C_2 point group has one two-fold axis of symmetry C_2 and E. Example of C_2 point group are:

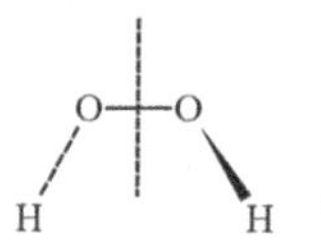

Gauche-H_2O_2

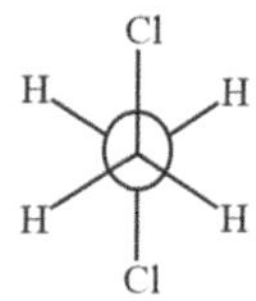

Staggered 1,2-dichloroethane

2.2.3.2 C_3 Group

C_3 point group has symmetry elements C_3, C_3^2, and C_3^3 (= E), for example, Molecules of this point group contain only one three-fold axis. Examples of C_3 point group are:

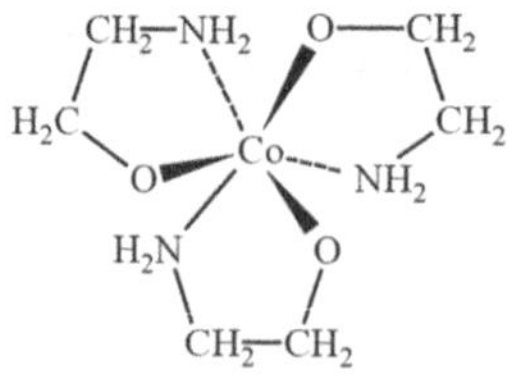

Tris(2-aminoethoxo)cobalt (III)

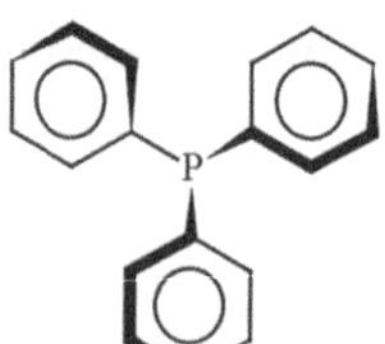

Thiphenyl phosphine.

These molecules are in this most stable conformations having C_3 symmetry.

2.2.3.3 $C_4 - C_6$ Groups

C_4 point group has C_4, C_4^2 (= C_2), C_4^3, and E symmetry elements. C_5 point group has C_5, C_5^2, C_5^3, C_5^4 and E and similarly C_6 point group has C_6, C_3, C_2, C_3^2, C_6^5 and E.

$$C_6^2 = C_3,\ C_6^3 = C_2,\ C_6^4 = C_3^2,\ C_6^6 = E$$

C_n belongs to cyclic point group and all the elements in the cyclic group commute with each other and hence, a cyclic group is always an Abelian group.

2.2.3.4 C_{nv} Group

Presence of n vertical planes of symmetry (σ_v) containing the rotation axis C_n gives C_{nv} point groups.

The symmetry operations in C_{nv} are $E + C_n + n\,\sigma_v$. In C_{nv}, n can vary from 2–6. Some example of C_{nv} are H_2O (C_{2v}), NH_3 (C_{3v}), trans-$[CoCN_3)_5Cl]^{2+}$ (C_{4v}), etc.

The point group of H_2O is C_{2v} as it has 1 C_2 + 2 σ_v symmetry elements.

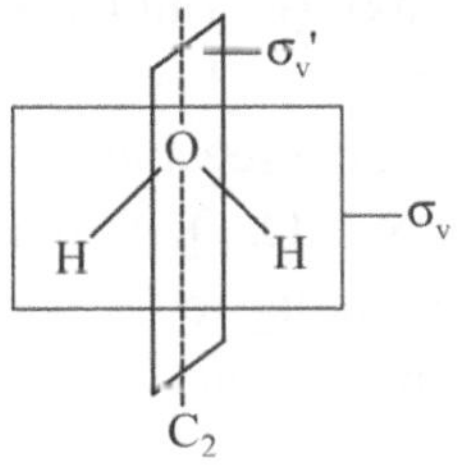

Similarly, NH_3 has a pyramidal geometry with 1 C_3 + 3 σ_v symmetry elements.

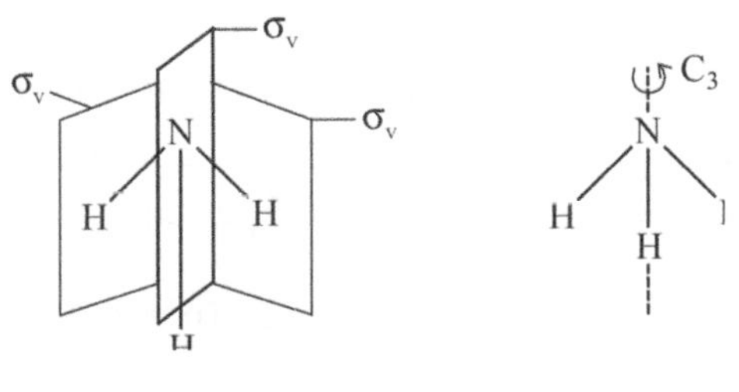

2.2.3.5 C_{nh} Group

If a horizontal plane perpendicular to principal axis (C_n) is present in a molecule, then the molecule is said to have C_{nh} symmetry. Such groups have $C_n + \sigma_h + E$ symmetry elements. Even if vertical planes are present in addition of C_{nh}, the point group remains C_{nh}. Examples of point group C_{nh} are trans-dichloroethylene (C_{2h}), B $(OH)_3$ (planar) (C_{3h}), etc.

C_{2h} – Trans-dichlorocthylene belongs to C_{2h} point group as it contains E, C_2 and σ_h apart from i.

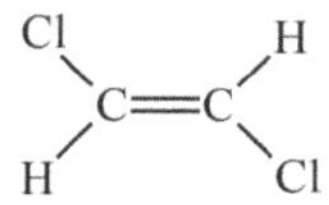

C_{3h} – Boric acid (planar shape) is an example of C_{3h} point group as it has E, C_3 and σ_h.

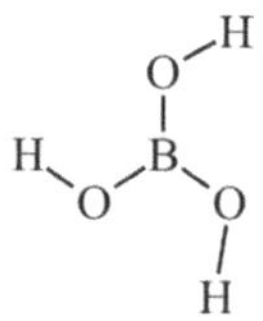

2.2.3.6 $C_{\infty v}$ Group

It has an infinite fold rotation axis (C_∞) and infinite number of vertical planes ($\infty\sigma_v$) passing through principal axis. Linear molecules can be rotated about its principal axis to any desired degree and have an infinite number of vertical planes (σ_v). e.g., HCl, HCN, NO, OCS, ICl, etc.

In general, molecules belonging to $C_{\infty v}$ point group are linear without center of symmetry (inversion element).

$$C_\infty + \infty\,\sigma_v = C_{\infty v}$$

2.2.4 *DIHEDRAL GROUPS*

Molecules having n two-fold axis (n C_2) perpendicular to the principal axis (C_n) belong to the dihedral groups, i.e., (n $C_2 \perp C_n$) and C_2 axis $\perp$ to C_n is called dihedral axis. The combination of D_n group is:

$$E + C_n + n\,C_2 \perp C_n = D_n \ (n = 2\text{–}6)$$

2.2.4.1 D_n Groups

Molecules with D_n group do not have plane of symmetry or mirror plane. Examples of D_2 point group is skew ethylene while tris(ethylenediamine) cobalt (III) cation belongs to D_3 point group.

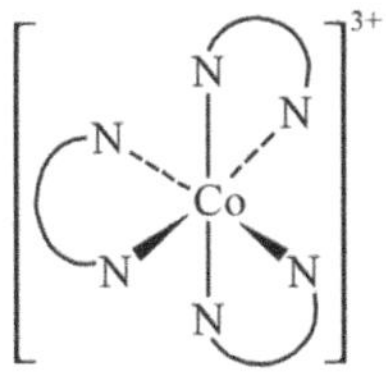

2.2.4.2 D_{nd} Group

When plane of symmetry contains the principal axis and bisects the angle between two adjacent C_2 axes, it is said to be a dihedral plane. The presence of σ_d (dihedral plane) operation in D_n group gives the D_{nd} point group. Examples of D_{nd} group are:

Staggered conformation of ethane (D_{3d}), staggered conformation of ferrocene (D_{5d}), S_8 molecule (D_{4d}), etc.

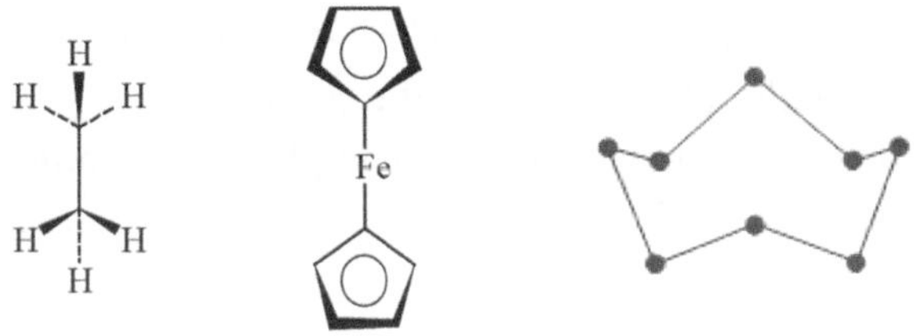

The combination of operation in this case is:

$$E + C_n + n\,C_2 \perp C_n + n\,\sigma_d = D_{nd}$$

2.2.4.3 D_{nh} Group

In D_{nd} point group, there are only D_{2d} and D_{3d} because from D_{4d} onwards to D_{6d}, σ_h is also present and therefore, the point group becomes D_{nh}. It means, when D_n group have σ_v plane in addition to σ_h, then these are considered under D_{nh} group. The combination of D_{nh} group is:

$$E + C_n + n\,C_2 \perp C_n + \sigma_h + \sigma_v = D_{nh}$$

Examples are BF_3, XeF_4, XeF_5, Benzene, etc.

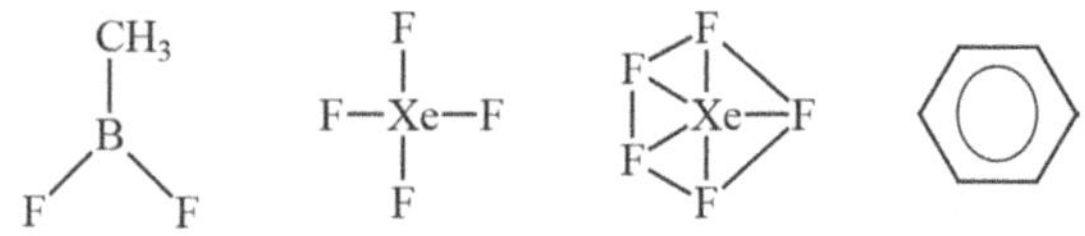

D_{3h} D_{4h} D_{5h} D_{6h}

2.2.4.4 $D_{\infty h}$ Group

Linear molecules, which have center of symmetry along with an infinite number of C_2 axes ⊥ to the principal axis and also horizontal plane, fall in point group $D_{\infty h}$. e.g., O_2, O = N = O, Br_2, BeF_2, H_2, etc.
The symmetry elements are $\infty C_\infty + \infty C_2 + \sigma_h + \infty\sigma_v + i + \infty S_\infty + E$.

2.2.5 GROUP WITH IMPROPER AXIS OF SYMMETRY (S_n)

All the groups discussed earlier are group with proper axis of symmetry. But there are molecules, which do not superimpose on their mirror image; these are referred as dissymmetric molecules. Such molecules posses improper axis of symmetry.

In S_n point group, n is the order of rotation axis. If S_n axis (n = even) exists in the molecule, then it implies the presence of $C_{n/2}$ axis in the molecule independently, collinear with S_n. The plane of symmetry (σ) perpendicular to $C_{n/2}$ or S_n axis will not be present independently. e.g., CH_4.

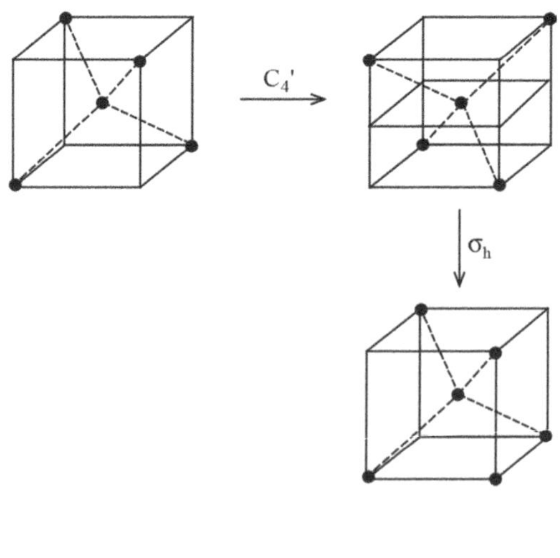

Methane molecule possess S_4 axis, which passes through C_2 axis and it also has another 3 C_2 axes. Thus, 3 S_4 axes exist in one CH_4 molecule. Symmetry operations for S_4 axis are:

$S_4^1 = C_4^1.\sigma_h$ (New operation)
$S_4^2 = C_4^2.\sigma_h^{\,2} = C_2^{\,1}.E = C_2$
$S_4^3 = C_4^3.\sigma_h^{\,3} = C_4^{\,3}.\sigma_h$ (New operation)
$S_4^4 = C_4^4.\sigma_h^{\,4} = E.E = E$

S_4^2 and S_4^4 operations do not result into new operation. Therefore, only S_4^1 and S_4^3 symmetry operations of S_4 are considered.

When S_n (n = odd) is present in the molecule, then horizontal plane of symmetry (σ_h) will also be present independently and total operation will be 2n. Examples of this point group are Gauche ethane, naphthalene, etc. Let us take example of BF_3 for finding symmetry operation of S_3 axis.

BF_3 molecule has S_3^1, S_3^2, S_3^3, S_3^4, S_3^5 and S_3^6 operations. Among these, only two symmetry operations are considered. They are S_3^1, and S_3^5 because $S_3^2 = C_3^2$, $S_3^3 = E$, $S_3^4 = C_3$, $S_3^6 = E$, which are already counted in symmetry operation of BF_3 molecule.

Thus, it can be concluded that as C_n and σ_h commute, and we obtain $S_n = C_n.\sigma_h$ for the general rotoreflection operations. When order of axis n (in S_n) is even, then S_n axis requires simultaneous independent existence of n $C_{n/2}$ rotation axis and inversion center. Whereas, in case of n = odd, rotoreflection axis requires independent existence of a C_n axis and a σ_h symmetry element.

Common point groups with their symmetry elements are:

Point group		Symmetry elements
C_1		E
C_s		$E + \sigma_h$
C_i		$E + i$
C_n		$E + C_n$
D_n	n — Even	$E + C_n + {}^{n}/_{2}\perp C_2^1, {}^{n}/_{2}\perp C_2^2$
	n — Odd	$E + C_n + n \perp C_2$
C_{nV}	n — Even	$E + C_n + {}^{n}/_{2}\sigma_v + {}^{n}/_{2}\sigma_d$
	n — Odd	$E + C_n + n\,\sigma_v$

CONTINUED

Point group		Symmetry elements
C_{nh}	n — Even	$E + C_n + \sigma_h + S_n + i$
	n — Odd	$E + C_n + \sigma_h + S_n$
D_{nh}	n — Even	$E + C_n + \sigma_h + {}^{n}/_{2} \perp C_2^{1} + {}^{n}/_{2} \perp C_2^{2} + S_n + {}^{n}/_{2}\sigma_v + {}^{n}/_{2}\sigma_d + i$
	n — Odd	$E + C_n + \sigma_h + n \perp C_2 + S_n + n\,\sigma_v$
D_{nd}	n — Even	$E + C_n + n\,C_2^{1} + S_n + n\,\sigma_d$
	n — Odd	$E + C_n + n \perp C_2 + S_{2n} + n\,\sigma_d + i$
S_n	n = Even only	$E + S_n + C_{n/2}$ and i, if ${}^{n}/_{2}$ is odd
T		$E + 4\,C_3 + 3\,C_2$
T_d		$E + 4\,C_3 + 3\,C_2 + 3\,S_4 + 6\,\sigma_d$
T_h/T_i		$E + 4\,C_3 + 3\,C_2 + 4\,S_n + i + 3\,\sigma_h$
O		$E + 3\,C_4 + 4\,C_3 + 6\,C_2$
O_h		$E + 3\,C_4 + 4\,C_3 + 6\,C_2 + 4\,S_6 + 3\,S_4 + i + 3\,\sigma_h + 6\,\sigma_d$
I		$E + 6\,C_5 + 10\,C_3 + 15\,C_2$
I_h		$E + 6\,C_5 + 10\,C_3 + 15\,C_2 + i + 6\,S_{10} + 10\,S_6 + 15\,\sigma$
K_h		E, ∞ number of all symmetry elements

2.3 DETERMINATION OF POINT GROUP

The following sequence of steps will decide the point group of a molecule.

Step 1: First step is to determine, whether the molecule belongs to any special group ($C_{\infty v}$, $D_{\infty h}$) or multiple high-order axis. It is simple to identify molecule, which belongs to $C_{\infty v}$ or $D_{\infty h}$ because only linear molecule belong to these groups. If molecule is not linear, then we must look for high symmetry groups, which include cubic group such as T, T_h, T_d, O and O_h. They require 4 C_3 axes, whereas I_h need 10 C_3 and 6 C_5. Therefore, these multiple C_3 and C_5 axes are the key things to look for I_h group, but only C_3 axes for other cubic groups.

All linear molecules belong to either $C_{\infty v}$ or $D_{\infty h}$ group depending upon whether a center of symmetry is present ($D_{\infty h}$ group) or absent ($C_{\infty v}$ group).

All the cubic groups T, T_h, T_d and O_h require four C_3 axes while I and I_h require ten C_3's and six C_5's. If the molecule appears to belong to T_d (CH_4),

O_h $[Fe(CN)_6]^{4-}$ and I_h [Boron compounds $(B_6H_{12}^{2-})$], it has to be confirmed as one is not sure that a molecule belongs to that particular group until one has verified it element by element that all the required symmetry elements are indeed present in the molecule.

Step 2: If the molecule does not belong to any of the special group as mentioned above, then, we look for whether any other axis of symmetry is present? If it has only proper or improper axis of symmetry, then the point group of the molecule is searched by Step 3. If no axis of symmetry is present, then we look for a plane of symmetry or an inversion center. If only a plane of symmetry is present, then the point group is C_s. If only a center of symmetry is present in the molecule, then it belongs to the group C_i. If no symmetry element exists, then the molecule belongs to the trivial group C_1, which contains only the identity operation.

Step 3: If an improper axis of even-order is present in the molecule (S_4, S_6 or S_8 is commonly present), but no plane of symmetry or any proper axis except a collinear one (or more); whose presence is automatically required by the presence of improper axis, then the point group of the molecule is S_4, S_6, S_8.... The presence of S_4 axis requires a C_2 axis and S_6 axis requires a C_3 axis and S_8 axis requires C_4 and C_2 axes. Here, it should be noted that S_n (n is even) group consists exclusively the operations generated by the S_n axis. If any additional operation is possible, then the molecule will belong to D_n, D_{nd} or D_{nh} group. Molecules belonging to these groups are relatively rare.

Step 4: If the molecule has more than one axes of rotation, then we have to locate some axis of highest order. This axis of highest order is termed as principal axis. Sometimes, there may not be any such unique principal axis of highest order. In such a case, we look, if one of them is geometrically unique in some sense; for example in allene, there are three C_2 axes. All the three axes are two fold axes and therefore, an axis of highest order cannot be decided. In such a case, it is observed that two C_2 axes pass through one carbon atom only while third C_2 axis passes through three carbon atoms. Hence, it is unique in nature as compared to other two C_2 axes that it passes through maximum number of atoms. Therefore, it can be designated as principal axis.

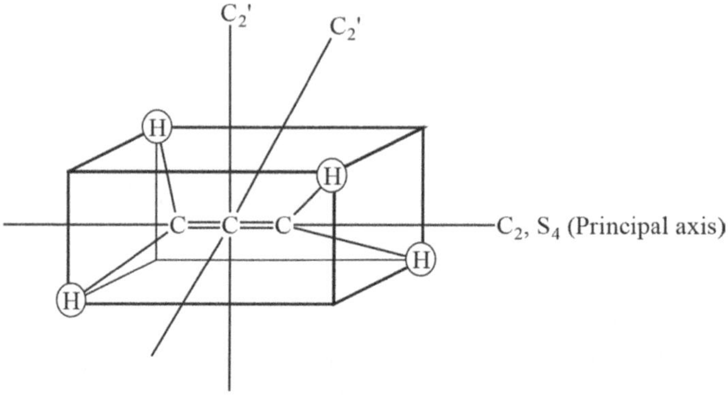

Now we look, whether a set of n C_2 axes perpendicular to the C_n axis is present or not? If yes, then proceed to Step 5. If not, then search for one of these groups C_n, C_{nv} or C_{nh}. If there are no symmetry elements other than the C_n axis, then the point group of the molecule is C_n.

Step 5: If in addition to the principal axis C_n, n C_2 axes are present perpendicular to the C_n axis, then the molecule belongs to one of the groups D_n, D_{nh} or D_{nd}. If there are no symmetry elements other than C_n and n C_2 axes, then the molecule belongs to D_n group. If there is a horizontal plane of symmetry in the molecule along with C_n and n C_2 axes, then the point group is D_{nh}. This D_{nh} group necessarily contains n vertical planes also. If n vertical planes (σ_v) are present along with C_n, then point group of molecule is C_{nv}. The presence of horizontal plane (σ_h) along with C_n, gives point group C_{nh}; however, it may possess vertical planes also.

The five-step procedure may be explained as follows:

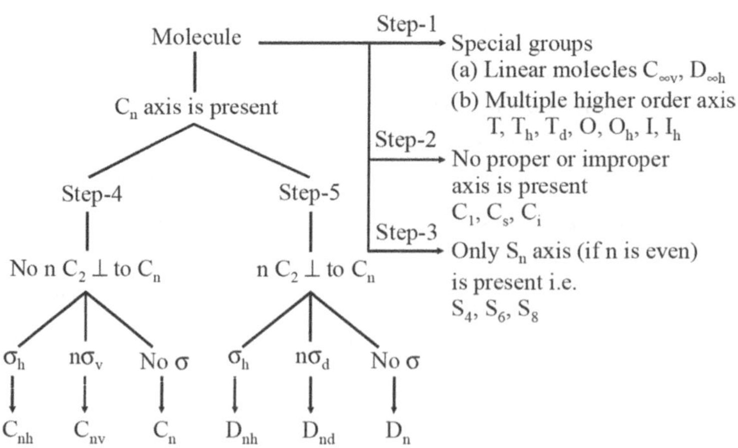

An alternate method of determination of point group is also there. One can easily find out the point group of a molecule by going in a sequence by replying certain simple questions, like whether a particular symmetry element is present or not?

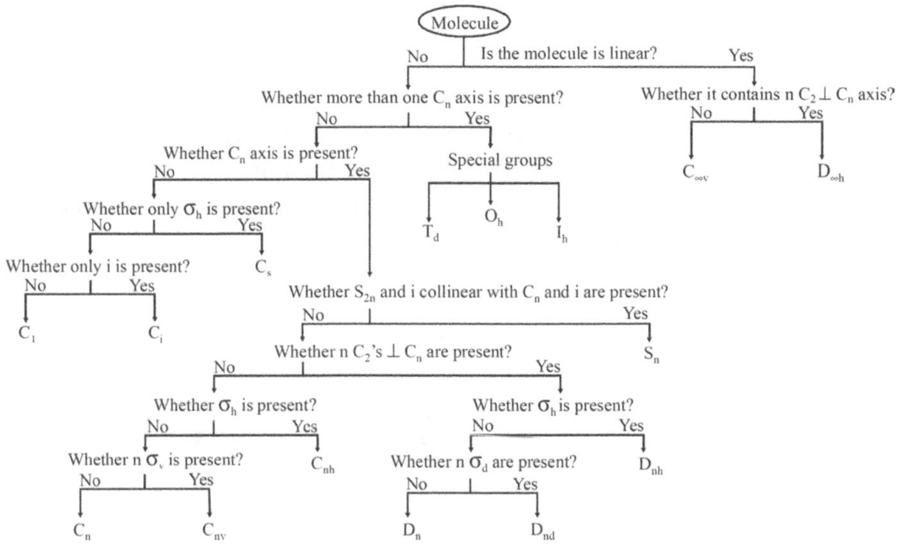

2.4 CHANGE IN POINT GROUP

The point group of a molecule changes on variation in its symmetry. If an atom B is replaced by another atom X one by one in a molecule belonging to a particular symmetry, then its point group will also change based on its new symmetry. Some examples will clarify it.

(i) If B atoms of a triangular planar molecule AB_3 (D_{3h} group) are exchanged by X one by one, then it changes its symmetry as well as point group as AB_2X and then ABX_2. These molecules belong to group C_{2v}.

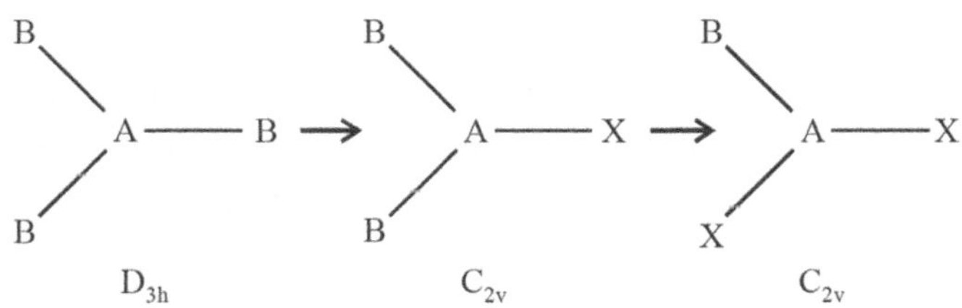

(ii) AB_3 molecule is pyramidal and it changes its point group for C_{3v} to C_s by replacing B atoms by X atom one by one.

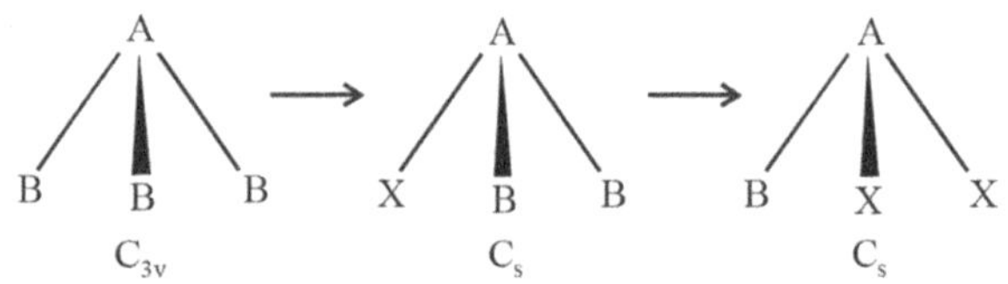

(iii) If B atoms of a tetrahedral molecule AB_4 (T_d group) are exchanged in sequence by X atoms, one by one; then, it changes its symmetry from T_d group to C_{3v} (in case of AB_3X and ABX_3) and then C_{2v} (in case of AB_2 X_2).

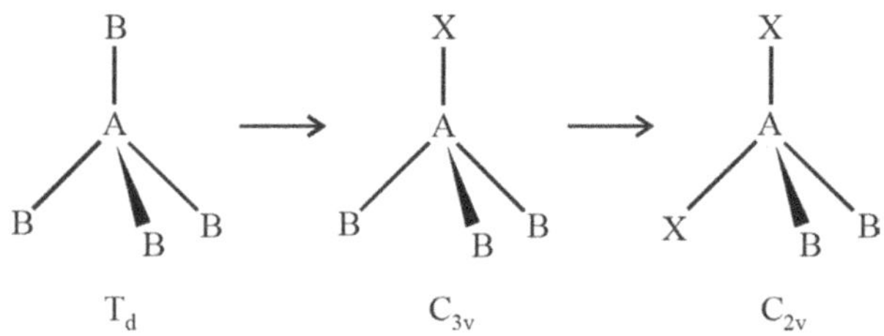

(iv) If B atoms of a square planar molecule AB_4 (D_{4h}) are exchanged by X atoms, one by one, then it changes its symmetry as well as point group from D_{4h} to C_{2v} and D_{2h} in different cases.

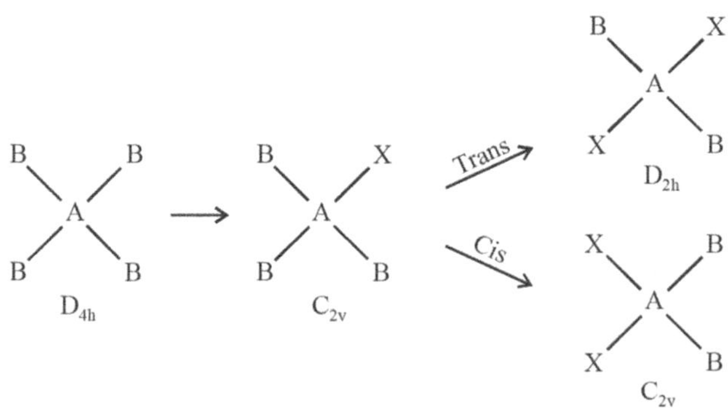

(v) If B atoms of a square pyramidal molecule AB_4 (C_{4v}) are exchanged by X atoms, one by one, then it changes its symmetry as well as point group from C_{4v} to C_s and C_{2v} in different cases.

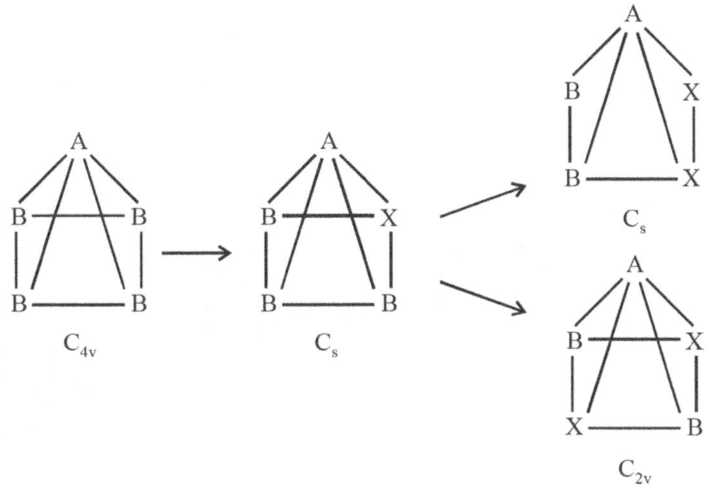

(vi) If B atoms of a square pyramidal molecule AB_5 (C_{4v}) are exchanged by X atoms, one by one, then it changes its symmetry as well as point group from C_{4v} to C_{2v} and C_s in different cases.

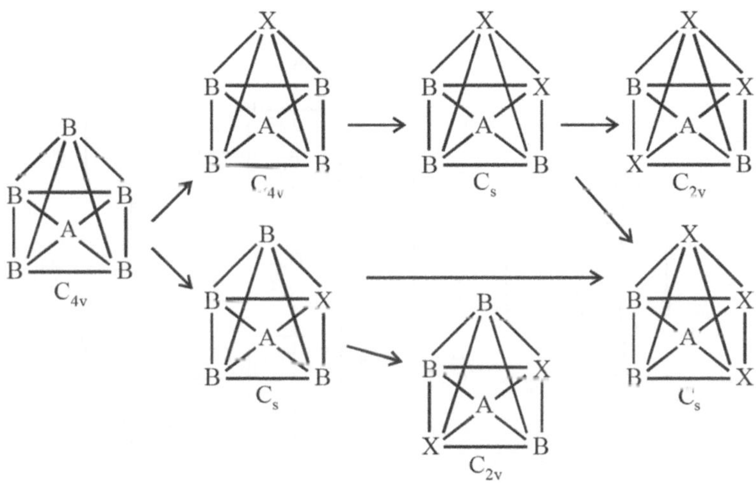

(vii) If B atoms of a trigonal bipyramidal molecule AB_5 (D_{3h}) are exchanged by X atoms, one by one, then it changes its symmetry as well as point group from D_{3h} to C_{3v}, C_{2v}, D_{3h}, C_s and D_{3h} in different cases.

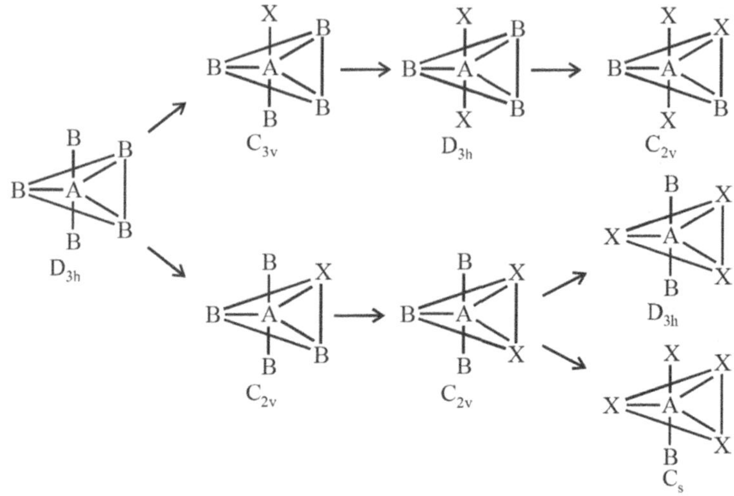

(viii) If B atoms of an octahedral molecule (AB_6) (O_h) are exchanged by X atoms, one by one, then the molecule AB_6 changes its symmetry as well as point group for O_h to C_{4h}, C_{2v}, C_{3v} and D_{4h} in different cases.

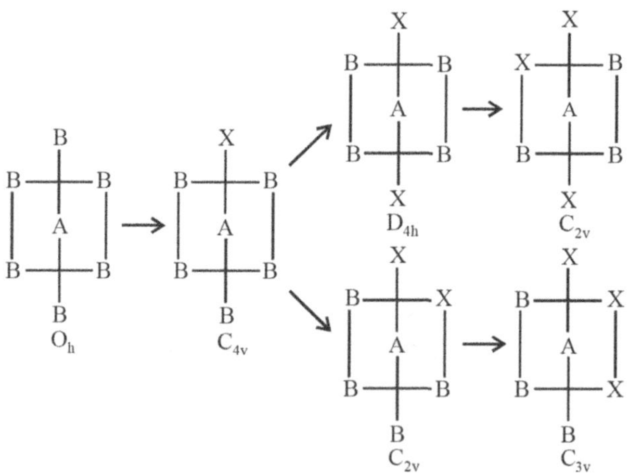

(ix) In case of benzene (D_{6h} group), if the H-atoms at different positions are substituted by X in various combinations, then it also changes its symmetry as well as point group from D_{6h} to $C_{2v,}$ D_{2h} and C_s points groups in different mono-, di- and tri- substituted benzenes.

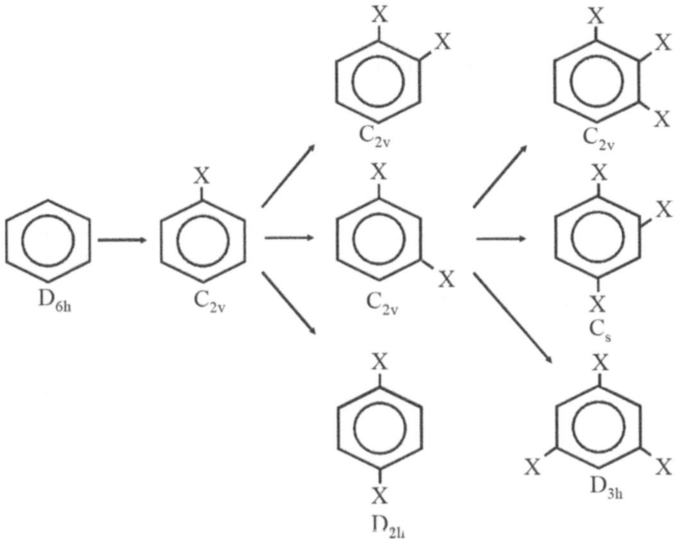

On the basis of this classification, point groups of different molecules/ions may be determined. Some examples of molecules with their point group, order of the group and symmetry elements are given in Table 2.1.

TABLE 2.1 Different Point Groups

Point group	Order of group	Symmetry elements	Example
C_1	1	E	$CH_3CH(OH)COOH$, CHFClBr, SiBrClFI, NbF_5, $TeCl_2Br_2$
C_2	2	$E + C_2$	Non-planar H_2O_2, F_2O_2, Gauche CH_2Cl-CH_2Cl, Cis-$[Co(en)_2Cl_2]^+$
C_3	3	$E + C_3^1 + C_3^2$	CH_3CF_3, PPh_3
C_s	2	$E + \sigma_h$	2-Bromonaphthalene, HOCl, $POBrCl_2$, HCOCl, CH_2ClBr, 1,2-Benzpyrene
C_i	2	$E + i$	CHFCl – CHFCl
C_{2v}	4	$E + C_2 + 2\sigma_v$ $\sigma_v(xz) + \sigma_v(yz)$	H_2O, SO_2, CH_2Cl_2, ClF_3, SO_2Cl_2, $SiCl_2Br_2$, Cis-CHCl = CHCl, $BClF_2$ Cyclohexane (Boat form), Pyridine, 2-Butene (Cis), C_6H_5X, $C_6H_4X_2$ (Ortho & meta), Cyclopentadiene, Cis-$[Pt(NH_3)_4Cl_2]^{2+}$, Cis-$[Pt(NH_3)_2$, Cis-$[Co(py)_2Cl_2]$, Cis-H_2O_2

TABLE 2.1 Continued

Point group	Order of group	Symmetry elements	Example
C_{3v}	4	$E + C_3 + 3\,\sigma_v$	NH_3, PH_3, PCl_3, $CHCl_3$, $POCl_3$, CH_3Cl, PF_4Cl
C_{4v}	8	$E + C_4 + 2\,\sigma_v + 2\,\sigma_d + C_2\,(C_4^{\,1} + C_4^{\,3})$	SF_5Cl, $Mn(CO)_5$, $[Co(NH_3)_4Cl(H_2O)]^{2+}$ Square pyramidal AB_4/AB_5, SbF_5, WOF_4
$C_{\infty v}$	∞	$E + C_{\infty v} + \infty\,\sigma_v$	HCl, CO, HCN, OCS, HBr, NO
C_{2h}	4	$E + C_2 + \sigma_h + i$	Trans-CHCl=CHCl, Trans-H_2O_2, Trans- −2-butene; 1,4-Difluoro. 2,5-dichlorobenzene, Glyoxal, P_2F_4, 1,1, 2,2-Tetrabromoethane (Staggered), $(Cu_2Cl_8)^{4-}$
C_{3h}	6	$E + C_3 + \tilde{A}_h + S_3\,(C_3^{\,1} + C_3^{\,2})$ $(S_3 + S_3^{\,5})$	H_3BO_3 (planar), Bicyclo [3.3.3] undecane
D_2	4	$E + C_2$	Skew ethylene, Skew biphenyl
D_3	6	$E + C_3 + 3C_2$	Gauche ethane
D_{2h}	8	$E + 3\,C_2 + 3\,\sigma + i$ [C_2 (*x*), C_2(*y*), C_2 (*z*)] [σ (xy), σ (xz), σ (yz)]	Naphthalene, Ethylene, Trans-$[Pt(NH_3)\,Cl_2]$ N_2O_4 (Planar), $C_6H_4X_2$ (para), C_2F_4, B_2H_6, 2, 2-Cyclophane
D_{3h}	12	$E + C_3 + 3\,C_2 \perp C_3 + 3\,\sigma_V + \sigma_h + 2\,S_3$ $(C_3^{\,1} + C_3^{\,2})$	BF_3, PF_5, C_2H_6 (Eclipsed), Tribromobenzene (Planar), Mesitylene, CO_3^{2-}, NO_3^-, Borazole, 1,4-Diazabicyclo [2.2.2] octane.
D_{4h}	16	$E + 2\,C_4 + 4\,C_2 \perp C_4 + 4\,\sigma_v + \sigma_h +$ C_2 and 2 S_4 (Colinear with C_4) + i $(2\,C_2' + 2\,C_2'')\,(2\,\sigma_v + 2\,\sigma_d)$	Cyclobutane, $PtCl_4^{2-}$, $Ni(CN)_4^{2-}$, Trans-$[Co(NH_3)_4Cl_2]^+$, Square planar AB_4 $[Ni\,(NH_3)_4]^{2+}$
D_{5h}	20	$E + 2\,C_5 + 5\,C_2 \perp C_5 + \sigma_h + 5\,\sigma_v +$ C_2 and S_5 (Contained within C_5) $(2\,C_5 + 2\,C_5^{\,2})\,(2\,S_5 + 2\,S_5^{\,3})$	Cyclopentane, Ferrocene (Eclipsed), Cyclopentadiemide ion
D_{6h}	24	$E + C_6 + 6\,C_2 \perp C_6 + 6\,\sigma_v + \sigma_h + C_2$ and S_6 (with C_6) + i $(2\,C_6)\,(2\,C_3)$ $(3\,C_2' + 3\,C_2'')\,(2\,S_3)\,(2\,S_6)\,(3\,\sigma_v +$ $3\,\sigma_d)$	Benzene, Dibenzenechromium (Eclipsed)
D_{2d}	8	$E + 3\,C_2$ (Mutually ⊥ + 2 S_4 (with one C_2) + 2 σ_d	Allene, Biphenyl, Cycloctatetraene twisted form.
D_{3d}	12	$E + 2\,C_3 + C_2 \perp C_3 + S_6$ (with C_3) + i + 3 σ_d (2 S_6)	Cyclohexane, Ethane (Staggered)

TABLE 2.1 Continued

Point group	Order of group	Symmetry elements	Example
$D_{\infty h}$	∞	$E + C_{\infty} + \infty\, C_2 \perp C_{\infty} + \infty\, \sigma_v + \sigma_h + i$	H_2, Br_2, $CH \equiv CH$, CO_2, $BeCl_2$, XeF_2
T_d	24	$E + 8\, C_3 + C_2 + 6\, S_4 + 6\, \sigma_d$	$Ni\,(CO)_4$, CH_4, CCl_4, $SiCl_4$ $[Zn(CN)_4]^{2-}$
O_h	48	$E + 8\, C_3 + 6\, C_2 + 6\, C_4 + 3\, C_2\, (= C_4^{\,2}) + i + 3\, \sigma_h + 6\, S_4 + 8\, S_6 + 6\, \sigma_d$	PCl_6^-, SF_6, $[Co(NH_3)_6]^{3+}$, $[PtCl_6]^{2-}$, $IrCl_6^{2-}$ Cubane (C_8H_8)
I_h	120	$E + 12\, C_5 + 12\, C_5^2 + 20\, C_3 + 15\, C_2 + i + 12\, S_{10} + 12\, S_{10}^3 + 20\, S_6 + 15\, \sigma$	Dodecaborane ($B_{12}H_{12}^{2-}$), Dodecahedrane $(CH)_{20}$

KEYWORDS

- **Cubic**
- **Dihedral**
- **Octahedral**
- **Point group**
- **Tetrahedral**

CHAPTER 3

GROUP THEORY

CONTENTS

Molecules having all the symmetry elements similar are placed in same group and these groups are termed as point group. For example, H_2O, SO_2, both the molecules are structurally different, but have same four symmetry elements, i.e., E, C_2, 2 σ_v. Therefore, these are assigned the same point group, i.e., C_{2v}. A complete set of symmetry operations follows some mathematical group rules. A mathematical group is a collection of elements, which are interrelated according to rules of group theory.

3.1 RULES OF GROUP THEORY

Group theory has four rules. These are: (1) identity rule; (ii) closure rule; (iii) associative law of multiplication; and (iv) inverse rule.

3.1.1 *IDENTITY RULE*

Group must have one element, which commute with all other elements of that group and leave them unchanged. Such an element is known as the identity element (E). Mathematically,

$$X.E \text{ or } E.X = X$$

The combination of E with X element in any order gives element X. For example C_2 operation followed by E operation or vice-versa gives:

$$E.C_2 = C_2 \text{ or } C_2.E = C_2$$

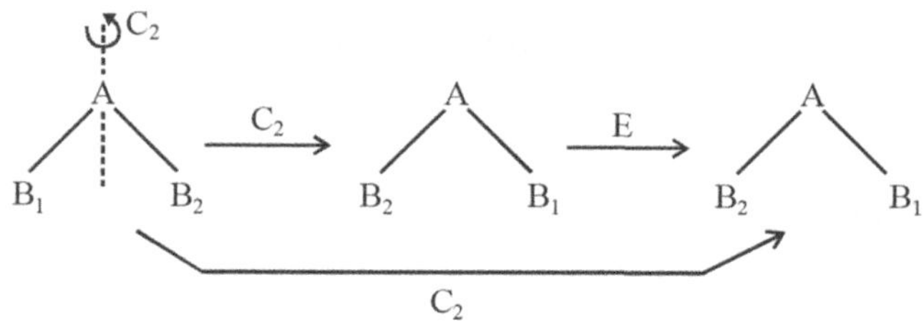

3.1.2 *CLOSURE RULE*

The combination (or multiplication) of any two elements in the group must result into an element (symmetry operation), which is also member of the same group.

AB_2 molecule is operated upon by C_2 and σ_v (xz) element (operation), then we get σ_v (yz) as the resultant product.

Operation

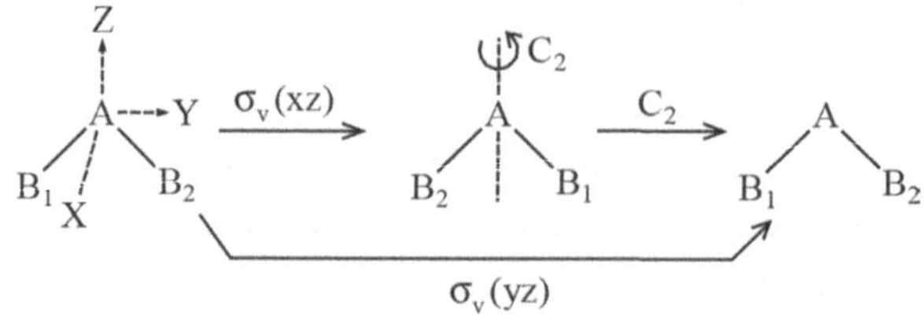

3.1.2.1 Cyclic Group

When all the elements of a group can be produced from one element, then the group is termed as a cyclic group of n order. The C_n and D_n point group are examples of such cyclic groups.

$$X, X^2, X^3, \ldots\ldots X_n^{\,n} = E$$

3.1.2.2 Abelian Group

A group is said to be Abelian, if all the elements commute with each other, i.e., follows commutative law. e.g., AB_2 has C_{2v} point group with E + 1 C_2 + 2 σ_v symmetry operations. We see that combination of these elements do commute, i.e., A.B = B.A.

$$C_2.\sigma_v(xz) = \sigma_v(xz)\, C_2$$

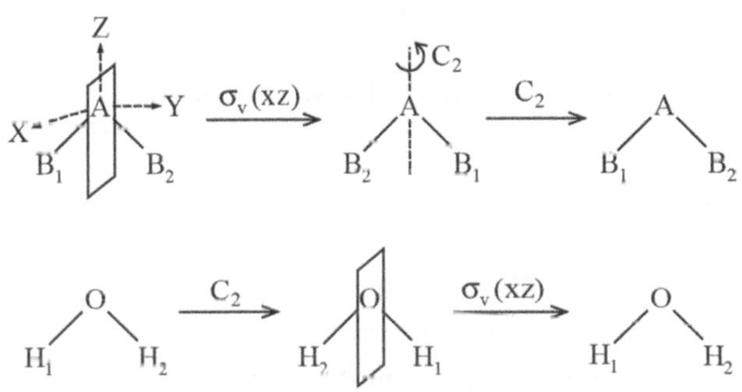

3.1.3 ASSOCIATIVE LAW OF MULTIPLICATION

The elements of a mathematical group should obey this law of association:

$$A.\,(BC) = (AB).\,C$$

$$C_2\,(\sigma_v(yz).\,\sigma_v(xz)) = (C_2.\,\sigma_v(yz).\,\sigma_v(xz))$$

L. H. S. =

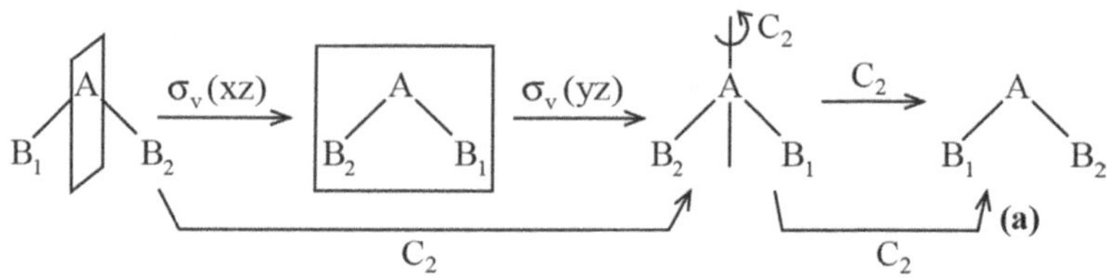

R. H. S. =

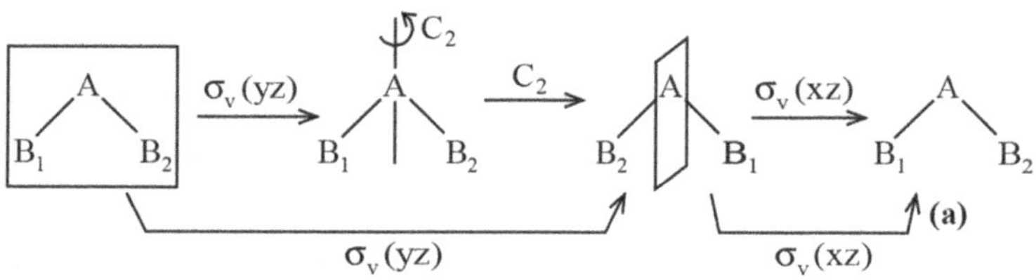

Thus, $C_2 (\sigma_v (yz). \sigma_v (xz)) = (C_2. \sigma_v (yz)). \sigma_v (xz)$

$$C_2.C_2 = \sigma_v (xz). \sigma_v (xz)$$

$$E = E$$

Resultant products on both the sides are identical.

3.1.4 INVERSE RULE

In a mathematical group, all the elements should have a reciprocal to itself, which is also a member of the same group. The combination of that element and its reciprocal results into the identity element (E) of the group.

Mathematically,

$$X. X^{-1} = X^{-1}. X = E$$

$$C_3^+. C_3^- = E$$

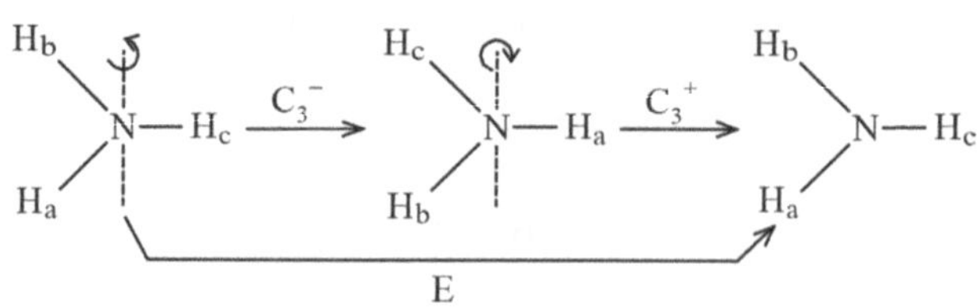

In case of two or more elements (operations), the reciprocal of elements are equal to product of the reciprocal in the reverse order.

$$(XYZ)^{-1} = Z^{-1}.\,Y^{-1}.\,X^{-1}$$

The number of elements in a finite group is known as the order of the group (h).

As a convention, multiplication of elements in a combination is done from right to left.

These rules can be applied on different molecules like water and ammonia molecules.

3.2 H_2O MOLECULE

H_2O molecule has four symmetry elements. These are:

$$C_{2v} = \{E, C_2, \sigma_v(xz), \sigma_v(yz)\}$$

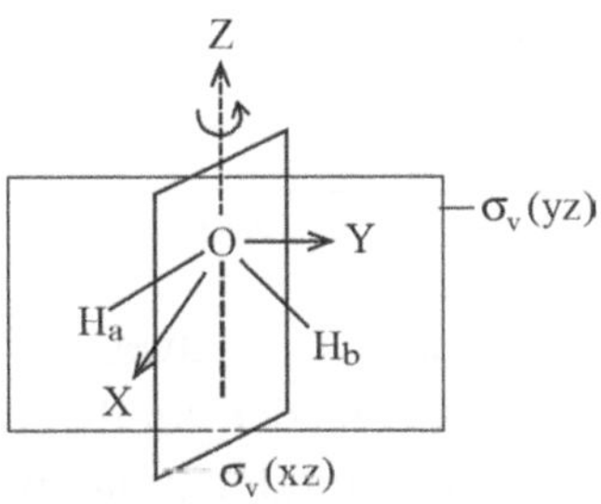

Now C_2 axis of symmetry generates one distinct symmetry operation, i.e., C_2 only, because C_2^2 equal to E.

Rule 1: Identity rule

The presence of E is very much there in point group C_{2v}. E is an element, which on combining (right or left) with any other element of the group leave them unchanged.

$$E.C_2 = C_2.E = C_2$$

$$E.\sigma_v(xz) = \sigma_v(xz).E = \sigma_v(xz)$$

$$E.\sigma_v(yz) = \sigma_v(yz).E = \sigma_v(yz)$$

$$E.E = E^2 = E$$

Hence, every element of the group obeys this rule.

Rule 2:Closure rule

Product of any two elements of a group must be an element of that group.

$$C_2 . \sigma_v (xz) = \sigma_v (yz)$$

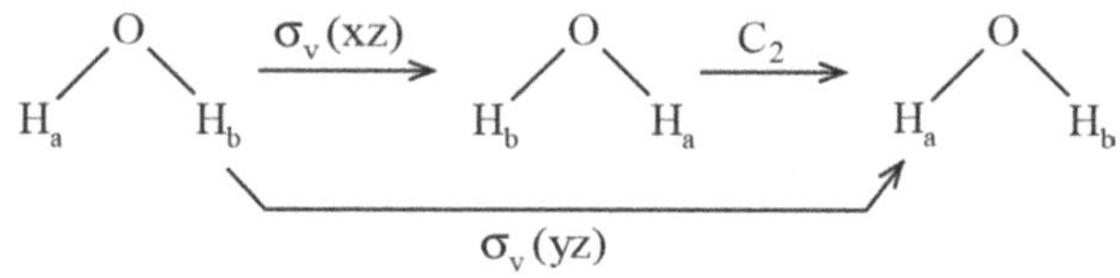

The resultant orientation by two successive symmetry operations $\sigma_v (xz)$ and C_2 can be directly obtained by a single operation of $\sigma_v (yz)$. Therefore, the result of the product $C_2 . \sigma_v (xz)$ is $\sigma_v (yz)$, which is also the element of this group. Same rule can be applied to any two elements of the group. Now let us take another combination of C_2 and $\sigma_v (yz)$.

$$C_2 . \sigma_v (yz) = \sigma_v (xz)$$

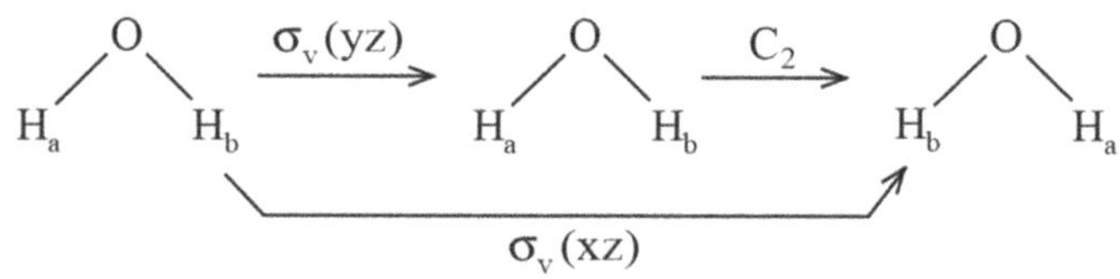

Thus, combination of C_2 and $\sigma_v (yz)$ gives $\sigma_v (xz)$, which is also an element of the C_{2v} group. Let us consider combination of planes but before applying it, one rule is to keep in mind that the plane designated initially in the configuration of the molecule will remain as such and it is not affected by the result of any operation. H_2O molecule has two planes, $\sigma_v (xz)$ and $\sigma_v (yz)$. So, on performing $\sigma_v (xz). \sigma_v (yz)$ combination, first $\sigma_v (yz)$ is performed on initial configuration followed by $\sigma_v (xz)$. The position of $\sigma_v (xz)$ remained same as designated initially in the molecule.

$$\sigma_v (xz). \sigma_v (yz) = C_2$$

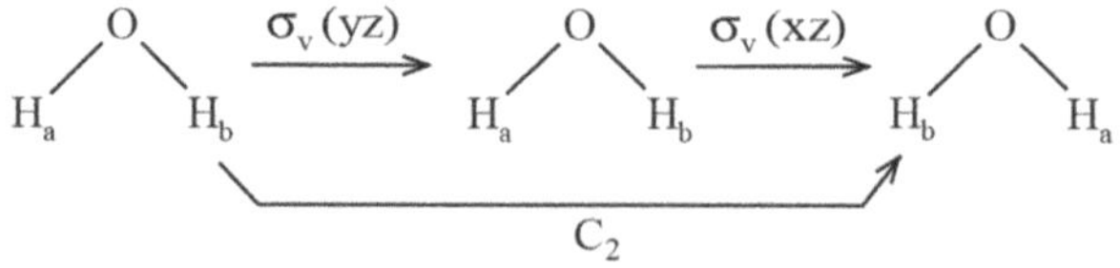

The combination of these two planes gives C_2 element, which is also element of the same group.

Thus, each elements of this point group obeys closure rule.

Rule 3:Associative law of multiplication

Hence $$C_2(\sigma_v(xz).\sigma_v(yz)) = (C_2.\sigma_v(xz)).\sigma_v(yz)$$
$$C_2.C_2 = \sigma_v(yz).\sigma_v(yz)$$
$$E = E$$

Hence, associative property is satisfied by the elements of this group.

Rule 4:Inverse rule

The product of any element and it inverse (reciprocal) is equal to identity element.

$$C_2.C_2^{-1} = C_2.C_2 = E$$
$$\sigma_v(xz).\sigma_v^{-1}(xz) = \sigma_v(xz).\sigma_v(xz) = E$$
$$\sigma_v(yz).\sigma_v^{-1}(yz) = \sigma_v(yz).\sigma_v(yz) = E$$

Hence, inverse rule is also followed by all the elements of this point group.

3.3 NH_3 MOLECULE

There are six symmetry operations present in ammonia molecule. These are E, C_3 (or C_3^+), C_3^- (or C_3^2), $\sigma_v(a)$, $\sigma_v(b)$, and $\sigma_v(c)$. Now, let us see, how these elements collectively from a mathematical group by obeying all the rules of a group theory?

Rule 1: Identity rule

$$E.E = E^2 = E$$
$$E.C_3^+ = C_3^+.E = C_3^+$$
$$E.C_3^- = C_3^-.E = C_3^-$$
$$E.\sigma_v(a) = \sigma_v(a).E = \sigma_v(a)$$
$$E.\sigma_v(b) = \sigma_v(b).E = \sigma_v(b)$$
$$E.\sigma_v(c) = \sigma_v(c).E = \sigma_v(c)$$

Hence, each element in combination with identity element E, remained unchanged. Thus, these obey identity rule.

Rule 2: Closure rule

$$C_3^+.\sigma_v(a) = \sigma_v(c)$$

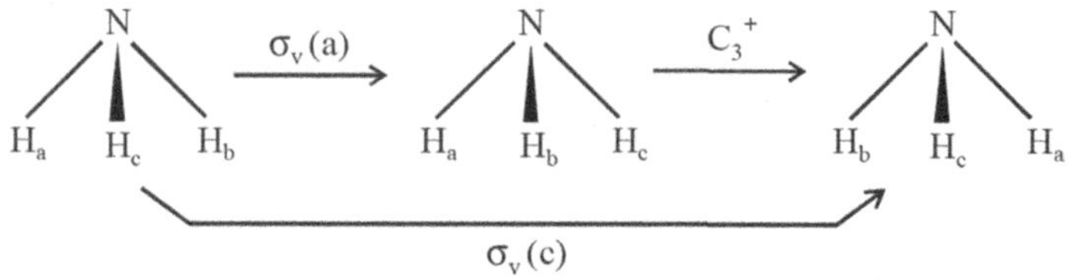

The configuration of NH_3 molecules obtained by two successive operations of C_3^+ and $\sigma_v(a)$, can be obtained directly by only $\sigma_v(c)$ operation. Hence, the combination of $C_3^+.\sigma_v(a)$ is $\sigma_v(c)$, which is also an element of the C_{3v} group.

Similarly $\sigma_v(a).\sigma_v(b) = C_3^+$ (or C_3^1)

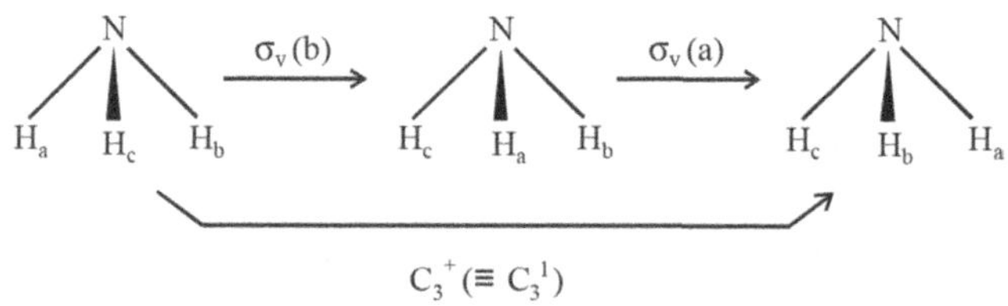

Therefore, $\sigma_v(a).\sigma_v(b)$ give C_3^1 or C_3^+, which is also element of this group. As mentioned earlier in case of planes in H_2O molecule, in point group C_{3v} also, $\sigma_v(b)$ is operated first, and then $\sigma_v(a)$ is performed, but in the same plane as designated initially $\sigma_v(a)$ in the molecule, and not the $\sigma_v(a)$ obtained after rotation.

Rule 3: Associative law of multiplication

Let us see, whether $\sigma_v(a)$, $\sigma_v(b)$ and $\sigma_v(c)$ followed associative law or not?

$$\sigma_v(a). (\sigma_v(b). \sigma_v(c)) = (\sigma_v(a). \sigma_v(b)). \sigma_v(c)$$

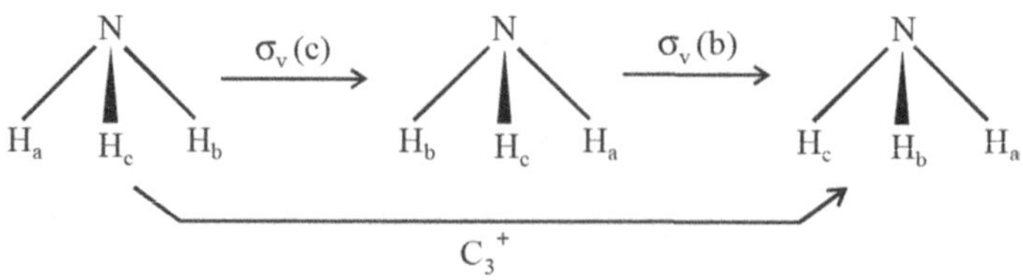

$$\sigma_v(a).\ (\sigma_v(b).\ \sigma_v(c))$$
$$= \sigma_v(a).\ C_3^+$$

After successive operation of $\sigma_v(b)$ and $\sigma_v(c)$, we obtain C_3^+. Now, this combination is used to obtain final result. First performing C_3^+ and then σ_v (a) on the molecule.

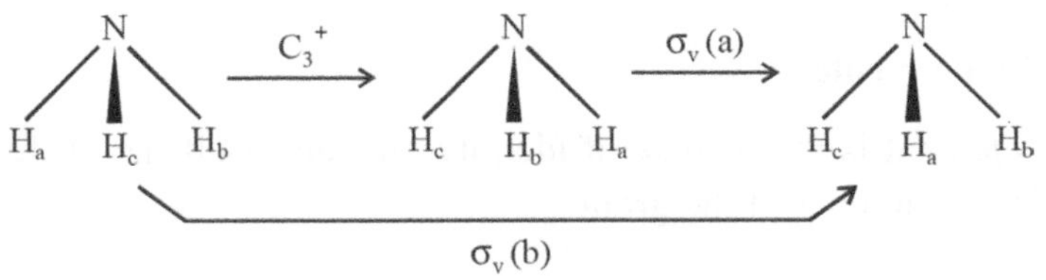

Over all, it may be written as:

$$\sigma_v(a)\ (\sigma_v(a).\ \sigma_v(c))$$
$$= \sigma_v(a).\ C_3^+$$
$$= \sigma_v(b)$$

On the other hand,

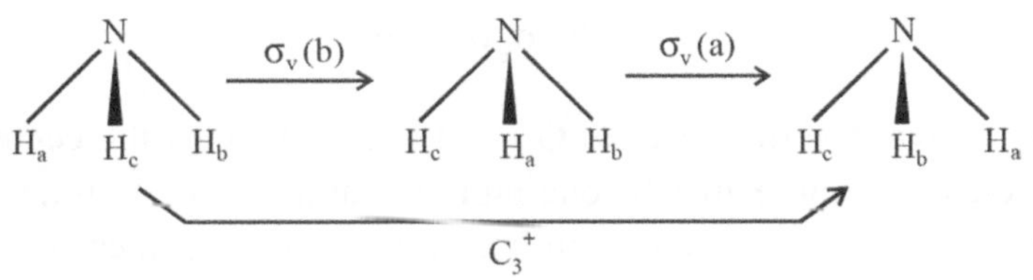

$$\sigma_v(a).\ \sigma_v(b) = C_3^+$$
$$(\sigma_v(a).\ \sigma_v(b)).\ \sigma_v(c)$$
$$= C_3^+.\ \sigma_v(c)$$

Now combination of $C_3^+.\ \sigma_v(c)$, is performed and we get,

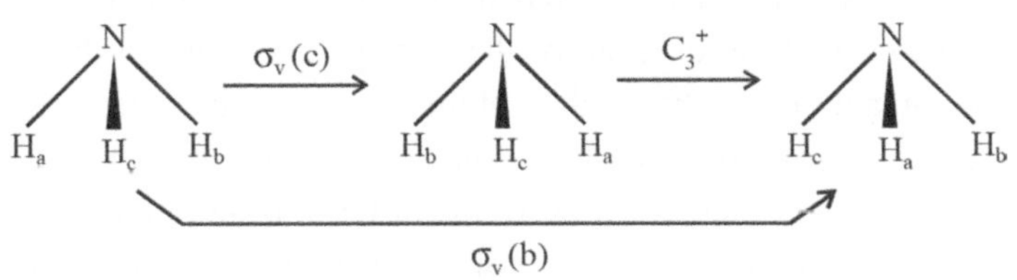

$$(\sigma_v(a).\ \sigma_v(b)).\ \sigma_v(c)$$
$$= C_3^+.\ \sigma_v(c)$$
$$= \sigma_v(b)$$

The over all result is same and therefore, elements of this group obey associative property.

Rule 4:Inverse rule

Identity element is the inverse of identity element itself, i.e., E.E = E, and thus, it is the member of the group.

$$C_3^-.C_3^+ = C_3^+.C_3^- = E$$

C_3^- is inverse of C_3^+. Their product gives element E in any order.

$\sigma_v(a)$, $\sigma_v(b)$ and $\sigma_v(c)$ are inverse of $\sigma_v(a)$, $\sigma_v(b)$ and $\sigma_v(c)$, respectively, and therefore, their multiplication will give identity. For example,

$$\sigma_v(a).\sigma_v(a) = E$$
$$\sigma_v(b).\sigma_v(b) = E$$
$$\sigma_v(c).\sigma_v(c) = E$$

From above description of H_2O (C_{2v}) and NH_3 (C_{3v}) molecules, it has become clear that molecules belonging to C_{2v} and C_{3v} point groups obey all rules of forming a mathematical group, In the same way, these rules can be applied to all other symmetry point groups.

3.4 GROUP MULTIPLICATION TABLES

If we have a complete and nonredundant list of h elements of a finite group, then the group is completely and uniquely defined. The forgoing information can be presented most conveniently in the form of the group multiplication table (GMT). This table consists of h rows and h columns. Each column as well as row is labeled with an element of that group. The entry in the table under a given column and row is the product of the element, which heads that column and that row. Here, multiplication is usually noncommutative, and therefore, we must have to follow certain rules for the order of multiplication. Arbitrarily, we shall take the factors in the order (column element) × (row element).

We can better explain these tables by rearrangement theorem. According to this theorem, "Each row and each column in the group multiplication table lists each of the group elements once and only once." In other words, we can say neither two rows nor two columns may be identical in this table. Thus, each row and column is a rearranged list of the group elements.

Let us explain these tables by constructing them one by one. Firstly, starting with smallest order of group, which is designated as G_1 and contains only one element, i.e., identity element E.

G_1	E
E	E

Then constructing the multiplication table for group G_2, which contains two elements, i.e., one is E and another is A.

G_2	E	A
E		
A		

A particular sequence must be maintained for writing these elements in the row and column. E must be the first element in both; the row and column, since it is a trivial element. Now multiplying in the order column × row and entering that product element at the cross section of the corresponding row and column. Thus, the multiplication table for G_2 group is represented as:

G_2	E	A
E	E	A
A	A	E

Any element remains unchanged, when multiplied with identity E and hence,

$$E.A = A.E = A$$

Element A is also inverse of itself.

$$A.A = E$$

There is no other way than this for writing a group multiplication table using the two elements E and A.

Similarly, a group with three elements E, A, and B will form group multiplication table of order G_3 in the following way:

G_3	E	A	B
E	E	A	B
A	A	B	E
B	B	E	A

If multiplication is commutative, then

$$A.B = B.A = E$$

These elements (A and B) do not have their inverse as themselves, but one element is the inverse of the other, so that:

$$A.A = B$$

and $$B.B = A$$

3.4.1 C_{2V} POINT GROUP

Let us take C_{2v} point group for explaining group multiplication table.

The top left corner has the symbol for the point group to which that molecule belongs. The rows and columns have all symmetry operations in sequence.

First operation to be per formed

Second operation to be performed

C_{2v}	E	C_2	σ_v (xz)	σ_v (yz)
E	*	*	*	*
C_2	*	*	*	*
σ_v (xz)	*	*	*	*
σ_v (yz)	*	*	*	*

If, we want to know the combination of $C_2.\sigma_v$ (xz), then first perform σ_v (xz) from the upper horizontal row followed by C_2 in the left vertical column. The product element (resultant) is placed at the junction of that row and column.

Similarly, all other combinations can be determined by putting the products obtained at crossing point (junction) of that row and column, i.e., column operation multiplied by row operation (column element × row element).

Let us explain this table by taking example of H_2O molecule having C_{2v} point group. C_{2v} point group possess four symmetry elements E, C_2, σ_v (xz), σ_v (yz). The number of symmetry elements and symmetry operations are same. Here, total number of symmetry elements is 4, which is also the order of this group. Now to construct the multiplication table, the symbol, C_{2v}, for point group is written in top left corner. Then on top row and first left column, all the four symmetry elements of the group are written.

C_{2v}	E	C_2	σ_v (xz)	σ_v (yz)
E	*	*	*	*
C_2	*	*	*	*
σ_v (xz)	*	*	*	*
σ_v (yz)	*	*	*	*

In the first row, the elements at star (*) are produced by multiplication of all symmetry elements by E on right side, i.e., E.E = E, C_2.E = C_2, σ_v (xz).E = σ_v (xz), σ_v (yz).E = σ_v (yz). These are represented in first row.

The products represented by star (*) are same in first column and these may be obtained by combination of any element followed by identity element (E). These are represented in first column.

C_{2v}	E	C_2	σ_v (xz)	σ_v (yz)
E	E	C_2	σ_v (xz)	σ_v (yz)
C_2	C_2	*	*	*
σ_v (xz)	σ_v (xz)	*	*	*
σ_v (yz)	σ_v (yz)	*	*	*

Other binary operations can be obtained in same manner as described earlier. Now, let us find product of the elements C_2 with C_2.

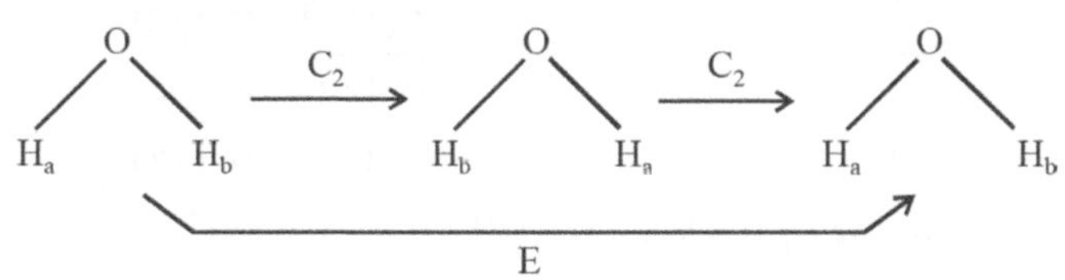

Therefore $C_2.C_2 = E$

Similarly, the product of $C_2.\sigma_v(xz)$ is:

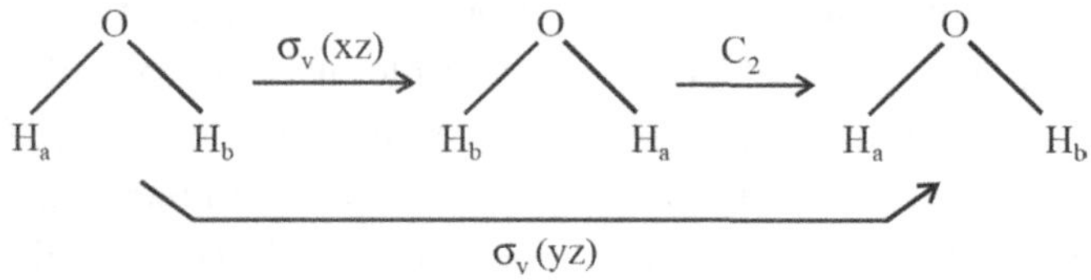

$C_2.\sigma_v(yz) = \sigma_v(xz)$

The product of $C_2.\sigma_v(yz)$ is:

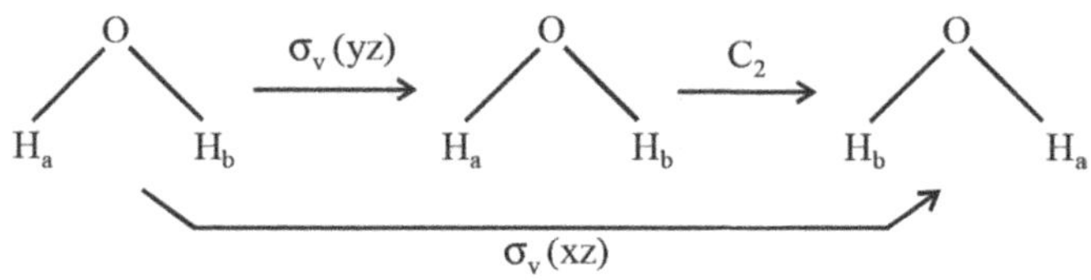

Hence, $C_2.\sigma_v(yz) = \sigma_v(xz)$

The product of $\sigma_v(yz).C_2$ is:

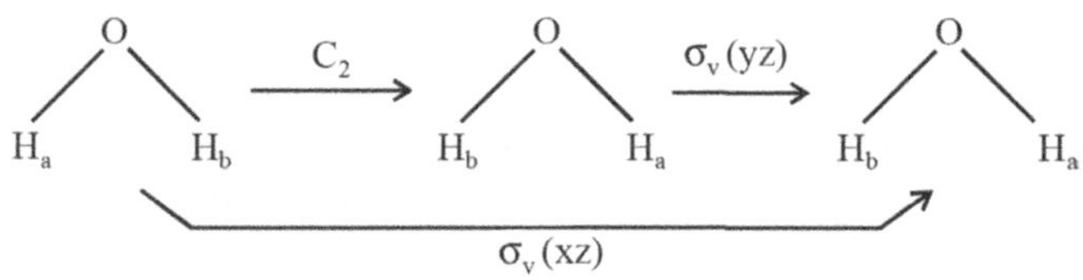

Hence, $\sigma_v(yz).C_2 = \sigma_v(xz)$

The product of $\sigma_v(xz).C_2$ is:

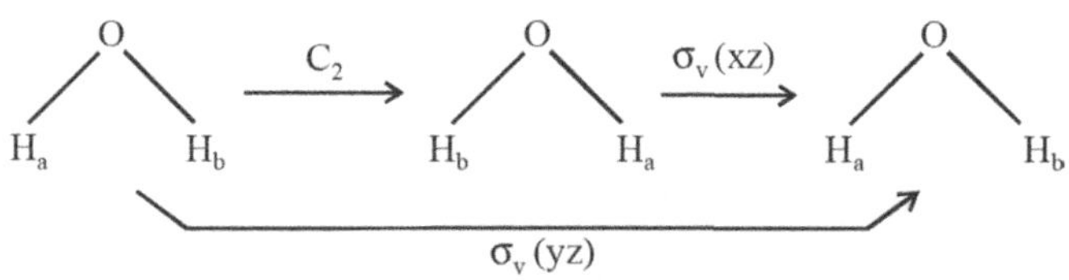

Hence, $\sigma_v(xz).C_2 = \sigma_v(yz)$

Putting these values in their respective places, we obtain:

C_{2v}	E	C_2	$\sigma_v(xz)$	$\sigma_v(yz)$
E	E	C_2	$\sigma_v(xz)$	$\sigma_v(yz)$
C_2	C_2	E	$\sigma_v(yz)$	$\sigma_v(xz)$
$\sigma_v(xz)$	$\sigma_v(xz)$	$\sigma_v(yz)$	*	*
$\sigma_v(yz)$	$\sigma_v(yz)$	$\sigma_v(xz)$	*	*

Similarly, products of remaining binary operations can also be determined as:

$$\sigma_v(xz).\sigma_v(xz) = E$$
$$\sigma_v(yz).\sigma_v(xz) = C_2$$
$$\sigma_v(xz).\sigma_v(yz) = C_2$$
$$\sigma_v(yz).\sigma_v(yz) = E$$

Thus, the complete multiplication table of point group C_{2v} is:

C_{2v}	E	C_2	$\sigma_v(xz)$	$\sigma_v(yz)$
E	E	C_2	$\sigma_v(xz)$	$\sigma_v(yz)$
C_2	C_2	E	$\sigma_v(yz)$	$\sigma_v(xz)$
$\sigma_v(xz)$	$\sigma_v(xz)$	$\sigma_v(yz)$	E	C_2
$\sigma_v(yz)$	$\sigma_v(yz)$	$\sigma_v(xz)$	C_2	E

This multiplication table is useful for understanding the properties of a group. Using this table, one can verify that the symmetry point group is also a mathematical group, as it satisfies all the four rules of group theory.

3.4.2 C_{3v} POINT GROUP

Now, let us construct multiplication table of point group C_{3v}. This point group consists of six elements {E, C_3^+, C_3^-, σ_v(a), σ_v(*b*), σ_v(*c*)}.

C_{3v}	E	C_3^+	C_3^-	σ_v(a)	σ_v(b)	σ_v(c)
E	*	*	*	*	*	*
C_3^+	*	*	*	*	*	*
C_3^-	*	*	*	*	*	*
σ_v(a)	*	*	*	*	*	*
σ_v(b)	*	*	*	*	*	*
σ_v(c)	*	*	*	*	*	*

At the top left corner, the symbol of point group of the molecule is written, C_{3v}. and in top most row and left column, all the six symmetry elements are written in sequence.

Now, we have to find out the products of all the binary combinations of the element of this group. The product of the combination is obtained by actually performing symmetry operation on the molecules. Therefore, the products in first row and column produced by operation on any element followed by identity and vice-versa are E.E = E, C_3^+.E = C_3^+, C_3^-.E = C_3^-, σ_v (a).E = σ_v (a), σ_v (*b*).E = σ_v (*b*), and σ_v (*c*).E = σ_v (*c*) in rows. Similarly, E.E = E, E.C_3^+ = C_3^+, E.C_3^- = C_3^-, E.σ_v (a) = σ_v (a), E.σ_v (*b*) = σ_v (*b*) and E. σ_v (*c*) = σ_v (*c*).

C_{3v}	E	C_3^+	C_3^-	$\sigma_v(a)$	$\sigma_v(b)$	$\sigma_v(c)$
E	E	C_3^+	C_3^-	$\sigma_v(a)$	$\sigma_v(b)$	$\sigma_v(c)$
C_3^+	C_3^+	*	*	*	*	*
C_3^-	C_3^-	*	*	*	*	*
$\sigma_v(a)$	$\sigma_v(a)$	*	*	*	*	*
$\sigma_v(b)$	$\sigma_v(b)$	*	*	*	*	*
$\sigma_v(c)$	$\sigma_v(c)$	*	*	*	*	*

Again, the vacant places of products shown by are filled by the results of binary combinations of column and row elements.

$$\left.\begin{aligned} C_3^+ . C_3^+ &= C_3^- \\ C_3^- . C_3^+ &= E \\ \sigma_v(a) . C_3^+ &= \sigma_v(c) \\ \sigma_v(b) . C_3^+ &= \sigma_v(a) \\ \sigma_v(c) . C_3^+ &= \sigma_v(b) \end{aligned}\right\}$$

$$\left.\begin{aligned} C_3^+ . C_3^- &= E \\ C_3^+ . \sigma_v(a) &= \sigma_v(b) \\ C_3^+ . \sigma_v(b) &= \sigma_v(c) \\ C_3^+ . \sigma_v(c) &= \sigma_v(a) \end{aligned}\right\}$$

Then

C_{3v}	E	C_3^+	C_3^-	$\sigma_v(a)$	$\sigma_v(b)$	$\sigma_v(c)$
E	E	C_3^+	C_3^-	$\sigma_v(a)$	$\sigma_v(b)$	$\sigma_v(c)$
C_3^+	C_3^+	C_3^-	E	$\sigma_v(c)$	$\sigma_v(a)$	$\sigma_v(b)$
C_3^-	C_3^-	E	*	*	*	*
$\sigma_v(a)$	$\sigma_v(a)$	$\sigma_v(b)$	*	*	*	*
$\sigma_v(b)$	$\sigma_v(b)$	$\sigma_v(c)$	*	*	*	*
$\sigma_v(c)$	$\sigma_v(c)$	$\sigma_v(a)$	*	*	*	*

Now, let us find further products:

$$\left.\begin{aligned} C_3^-.C_3^- &= C_3^+ \\ \sigma_v(a).C_3^- &= \sigma_v(b) \\ \sigma_v(b).C_3^- &= \sigma_v(c) \\ \sigma_v(c).C_3^- &= \sigma_v(a) \end{aligned}\right\}$$

$$\left.\begin{aligned} C_3^+.\sigma_v(a) &= \sigma_v(c) \\ C_3^+.\sigma_v(b) &= \sigma_v(a) \\ C_3^+.\sigma_v(c) &= \sigma_v(b) \end{aligned}\right\}$$

Putting these values in multiplication table of point group C_{3v}, we get:

C_{3v}	E	C_3^+	C_3^-	$\sigma_v(a)$	$\sigma_v(b)$	$\sigma_v(c)$
E	E	C_3^+	C_3^-	$\sigma_v(a)$	$\sigma_v(b)$	$\sigma_v(c)$
C_3^+	C_3^+	C_3^-	E	$\sigma_v(c)$	$\sigma_v(a)$	$\sigma_v(b)$
C_3^-	C_3^-	E	C_3^+	$\sigma_v(b)$	$\sigma_v(c)$	$\sigma_v(a)$
$\sigma_v(a)$	$\sigma_v(a)$	$\sigma_v(b)$	$\sigma_v(c)$	*	*	*
$\sigma_v(b)$	$\sigma_v(b)$	$\sigma_v(c)$	$\sigma_v(a)$	*	*	*
$\sigma_v(c)$	$\sigma_v(c)$	$\sigma_v(a)$	$\sigma_v(b)$	*	*	*

Similarly, rest all other binary combinations can be determined and final multiplication table of point group C_{3v} can be constructed.

C_{3v}	E	C_3^+	C_3^-	$\sigma_v(a)$	$\sigma_v(b)$	$\sigma_v(c)$
E	E	C_3^+	C_3^-	$\sigma_v(a)$	$\sigma_v(b)$	$\sigma_v(c)$
C_3^+	C_3^+	C_3^+	E	$\sigma_v(c)$	$\sigma_v(a)$	$\sigma_v(b)$
C_3^-	C_3^-	E	C_3^+	$\sigma_v(b)$	$\sigma_v(c)$	$\sigma_v(a)$
$\sigma_v(a)$	$\sigma_v(a)$	$\sigma_v(b)$	$\sigma_v(c)$	E	C_3^+	C_3^-
$\sigma_v(b)$	$\sigma_v(b)$	$\sigma_v(c)$	$\sigma_v(a)$	C_3^-	E	C_3^+
$\sigma_v(c)$	$\sigma_v(c)$	$\sigma_v(a)$	$\sigma_v(b)$	C_3^+	C_3^-	E

It is very clear from the observation that each row and each column does not repeat any element in the multiplication tables.

3.5 SUBGROUPS

A group consists of number of symmetry elements (operations). Total number of elements in that group is called its order, and is denoted by h. Generally, a smaller set of symmetry elements exists in the total symmetry elements, which also form a symmetry group (point group), following all the four rules of a mathematical group, i.e., closure, association, identity and inverse rules. Thus, such smaller groups existing within a larger group are called subgroups. These are of two types and these are:

(i) Trivial Subgroup

When subgroup of a group consists of only one symmetry element, i.e., identity element (E), then it is known as trivial subgroup. Identity element is itself a group of order 1.

(ii) Non-Trivial Subgroup

When subgroup does not solely of the identity element but also contain more elements other than E in the group, then it is known as non-trivial subgroup.

E, A; E, B; E, C; Order of subgroup (g) = 2
E, D, F; Order of subgroup (g) = 3

There is another group also, which is recognized as a cyclic group G_3, in which $D^2 = F$, $D^3 = DF = FD = E$.

3.5.1 RELATIONSHIP BETWEEN ORDER OF GROUP (H) AND SUBGROUP (G)

If order of any group is h and order of its subgroup is g, then both these orders are related to each other by the relation.

$$h = kg \text{ (k = Integer greater than one)}$$

$$= k \tag{3.1}$$

It follows the Lagrange's theorem, i.e., the order of any subgroup (g) of a group of order (h) must be a divisor of h.

Although, it has been shown that the order of any subgroup (g) must be a divisor of h, but reverse may generally not true that there are subgroups of all orders that are divisors of h.

Let us take an example of C_{2v} point group, containing the following symmetry elements.

$$C_{2v} = E, 1\ C_2, 2\ \sigma_v$$

The order of this group (h) is 4. As the order of subgroup (g) must be an integral divisor of h. Therefore, this group has two possible subgroups with order (g), 1 and 2. Using multiplication table, following subgroups can be easily recognized.

$C_1 = \{E\}$	where	(g = 1)
$C_2 = \{E, C_2\}$		(g = 2)
$C_s = \{E, \sigma_v(xz)\}$		(g = 2)
$C_s = \{E, \sigma_v(yz)\}$		(g = 2)

It can be seen that value of h can be divided by 1 (as an integer) and it contain only symmetry element E and hence, it will be a subgroup of any other group.

Hence, there are 4 subgroups of C_{2v} (C_1, C_2, C_s and C_s). We can form group multiplication tables for these subgroups as:

C_1	E
E	E

C_2	E	C_2
E	E	C_2
C_2	C_2	E

C_3	E	σ_v
E	E	σ_v
σ_v	σ_v	E

Some other sets like $\{E, C_2, \sigma_v\}$ or $\{E, \sigma_v(xz), \sigma_v(yz)\}$ have order 3, which is not integral divisor of the group order 4. It means these two possible sets are not the subgroup of C_{2v} group.

In case of C_{3v} point group, E, C_3^+, C_3^- $\sigma_v(a)$, $\sigma_v(b)$, $\sigma_v(c)$ symmetry elements are present, which gives it an order of 6, one can easily identify as many as five subgroups in this point group having order 3, 2 and 1. Therefore,

$C_1 = \{E\}$	(g = 1)
$C_3 = \{E, C_3^+, C_3^-\}$	(g = 3)
$C_s^a = \{E, \sigma_v(a)\}$	(g = 2)
$C_s^b = \{E, \sigma_v(b)\}$	(g = 2)
$C_s^c = \{E, \sigma_v(c)\}$	(g = 2)

Hence, there are four subgroup in C_{3v} point group (C_1, C_3, C_s^a, C_s^b, C_s^c). We can form group multiplication tables for these subgroups as:

C_1	E
E	E

C_3	E	C_3^+	C_3^-
E	E	C_3	C_3^-
C_3^+	C_3^+	C_3^-	E
C_3^-	C_3^-	E	C_3^+

C_s^a	E	$\sigma_v(a)$
E	E	$\sigma_v(a)$
$\sigma_v(a)$	$\sigma_v(a)$	E

C_s^b	E	$\sigma_v(b)$
E	E	$\sigma_v(b)$,
$\sigma_v(b)$	$\sigma_v(b)$	E

C_s^c	E	$\sigma_v(c)$
E	E	$\sigma_v(c)$
$\sigma_v(c)$	$\sigma_v(c)$	E

D_{6h} point group has the order 24, and it consists of subgroups with order 6, for example, $C_6 = \{E, 5\ C_5\}$ and O_h point group (h = 18). O is the subgroup with only rotational operation and E.

$O = \{E, 9\ C_4, 8\ C_3, 6\ C_2\}$ (g = 24)

There can be more than one subgroup of a given order. As subgroups follow all the rules of a mathematical group and it must, therefore, contain the identity element (E). The subgroups are always Abelian like because the elements of a group do commute essentially. On the contrary, the elements of main group need not necessarily commute.

Besides subgroups, there is one more way by which symmetry elements of group may be separated into smaller sets known as classes. In simple words, the geometrically equivalent symmetry elements are placed in one class. For examples AB_4 molecule having square planar geometry possesses 4 C_2 axes in the plane of the molecule. Among these 4 C_2 axis, two $C_{2'}$ axes belong to same class due to geometrical equivalence and other two C_2'' axes belong to other class because of the same reason. It means, 2 $C_{2'}$ and 2 C_2'' belong to two different classes.

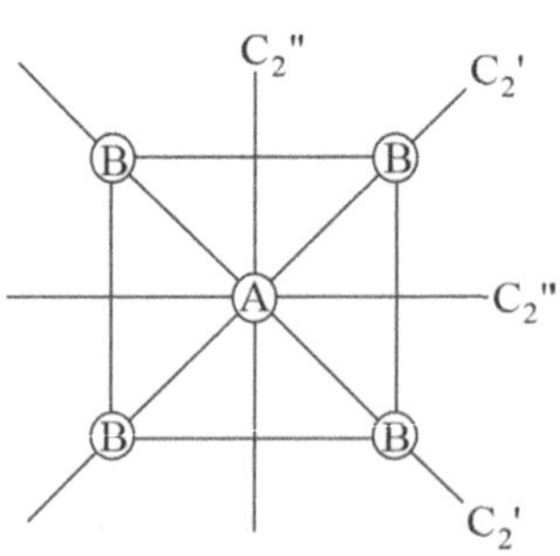

In order to understand classes present in a group in detail, a mathematical process known as similarity transformation of elements must be defined.

We know that there are some smaller groups or set of elements present in a large group. One way to separate these smaller sets of elements is to form subgroups or in another way, the elements of group may be separated into smaller sets known as classes. Before defining a class, one must consider a mathematical operation known as similarity transformation.

3.6 SIMILARITY TRANSFORMATION

Similarity transformation is an operation, which defines the classes in more general and mathematical way. Suppose a group is having X, A, and B elements, then according to the similarity transformation.

$$B = X^{-1}.\,A.\,X$$

where X^{-1} is the reciprocal (inverse) of X. Then B is said to be similarity transform of A by X. It may also be said that A and B are conjugate to each other. B is obtained, when operations are performed in the order as X, then A followed by X^{-1}. Such conjugate elements will form a class.

Conjugate elements have three properties:

(i) Every element is conjugate with itself, i.e., if we select any particular element, let A, then there must be an element X in that group, such that:

$$A = X^{-1}.\,A.\,X$$

On multiplying both the sides by A^{-1}, we get:

$$A^{-1}.\,A = E = A^{-1}.\,X^{-1}.\,A.\,X$$
$$= (XA)^{-1}.\,(AX)$$

This relationship holds good only, when A and X commute with each other. The X may be E or it may be any other element, which commute with element A.

(ii) If A is conjugate with B, then B must also be conjugate with A.

$$A = X^{-1}.\,B.\,X$$

Then there must be some another element Y present in the group, such that

$$B = Y^{-1}. A. Y$$

It can be proved by appropriate multiplication that

$$X. A. X^{-1} = X. X^{-1}.B.X.X^{-1} = B$$

Thus, if $Y = X^{-1}$ (or $Y^{-1} = Y$), then

$$B = Y^{-1}. A. Y$$

It is possible only, when any element (X) has an inverse element (Y).

(iii) If A is conjugate with B and C, then latter two, i.e., B and C are also conjugate to each other.

Therefore, a class is a set of conjugate elements and these conjugate elements are related by similarity transformation.

In order to determine class within a group, we begin with one element and work out all its transform, using all elements present in that group, including itself. Then select any other element, which is not among those found to be conjugate to the first. All the transforms corresponding to each element of the group are determined in this way, until all elements in that group have been placed in one class or another.

We can determine by this procedure, whether a particular element belongs to class or not?

(i) E forms its own independent class and no other elements are present in this class.
(ii) Axes form their own class and no other elements are included in it.
(iii) Planes have their own class and no other elements are present in their class.
(iv) Inversion center has its own separate class like identity E.

Let us consider the group G_3. Each element is tried one by one, starting from E.

$$E^{-1}.E.E = E.E = E$$
$$A^{-1}.E. A = A^{-1}. A. = E$$
$$B^{-1}.E. B = B^{-1}. B. = E$$

It means that E is not conjugate with any other element except itself. Thus, E is a class of order 1. Now let us take other elements of same group, like A. B, C and others. If the following relations hold good.

$$E^{-1}. A.E = A$$
$$A^{-1}. A. A = A$$
$$B^{-1}. A. B = C$$
$$C^{-1}. A. C = B$$
$$D^{-1}. A. D = B$$
$$F^{-1}. A. F = C$$

Thus, the element, A, B and C are conjugate to each other and therefore, belong to the same class. Now consider remaining element, for example D, which can be expressed as:

$$E^{-1}. D.E = D$$
$$A^{-1}. D. A = F$$
$$B^{-1}. D. B = F$$
$$C^{-1}. D. C = F$$
$$D^{-1}. D. D = D$$
$$F^{-1}. D. F = D$$

Similarly, every transformation of F gives either D or F element. Therefore, D and F constitute a class having order Z. Here, it can be noticed that the order of all classes must be integral factors of the order of the group.

3.7 DETERMINATION OF CLASSES

Using similarity transformation, classes can be determined in different point groups. Let us see that whether the 2 σ_v planes in C_{2v} group form a class or not?

There are four symmetry elements in this point group {E, 1 C_2, 2 σ_v}. E and C_2 form their own classes all alone with the order 1. Now the class of planes may be determined.

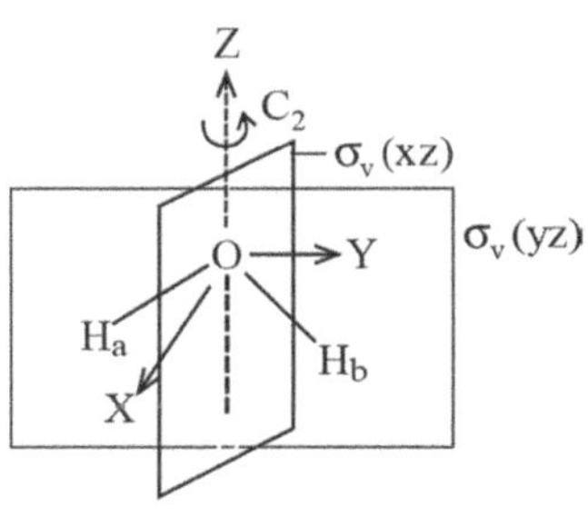

If we observe the class of σ_v (xz) by right multiply it by C_2 and left multiplying by C_2^{-1} (or C_2), then mathematically,

$$C_2^{-1} \text{ (or } C_2). \sigma_v (xz). C_2$$

The symmetry operations are performed on H_2O molecule, one by one from right to left.

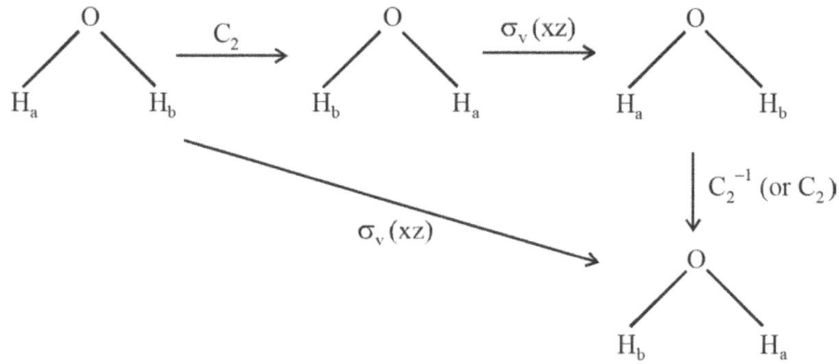

$$\sigma_v (xz) = C_2^{-1}. \sigma_v (xz). C_2$$
$$\sigma_v (xz) = \sigma_v (xz)$$

It means σ_v (yz) and σ_v (xz) are not conjugate to each other and the resultant product is σ_v (xz). Therefore, σ_v (xz) is an independent class of its own.

Similarly, considering σ_v (yz) and applying similarity transformation with C_2, Mathematically,

$$C_2^{-1}. \sigma_v (yz). C_2$$

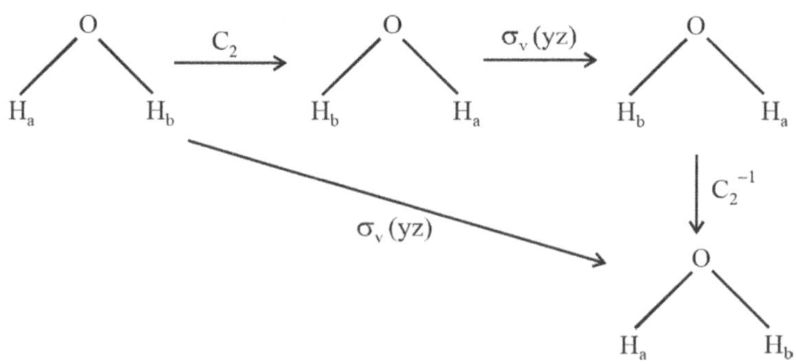

$$\sigma_v(yz) = C_2^{-1}.\ \sigma_v(yz).\ C_2$$
$$\sigma_v(yz) = \sigma_v(yz)$$

Here also $\sigma_v(yz)$ is conjugate to itself and it is not conjugate to $\sigma_v(xz)$.

Thus, $\sigma_v(xz)$ and $\sigma_v(yz)$ are not conjugate elements because both are not related by similarity transformation relation. They are conjugate to themselves, so $\sigma_v(xz)$ and $\sigma_v(yz)$ planes will form two separate classes of their own. Therefore, C_{2v} possess four classes.

$$1\ E,\ 1\ C_2,\ 1\ \sigma_v(xz),\ 1\ \sigma_v(yz)$$

It is quite interesting to note that two different planes form two different classes in C_{2v} group, but C_{3v} group has three planes, i.e., $\sigma_v(a)$, $\sigma_v(b)$ and $\sigma_v(c)$ and all these three planes form a class.

C_{3v} point group have six symmetry elements (E, C_3^+, C_3^-, $\sigma_v(a)$, $\sigma_v(b)$, $\sigma_v(c)$. E always remains all-alone as a class. Now, let us find out the conjugate of C_3^+, if any? For this, applying $\sigma_v(a)$ on right side and its inverse $\sigma_v(a)^{-1}$ on left side of C_3^+, which is the element under observation. Mathematically, it is:

$$\sigma_v(a)^{-1}.\ C_3^+.\ \sigma_v^-(a)$$

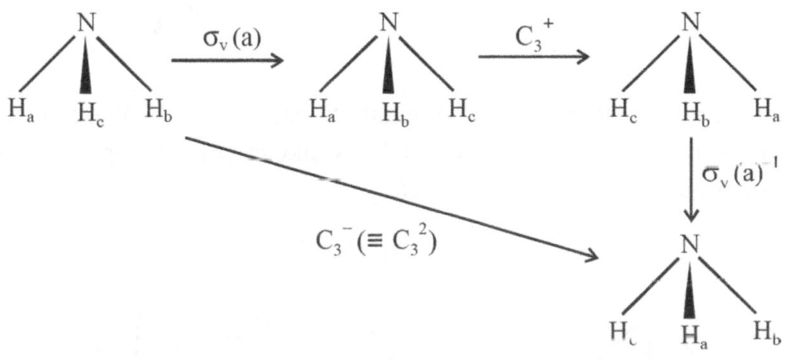

The resultant is C_3^-.

Similarly, $$C_3^2 \text{ or } C_3^- = \sigma_v(a)^{-1}.\ C_3^+.\ \sigma_v(a)$$

Hence, C_3^+ and C_3^- are conjugate to each other and belong to same class and it can be proved that:

$$C_3^+ = \sigma_v(b)^{-1}.\ C_3^-.\ \sigma_v(b)$$

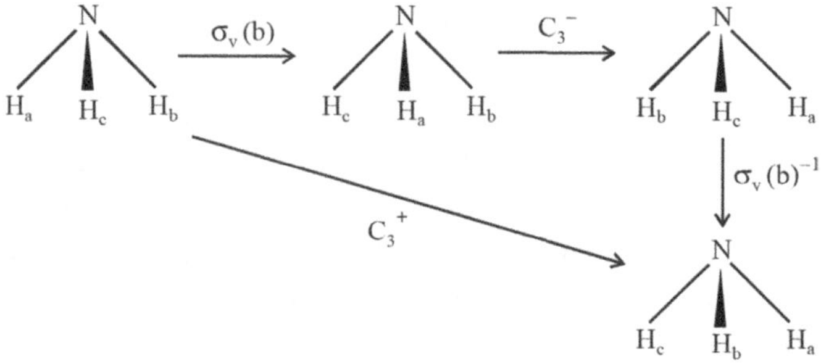

Now let us determine the class of any plane, i.e., $\sigma_v(a)$. For this purpose, we have to multiply σ_v (a) by C_3^+ on right side and its inverse $C_3^- = C_3^2$ on the left side.

$$\sigma_v(c) = C_3^-.\ \sigma_v(a).\ C_3^+$$

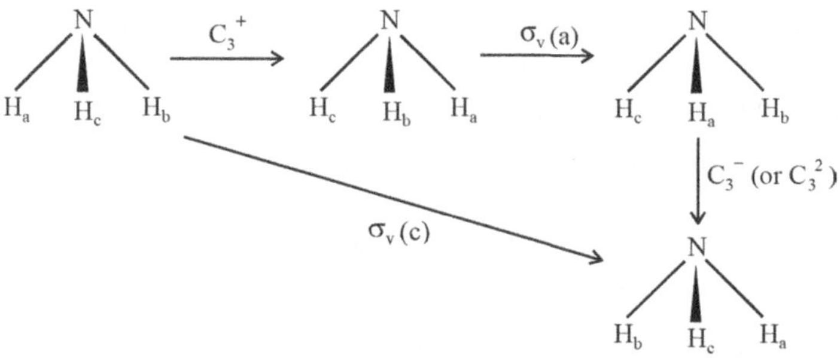

Hence, σ_v (a) and $\sigma_v(c)$ are conjugate to each other. As these are conjugate elements, they will form a class. Now take $\sigma_v(b)$ and applying similarity transformation on it. Mathematically.

$$C_3^{-1}.\ \sigma_v(b).\ C_3^+$$

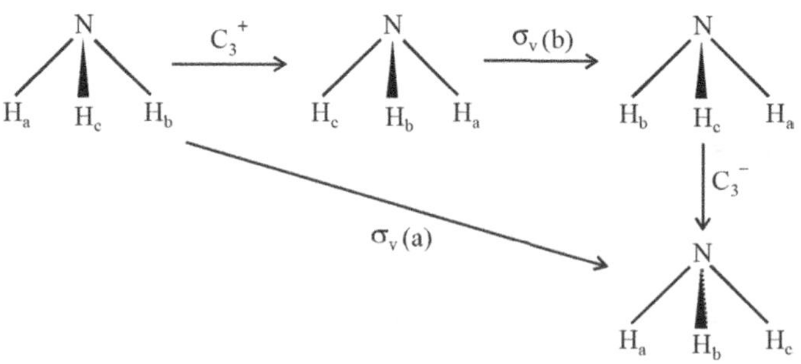

$$\sigma_v(a) = C_3^-.\ \sigma_v(b).\ C_3^+$$
$$\sigma_v(b) = C_3^-.\ \sigma_v(c).\ C_3^+$$

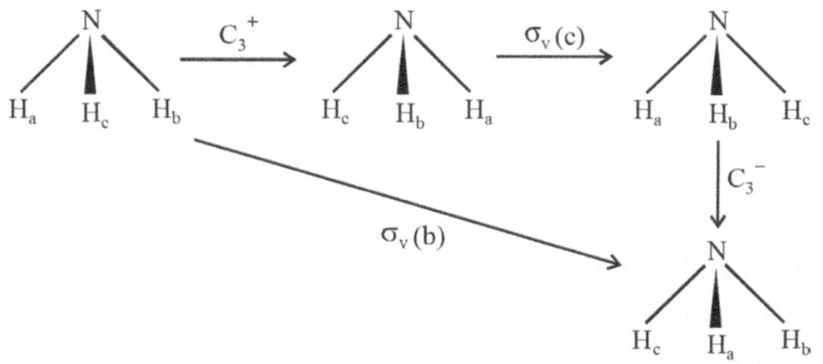

Hence, $\sigma_v(b)$ and $\sigma_v(a)$ are also conjugate to each other. Thus, as per the property of conjugate element, it is clear that when $\sigma_v(a)$ is conjugate with $\sigma_v(c)$ and $\sigma_v(a)$ is conjugate with $\sigma_v(b)$, then $\sigma_v(b)$, $\sigma_v(c)$ and $\sigma_v(a)$ are also conjugate to each other.

So, $\sigma_v(a)$, $\sigma_v(b)$ and $\sigma_v(c)$ will form a class and written as 3 σ_v in the character table of the symmetry point group C_{3v}

Hence, other possibility on similarity transformation on C_3 gives:

$$C_3^- = \sigma_v(c).\ C_3^+.\ \sigma_v(c),$$
$$\sigma_v^{-1}.\ C_3^-.\ \sigma_v = C_3^+$$
$$C_3^- = \sigma_v(b).\ C_3^+.\ \sigma_v(b)$$
$$C_3^+ = C_3^+.\ C_3^+.\ C_3$$
$$C_3^+ = C_3^-.\ C_3^+.\ C_3^+$$
$$C_3^+ = E.\ C_3^+.\ E$$

C_3^+ and C_3^- are conjugate elements. Hence in C_{3v} point group, there will be three classes, i.e., 1 E, 2 C_3, and 3 σ_v and over all order of the group is six.

It is quite interesting to note that three planes of C_{3v} point group, $\sigma_v(b)$, σ_v (b) and $\sigma_v(c)$ form a class 3 σ_v, where as only two planes in C_{2v} point group do not form a class and cannot be represented as 2 σ_v, but separate classes as σ_v(xz) and σ_v(yz).

KEYWORDS

- **Associative**
- **Class**
- **Closure**
- **Identity**
- **Inverse**
- **Multiplication table**
- **Similarity transformation**
- **Subgroup**

CHAPTER 4

MATRICES

CONTENTS

4.1 MATRICES

Group element corresponds to symmetry element, which are carried out on spatial coordinates resulting into change in configuration of the molecule. But in order to show its effect on mathematical basis, some numerical representation is required. These are represented in the form of matrices. When these operations are represented as linear transformation with respect to the Cartesian coordinate system, matrices are obtained. They follow general rules of matrix to the group of symmetry operations. In essence, matrices are useful in describing the symmetry operation mathematically, which is otherwise a tedious job because it requires many equations of relation between the set of coordinates on which each atom of a molecule is present.

Thus, matrices are representation of the symmetry group with each element corresponding to a particular matrix. The matrices for symmetry operations are derived from vectors and the vectors are basic of the representation.

Matrix is a rectangular array, in which combinations of numbers are arranged and it is combined with another matrix following a definite rules of matrices. Generally, matrix is given in the form of square brackets. Matrix made up of a numbers of rows (m) and number of columns (m) as:

$$\begin{bmatrix} x_{11} & x_{12} & x_{13} & \cdots & x_{1n} \\ x_{21} & x_{22} & x_{23} & \cdots & x_{2n} \\ x_{31} & x_{32} & x_{33} & \cdots & x_{3n} \\ x_{m1} & x_{m2} & x_{m3} & \cdots & x_{mn} \end{bmatrix}_{m \times n}$$

The number of rows and number of columns is termed as dimension of matrix. It means that the dimension of such a matrix is $m \times n$, which is written on right bottom corner of the matrix. Matrix can be symbolized as [x] or $[x_{ij}]$ in which 'x' is the element of the matrix, which is present in i^{th} row and j^{th} column.

4.2 TYPES OF MATRICES

There are different type of matrices such as square matrix, unit matrix, diagonal matrix, symmetric matrix, transpose matrix, vector matrix and many more.

4.2.1 SQUARE MATRIX

It includes matrix with equal number of rows and columns ($m = n$). It is useful for symmetry and group theory considerations.

4.2.2 UNIT MATRIX

In this matrix, diagonal element (x_{ii}) is equal to one and rest all elements (diagonal element) (x_{ij}, when $i \neq j$) are equal to zero. It is represented as E, I or 1. It is also called unitary matrix.

Example of unit matrix is:

$$\begin{bmatrix} 1 & 0 & 0 \\ 0 & 1 & 0 \\ 0 & 0 & 1 \end{bmatrix}_{3 \times 3}$$

$a_{ij} = \delta_{ij}$ (Represent unit matrix)

δ_{ij} is a Kronecker delta $\rightarrow$ 1 (i = j); $\rightarrow$ 2 (i ≠ j)

4.2.3 DIAGONAL MATRIX

In this matrix, all diagonal elements (x_{ij}) are different and all other elements are zero.

Example of a diagonal matrix is:

$$\begin{bmatrix} 2 & 0 & 0 & 0 \\ 0 & 6 & 0 & 0 \\ 0 & 0 & 4 & 0 \\ 0 & 0 & 0 & 1 \end{bmatrix}_{4\times 4}$$

4.2.4 TRANSPOSE MATRIX

It involves interchange of the elements across the diagonal.

$$\begin{bmatrix} 1 & 2 & 3 & 4 \\ 5 & 6 & 7 & 8 \\ 9 & 1 & 2 & 4 \\ 6 & 7 & 8 & 9 \end{bmatrix}_{4\times 4} \longrightarrow \begin{bmatrix} 1 & 5 & 3 & 4 \\ 2 & 6 & 1 & 8 \\ 9 & 7 & 2 & 8 \\ 6 & 7 & 4 & 9 \end{bmatrix}_{4\times 4}$$

Transposed matrix

4.2.5 SYMMETRIC MATRIX

In this matrix, equal elements are symmetrically disposed along the diagonal, which has $x_{ij} = x_{ji}$ relationship.

Example of a symmetric matrix is:

$$\begin{bmatrix} 5 & 1 & 2 & 6 \\ 1 & 2 & 4 & 3 \\ 2 & 4 & 3 & 7 \\ 4 & 3 & 7 & 6 \end{bmatrix}_{4\times 4}$$

4.2.6 VECTOR MATRIX

Matrix with single row or single column is called a vector matrix. Example of vector matrix is:

$$\begin{bmatrix} 1 \\ 2 \\ 3 \end{bmatrix}_{3\times 4} \qquad \begin{bmatrix} 1 & 2 & 3 \end{bmatrix}_{1\times 3}$$

Column matrix   Row matrix

Suppose, vector $\overrightarrow{OA}$ is in three-dimensional space with point A having x, y, and z Cartesian coordinates.

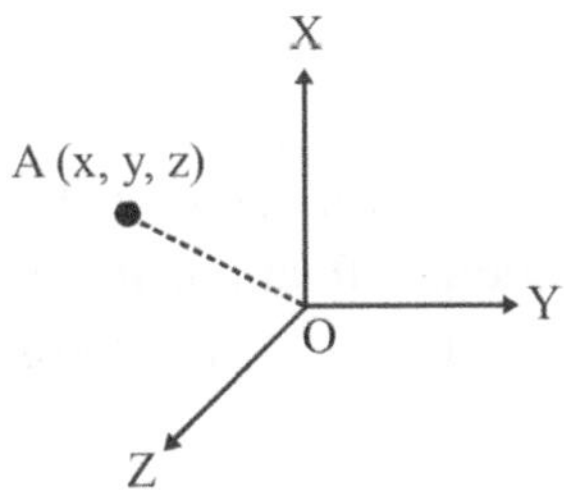

Then these three coordinates can be represented as row and column matrix, respectively to show $\overrightarrow{OA}$ vector.

$$[xyz] \text{ and } \begin{bmatrix} x \\ y \\ z \end{bmatrix}$$

4.2.7 SCALAR MATRIX

When all the diagonal elements are equal and other elements are zero, then the matrix is called scalar matrix.

$$\begin{bmatrix} 2 & 0 & 0 \\ 0 & 2 & 0 \\ 0 & 0 & 2 \end{bmatrix}_{3\times 3}$$

4.2.8 NULL OR ZERO MATRIX

Null or Zero matrix is a matrix consisting of all zero elements.

$$\begin{bmatrix} 0 & 0 & 0 \\ 0 & 0 & 0 \\ 0 & 0 & 0 \end{bmatrix}_{3\times 3}$$

4.3 CHARACTERS OF CONJUGATE MATRICES

Character is the sum of the diagonal element of a square matrix. It is an important property of a square matrix and denoted by χ (Greek chi).

$$\chi = \Sigma x_{ii}$$

Different symmetry operations have different characters of matrix. For example, identity rotation, reflection, inversion and improper rotation operations have 3, $2\cos\theta + 1$, 1, –3, and $2\cos\theta - 1$ character of matrix, respectively.

4.4 EQUALITY, ADDITION AND SUBTRACTION

Mathematical matrices should have identical dimensions for equality, addition and subtraction. Two matrices, A and B, which are of $m \times n$ dimensions; then each one of these matrices is said to be equal if $a_{ij} = b_{ij}$; $i = 1$ to m and $j = 1$ to n.

Addition or subtraction of A and B matrices consists of addition or subtraction of their corresponding elements, i.e.,

$$A \pm B = C \text{ means } a_{ij} \pm b_{ij} = c_{ij}$$

For example, if $A = \begin{bmatrix} 2 & 1 \\ 6 & 8 \end{bmatrix}$ and $B = \begin{bmatrix} 4 & 7 \\ 3 & 5 \end{bmatrix}$

Then $A + B = \begin{bmatrix} 2+4 & 1+7 \\ 6+3 & 8+5 \end{bmatrix} = \begin{bmatrix} 6 & 8 \\ 9 & 13 \end{bmatrix}$

$$\text{and } A - B = \begin{bmatrix} 2-4 & 1-7 \\ 6-3 & 8-5 \end{bmatrix} = \begin{bmatrix} -2 & -6 \\ 3 & 3 \end{bmatrix}$$

4.5 MULTIPLICATION

For multiplication, the matrices should be confirmable, i.e., if $A \times B = C$; then number of columns in A should be equal to number of rows in B. Hence, if dimensions of A is $m \times p$, then that of B should be $p \times n$.

Dimensions of C will be $m \times n$. Thus,

$$c_{ik} = a_{i1}b_{1k} + a_{i2}b_{2k} + a_{i3}b_{3k} + \ldots + a_{ip}b_{pk}$$

For example,

$$A = \begin{bmatrix} 2 & 3 \\ 1 & 8 \\ 5 & 6 \end{bmatrix} \text{ and } B = \begin{bmatrix} 1 & 3 & 0 \\ -4 & 2 & 1 \end{bmatrix}$$

$$A \times B = \begin{bmatrix} 2\times1+3\times(-4) & 2\times3+3\times2 & 2\times0+3\times1 \\ 1\times1+8\times(-4) & 1\times3+8\times2 & 1\times0+8\times1 \\ 5\times1+6\times(-4) & 5\times3+6\times2 & 5\times0+6\times1 \end{bmatrix}$$

or

$$C = \begin{bmatrix} -10 & 12 & 3 \\ -31 & 19 & -8 \\ -19 & 27 & 6 \end{bmatrix}$$

That is, multiply elements of ith row of A by corresponding elements of kth column of B and add these to obtain c_{ik}.

The following points should be noted in multiplication of matrices.

(i) Multiplication of any matrix by unit matrix (identity matrix) of appropriate dimensions leaves the matrix unchanged. Thus,

$$A \times E = A$$

(ii) Generally, matrix multiplication is non-commutative, i.e.,

$$A \times B \neq B \times A$$

(iii) An inverse of a matrix is defined as:

$$A \times B = E$$

In this case, B is the inverse of A, i.e., B = A^{-1} and therefore, $A \times B = A \times A^{-1} = E$.

It is to be noted here that only square matrices can have an inverse matrix. Inverse of a matrix involves the division of a cofactor by the determinant of the matrix, i.e., |A|, |A|≠0, this cofactor should not be singular. It has to be non-singular, i.e., non-zero. As only square determinants can be non-zero and hence, only square matrices can have the corresponding inverse matrices.

Since we shall be dealing mostly with square matrices here, it is important to note their relevant properties. These are:

(i) An important property is its character or (Trace), which is represented by χ (chi) = Σa_{ij}, i.e., sum of the diagonal elements.
(ii) If the product of two matrices A and B are AB = C and BA = D, then their characters $\chi(C) = \chi(D)$, are equal.
(iii) Conjugate matrices have identical characters. Conjugate matrices, like conjugate elements of group are related by similarity transformation and then there is a third matrix X such that $A = X^{-1}. B. X$.

These properties of matrices closely parallel those of group elements (symmetry operators/operations).

4.6 INVERSE OF A MATRIX

This concept will be useful in evaluating the coefficients a_{ij} of equivalent hybrid orbital. Consider linear homogeneous equations:

$$x' = a_1x + b_1y + c_1z \quad (4.1)$$

$$y' = a_2x + b_2y + c_2z \quad (4.2)$$

$$z' = a_3x + b_3y + c_3z \quad (4.3)$$

analogous to LCAOψ_s in hybridization.

$$\psi_1 = a_1\phi_1 + b_1\phi_2 + c_1\phi_3 \quad (4.4)$$

$$\psi_2 = a_2\phi_1 + b_2\phi_2 + c_2\phi_3 \quad (4.5)$$

$$\psi_3 = a_3\phi_1 + b_3\phi_2 + c_3\phi_3 \tag{4.6}$$

This can be put in matrix form as:

$$\underset{R'}{\begin{bmatrix} x' \\ y' \\ z' \end{bmatrix}} = \underset{T_r}{\begin{bmatrix} a_1 & b_1 & c_1 \\ a_2 & b_2 & c_2 \\ a_3 & b_3 & c_3 \end{bmatrix}} \underset{R}{\begin{bmatrix} x \\ y \\ z \end{bmatrix}}$$

Mathematically, R' = T_r. R

To obtain x, y, z (or R) in terms of primed variables x,' y,' z' (or R'), we can use inverse of the transformation matrix, T_r^{-1}:

$$\underset{R}{\begin{bmatrix} x \\ y \\ z \end{bmatrix}} = \underset{T_r^{-1}}{\begin{bmatrix} a_1' & b_1' & c_1' \\ a_2' & b_2' & c_2' \\ a_3' & b_3' & c_3' \end{bmatrix}} \underset{R}{\begin{bmatrix} x' \\ y' \\ z' \end{bmatrix}}$$

Mathematically, R = T_r^{-1} R'

To obtain the elements of T_r^{-1} (primed letters), T_r is treated as a determinant. Thus, its value $|T_r|$ is obtained. Elements of T_r are now replaced by their cofactors divided by determinant of T_r, i.e., $|T_r|$. The cofactor of elements ij of matrix T_r (a_1, b_1, etc.) is equal to the minor of element ij multiplied by (1)$^{i+j}$. The minor of the element ij is the determinant obtained by striking out the ith row and jth column in the matrix T_r. This matrix of cofactors is:

$$T_r = \begin{bmatrix} \dfrac{\begin{vmatrix} b_2 & c_2 \\ b_3 & c_3 \end{vmatrix}}{|T_r|} & -\dfrac{\begin{vmatrix} a_2 & c_2 \\ a_3 & c_3 \end{vmatrix}}{|T_r|} & +\dfrac{\begin{vmatrix} a_2 & b_2 \\ a_3 & b_3 \end{vmatrix}}{|T_r|} \\[2ex] -\dfrac{\begin{vmatrix} b_1 & c_1 \\ b_3 & c_3 \end{vmatrix}}{|T_r|} & \dfrac{\begin{vmatrix} a_1 & c_1 \\ a_3 & c_3 \end{vmatrix}}{|T_r|} & -\dfrac{\begin{vmatrix} a_1 & b_1 \\ a_3 & b_3 \end{vmatrix}}{|T_r|} \\[2ex] \dfrac{\begin{vmatrix} b_1 & c_1 \\ b_2 & c_2 \end{vmatrix}}{|T_r|} & -\dfrac{\begin{vmatrix} a_1 & c_1 \\ a_2 & c_2 \end{vmatrix}}{|T_r|} & \dfrac{\begin{vmatrix} a_1 & b_1 \\ a_2 & b_2 \end{vmatrix}}{|T_r|} \end{bmatrix}$$

The inverse of the transformation matrix will be:

$$\begin{bmatrix} \dfrac{\begin{vmatrix} b_2 & c_2 \\ b_3 & c_3 \end{vmatrix}}{|T_r|} & -\dfrac{\begin{vmatrix} b_1 & c_1 \\ b_3 & c_3 \end{vmatrix}}{|T_r|} & \dfrac{\begin{vmatrix} b_1 & c_1 \\ b_2 & c_2 \end{vmatrix}}{|T_r|} \\ -\dfrac{\begin{vmatrix} a_2 & c_2 \\ a_3 & c_3 \end{vmatrix}}{|T_r|} & \dfrac{\begin{vmatrix} a_1 & c_1 \\ a_3 & c_3 \end{vmatrix}}{|T_r|} & -\dfrac{\begin{vmatrix} a_1 & c_1 \\ a_2 & c_2 \end{vmatrix}}{|T_r|} \\ \dfrac{\begin{vmatrix} a_2 & b_2 \\ a_3 & b_3 \end{vmatrix}}{|T_r|} & -\dfrac{\begin{vmatrix} a_1 & c_1 \\ a_2 & c_2 \end{vmatrix}}{|T_r|} & \dfrac{\begin{vmatrix} a_1 & b_1 \\ a_2 & b_2 \end{vmatrix}}{|T_r|} \end{bmatrix}$$

Example: Consider the matrix

$$A = \begin{bmatrix} 1 & 1 & 0 \\ 1 & -1 & 1 \\ 0 & 1 & -1 \end{bmatrix} = T_r$$

The value of $|A| = +1$. Then its inverse transformation matrix will be:

$$T_r^{-1} = \begin{bmatrix} \dfrac{\begin{vmatrix} -1 & 1 \\ 1 & -1 \end{vmatrix}}{+1} & -\dfrac{\begin{vmatrix} 1 & 1 \\ 0 & -1 \end{vmatrix}}{+1} & \dfrac{\begin{vmatrix} 1 & -1 \\ 0 & -1 \end{vmatrix}}{+1} \\ -\dfrac{\begin{vmatrix} 1 & 0 \\ 1 & -1 \end{vmatrix}}{+1} & \dfrac{\begin{vmatrix} 1 & 0 \\ 1 & -1 \end{vmatrix}}{+1} & -\dfrac{\begin{vmatrix} 1 & 1 \\ 0 & 1 \end{vmatrix}}{+1} \\ \dfrac{\begin{vmatrix} 1 & 0 \\ -1 & 1 \end{vmatrix}}{+1} & -\dfrac{\begin{vmatrix} 1 & 0 \\ 1 & 1 \end{vmatrix}}{+1} & \dfrac{\begin{vmatrix} 1 & 1 \\ 1 & -1 \end{vmatrix}}{+1} \end{bmatrix}$$

$$T_r^{-1} = \begin{bmatrix} 0 & 1 & -1 \\ 1 & -1 & -1 \\ 1 & -1 & 1 \end{bmatrix}$$

A is a symmetric matrix and its inverse is also symmetric.

In hybridization, we deal with similar linear homogeneous equation obtained by LCAO method:

$$\psi_i = \sum_{j=1}^{n} a_{ij}\, \varphi_j$$

Since ψ_i are orthogonal wave functions, the transformation matrices obtained by collection of their coefficients a_{ij} (the elements) are also orthogonal matrices. Inverse of such orthogonal matrices are their transposed matrices. The elements of orthogonal matrices follow the relation:

$$\sum_{j=1}^{n} a_{ij}\, a_{ki} = a_{ij}\, a_{ik} = \delta_{jk}\,(\text{Kronecker delta}) = 1, \text{ if } j = k$$

4.7 REPRESENTATION OF GROUP BY MATRIX

One of the very convenient methods of representation of a point group is to attach one or more vectors to it, i.e., lines of specified length and direction to the molecule and to see the objects after the operations by the vectors attached.

We shall show that each of the five types of symmetry operations of molecules can be described by a matrix.

4.7.1 IDENTITY OPERATION (E)

Identity operation is a trivial operation even though it has much importance in group theory. When a point on the vector $\vec{r}$(OA) with coordinates x_1, y_1 and z_1, is subjected to identity operation, then it gives new coordinates (x_2, y_2, z_2), which is same as the initial one, i.e., E keep vectors unaltered. Therefore, linear homogeneous equation can be expressed as:

$$x_2 = 1\,x_1 + 0\,y_1 + 0\,z_1$$

$$y_2 = 0\,x_1 + 1\,y_1 + 0\,z_1$$

$$z_2 = 0\,x_1 + 0\,y_1 + 1\,z_1$$

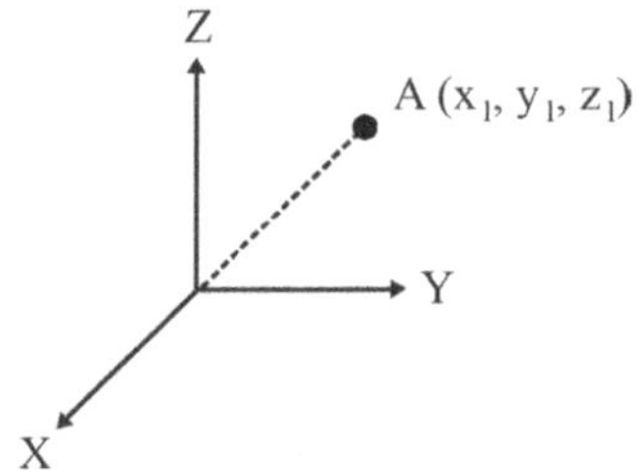

This may be expressed in the form of a matrix as:

$$\begin{bmatrix} x_2 \\ y_2 \\ z_2 \end{bmatrix} = \begin{bmatrix} 1 & 0 & 0 \\ 0 & 1 & 0 \\ 0 & 0 & 1 \end{bmatrix} \begin{bmatrix} x_1 \\ y_1 \\ z_1 \end{bmatrix}$$

or $\vec{r} = E.\ r.$

Identity matrix is a unit matrix, where diagonal elements are equal to 1 and all other elements are zero. It is a square matrix of order 3 × 3. The matrix of identity element can also represented as:

$$E = \begin{bmatrix} 1 & 0 & 0 \\ 0 & 1 & 0 \\ 0 & 0 & 1 \end{bmatrix}$$

Thus, identity is represented by unit matrix of order 3 × 3, in which x_1, y_1, and z_1 set of coordinates resulted into another set of coordinates x_2, y_2, and z_2 on applying the identity operation.

4.7.2 AXIS OF SYMMETRY (PROPER ROTATION) (C_n)

Consider that a vector r is having coordinates *x*, *y*, and *z* with an angle α on X-axis.

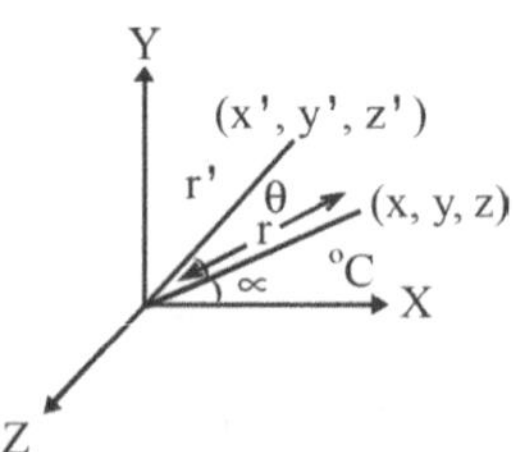

Rotation of this vector through an angle θ (anticlockwise) will give new location of the vector with coordinate such as *x*,’ *y*,’ and *z*.’

As $\sin\alpha = \dfrac{\text{Perpendicular}}{\text{Hypotenuse}} = \dfrac{y}{r}$, and therefore,

$$y = r \sin\alpha \tag{4.7}$$

As $\cos\alpha = \dfrac{\text{Base}}{\text{Hypotenuse}} = \dfrac{x}{r}$ and therefore,

$$x = r \cos\alpha \tag{4.8}$$

After rotation through an angle θ,

$$\sin(\alpha + \theta) = \frac{y'}{r} \text{ and therefore}$$

$$y' = r \sin(\alpha + \theta) \tag{4.9}$$

We know from trigonometry that sin (A + B) = sin A cos B + sin B cos A

Hence
$$y' = r \sin\alpha \cos\theta + r \sin\theta \cos\alpha \tag{4.10}$$

Putting the values of *x*, and *y* from equation (4.7) and (4.8) in equation (4.10), wc get

$$y' = y\cos\theta + x \sin\theta \tag{4.11}$$

Similarily,

$$\cos(\alpha + \theta) = \frac{x'}{r} \text{ and therefore}$$

$$x' = r \cos(\alpha + \theta)$$

As from trigonometry, cos (A + B) = cos A cos B – sin A sin B.

$$x' = r \cos\alpha \cos\theta - r \sin\alpha \sin\theta \tag{4.12}$$

Putting the values of *x* and *y* from equation (4.7) and (4.8) into (4.12)

$$x' = x \cos\theta - y \sin\theta \tag{4.13}$$

Since rotation occurs in xy plane; z vector remains unchanged after rotation. In other words, we can say $z' = z$

$$C_n(z).r = \begin{bmatrix} x' \\ y' \\ z' \end{bmatrix} = \begin{bmatrix} \cos\theta & -\sin\theta & 0 \\ \sin\theta & \cos\theta & 0 \\ 0 & 0 & 1 \end{bmatrix} \begin{bmatrix} x \\ y \\ z \end{bmatrix}$$

Operation is always done in a clockwise direction. Hence, for conversion of this relation into clockwise direction, we replace θ by (-θ) and then, the matrix becomes:

$$C_n(z).r = \begin{bmatrix} \cos(-\theta) & -\sin(-\theta) & 0 \\ \sin(-\theta) & \cos(-\theta) & 0 \\ 0 & 0 & 1 \end{bmatrix} \begin{bmatrix} x \\ y \\ z \end{bmatrix}$$

Because $\cos(-\theta) = \cos\theta$ and $\sin(-\theta) = -\sin\theta$

$$C_n(z).r = \begin{bmatrix} \cos\theta & -\sin\theta & 0 \\ \sin\theta & \cos\theta & 0 \\ 0 & 0 & 1 \end{bmatrix} \begin{bmatrix} x \\ y \\ z \end{bmatrix}$$

4.7.3 PLANE OF SYMMETRY (REFLECTION)

If a plane of symmetry is selected along or coplanar with one of the principal Cartesian plane, i.e., xy, xz or yz plane, respective planes are designated as σ_{xy}, σ_{xz}, or σ_{yz}. The reflection plane changes the sign of the coordinate lying perpendicular to that plane, while the two coordinates remain unchanged, whose axes are in the plane.

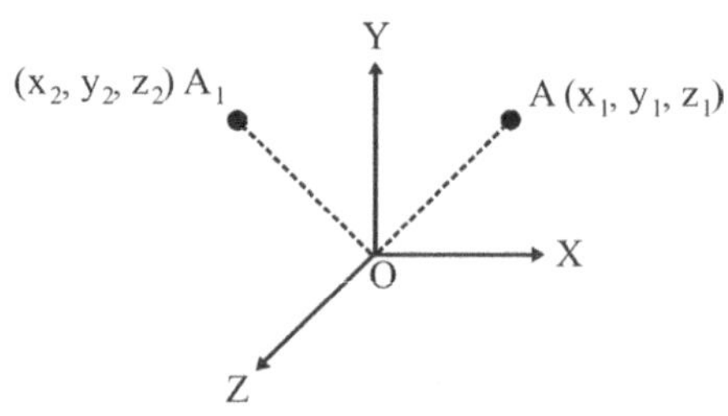

Relation between x_1, y_1 and z_1 and x_2, y_2 and z_2, for reflection in yz plane is represented by the equations:

$$x_2 = (-1)\, x_1 + 0\, y_1 + 0\, z_1$$

$$y_2 = 0\, x_1 + 1\, y_1 + 0\, z_1$$

$$z_2 = 0\, x_1 + 0\, y_1 + 1\, z_1$$

The equation can be represented in form of a matrix as:

$$\begin{bmatrix} x_2 \\ y_2 \\ z_2 \end{bmatrix} = \begin{bmatrix} -1 & 0 & 0 \\ 0 & 1 & 0 \\ 0 & 0 & 1 \end{bmatrix} \begin{bmatrix} x_1 \\ y_1 \\ z_1 \end{bmatrix}$$

$$r' = \sigma_{yz.}\, r$$

or $$\sigma_{yz} = \begin{bmatrix} -1 & 0 & 0 \\ 0 & 1 & 0 \\ 0 & 0 & 1 \end{bmatrix}$$

Thus, reflection in other two principal planes will be–

$$\sigma_{xy} - \begin{bmatrix} 1 & 0 & 0 \\ 0 & 1 & 0 \\ 0 & 0 & -1 \end{bmatrix} \text{ and } \sigma_{xz} = \begin{bmatrix} 1 & 0 & 0 \\ 0 & -1 & 0 \\ 0 & 0 & 1 \end{bmatrix}$$

It is to be noted that the reflection in σ_{yz} plane changes the sign of the coordinate *x*. in the same manner, the reflection in σ_{xy} and σ_{xz} planes changes the sign of the coordinate *z* and *y*, respectively.

4.7.4 *CENTRE OF SYMMETRY (INVERSION)*

Inversion changes the sign of all the coordinates. Therefore, *x*, *y* and *z* coordinates of vector r are transformed into their respective negative coordinates (–*x*, –*y*, and –*z*). Thus, for inversion operation, the equations will be:

$$x_2 = (-1)\, x_1 + 0\, y_1 + 0\, z_1$$

$$y_2 = 0\,x_1 + (-1)\,y_1 + 0\,z_1$$

$$z_2 = 0\,x_1 + 0\,y_1 + (-1)\,z_1$$

The matrix representation for these equations may be expressed as:

$$\begin{bmatrix} x_2 \\ y_2 \\ z_2 \end{bmatrix} = \begin{bmatrix} -1 & 0 & 0 \\ 0 & -1 & 0 \\ 0 & 0 & -1 \end{bmatrix} \begin{bmatrix} x_1 \\ y_1 \\ z_1 \end{bmatrix}$$

Hence, transformation matrix for inversion will be:

$$i = \begin{bmatrix} -1 & 0 & 0 \\ 0 & -1 & 0 \\ 0 & 0 & -1 \end{bmatrix}$$

Thus, the center of symmetry matrix representation is a negative unit matrix.

4.7.5 IMPROPER ROTATION (S_n)

Any improper rotation through an angle θ involves two successive symmetry operations, i.e., rotation followed by reflection in perpendicular plane, which may be represented as the product of the two matrices. Thus,

$$S_n\,(z) = C_n\,(z).\,\sigma_h$$

$$S_n\,(z) = \begin{bmatrix} \cos\theta & \sin\theta & 0 \\ -\sin\theta & \cos\theta & 0 \\ 0 & 0 & 1 \end{bmatrix} \times \begin{bmatrix} 1 & 0 & 0 \\ 0 & 1 & 0 \\ 0 & 0 & 1 \end{bmatrix}$$

Thus, matrix representation of $S_n(z)$ will be

$$= \begin{bmatrix} \cos\theta & \sin\theta & 0 \\ -\sin\theta & \cos\theta & 0 \\ 0 & 0 & 1 \end{bmatrix}$$

This equation represents the matrix for an improper rotation about the Z-axis. As the operation is σ_h, i.e., σ_{xy}, and therefore, only the sign of coordinate *z* changes.

It should also be noted that the set of matrices describing symmetry operations of a point group can be multiplied (combined) together and the product matrix operations also belong to that group. Thus, a point group may be also represented in terms of a set of matrices corresponding to each operation. The collection of matrices obeys the definition of a group.

4.8 C_{2h} MOLECULE

Now let us consider the group C_{2h} as an example. The multiplication table for the point group C_{2h} is:

C_{2h}	E	$C_2(z)$	i	σ_h
E	E	C_2	i	σ_h
$C_2(z)$	C_2	E	σ_h	i
i	i	σ_h	E	C_2
σ_h	σ_h	i	C_2	E

We shall make use of a vector having three Cartesian components as a base set. We get the following matrices:

$$E = \begin{bmatrix} 1 & 0 & 0 \\ 0 & 1 & 0 \\ 0 & 0 & 1 \end{bmatrix}$$

$$C_2 = \begin{bmatrix} \cos\theta & \sin\theta & 0 \\ \sin\theta & \cos\theta & 0 \\ 0 & 0 & 1 \end{bmatrix}$$

$$i = \begin{bmatrix} -0 & 0 & 0 \\ 0 & -1 & 0 \\ 0 & 0 & -1 \end{bmatrix}$$

$$\sigma_h = \begin{bmatrix} -1 & 0 & 0 \\ 0 & 1 & 0 \\ 0 & 0 & -1 \end{bmatrix}$$

As we know that cos 180° = –1 and sin 180° = 0, putting these value in matrix for operation C_2. Therefore, matrix for C_2 will be:

$$C_2 = \begin{bmatrix} -1 & 0 & 0 \\ 0 & -1 & 0 \\ 0 & 0 & 1 \end{bmatrix}$$

If all the four characteristics of a group are applied to the above set of matrices.

4.8.1 IDENTITY RULE OR LAW OF COMMUTATION

One element in the group must be such that it should leave all the other elements unchanged or in other words, it is commutative will all other elements in the group, i.e.,

$$E.C_2 = C_2.E = C_2$$

$$\underset{E}{\begin{bmatrix} 1 & 0 & 0 \\ 0 & 1 & 0 \\ 0 & 0 & 1 \end{bmatrix}} \cdot \underset{C_2}{\begin{bmatrix} -1 & 0 & 0 \\ 0 & -1 & 0 \\ 0 & 0 & 1 \end{bmatrix}} = \underset{C_2}{\begin{bmatrix} -1 & 0 & 0 \\ 0 & -1 & 0 \\ 0 & 0 & 1 \end{bmatrix}}$$

$$\underset{C_2}{\begin{bmatrix} -1 & 0 & 0 \\ 0 & -1 & 0 \\ 0 & 0 & 1 \end{bmatrix}} \cdot \underset{E}{\begin{bmatrix} 1 & 0 & 0 \\ 0 & 1 & 0 \\ 0 & 0 & -1 \end{bmatrix}} = \underset{C_2}{\begin{bmatrix} -1 & 0 & 0 \\ 0 & -1 & 0 \\ 0 & 0 & 1 \end{bmatrix}}$$

4.8.2 LAW OF COMBINATION

The product of any two elements must be also an element of the group. Let us consider the product $i.C_2 = \sigma_h$.

$$\overset{i}{\begin{bmatrix} -1 & 0 & 0 \\ 0 & -1 & 0 \\ 0 & 0 & -1 \end{bmatrix}} \cdot \overset{C_2}{\begin{bmatrix} -1 & 0 & 0 \\ 0 & -1 & 0 \\ 0 & 0 & 1 \end{bmatrix}} = \overset{\sigma_h}{\begin{bmatrix} 1 & 0 & 0 \\ 0 & 1 & 0 \\ 0 & 0 & -1 \end{bmatrix}}$$

The product of the two matrices i and C_2 is the matrix, which is represented by the operations σ_h. Likewise the product $C_2.\sigma_h = i$

$$\overset{C_2}{\begin{bmatrix} -1 & 0 & 0 \\ 0 & -1 & 0 \\ 0 & 0 & 1 \end{bmatrix}} \cdot \overset{\sigma_h}{\begin{bmatrix} 1 & 0 & 0 \\ 0 & 1 & 0 \\ 0 & 0 & -1 \end{bmatrix}} = \overset{i}{\begin{bmatrix} -1 & 0 & 0 \\ 0 & -1 & 0 \\ 0 & 0 & -1 \end{bmatrix}}$$

These results show that the combination of any two matrices (product) obey the rule of combination.

4.8.3 LAW OF ASSOCIATION

The associative law must also hold good.

$$C_2.(\sigma_h.i) = (C_2.\sigma_h).i$$

$$C_2.C_2 = i.i$$

$$E = E$$

The L.H.S.: $C_2.(\sigma_h.i)$ is:

$$\overset{i}{\begin{bmatrix} -1 & 0 & 0 \\ 0 & -1 & 0 \\ 0 & 0 & -1 \end{bmatrix}} \cdot \overset{\sigma_h}{\begin{bmatrix} 1 & 0 & 0 \\ 0 & 1 & 0 \\ 0 & 0 & -1 \end{bmatrix}} = \overset{C_2}{\begin{bmatrix} -1 & 0 & 0 \\ 0 & -1 & 0 \\ 0 & 0 & 1 \end{bmatrix}}$$

Then

$$\overset{C_2}{\begin{bmatrix} -1 & 0 & 0 \\ 0 & -1 & 0 \\ 0 & 0 & 1 \end{bmatrix}} \cdot \overset{C_2}{\begin{bmatrix} -1 & 0 & 0 \\ 0 & -1 & 0 \\ 0 & 0 & 1 \end{bmatrix}} = \overset{E}{\begin{bmatrix} 1 & 0 & 0 \\ 0 & 1 & 0 \\ 0 & 0 & 1 \end{bmatrix}}$$

The R.H.S. $(C_2.\sigma_h).i$ can be solved as

$$\overset{C_2}{\begin{bmatrix} -1 & 0 & 0 \\ 0 & -1 & 0 \\ 0 & 0 & 1 \end{bmatrix}} \cdot \overset{\sigma_h}{\begin{bmatrix} 1 & 0 & 0 \\ 0 & 1 & 0 \\ 0 & 0 & -1 \end{bmatrix}} = \overset{i}{\begin{bmatrix} -1 & 0 & 0 \\ 0 & -1 & 0 \\ 0 & 0 & -1 \end{bmatrix}}$$

$$\overset{i}{\begin{bmatrix} -1 & 0 & 0 \\ 0 & -1 & 0 \\ 0 & 0 & -1 \end{bmatrix}} \cdot \overset{i}{\begin{bmatrix} -1 & 0 & 0 \\ 0 & -1 & 0 \\ 0 & 0 & -1 \end{bmatrix}} = \overset{E}{\begin{bmatrix} 1 & 0 & 0 \\ 0 & 1 & 0 \\ 0 & 0 & 1 \end{bmatrix}}$$

Hence, R.H.S. is equal to the L.H.S. These results obey the associative law.

4.8.4 LAW OF INVERSE

Every elements of the group must have its reciprocal (inverse) also as a member of that group.

If the multiplication of any element with another element gives the element identity, then it is called its reciprocal. In the group C_{2h}, each element is its own reciprocal or inverse. It can be checked by determining the products of the elements with themselves.

$$\overset{C_2}{\begin{bmatrix} -1 & 0 & 0 \\ 0 & -1 & 0 \\ 0 & 0 & 1 \end{bmatrix}} \cdot \overset{C_2}{\begin{bmatrix} -1 & 0 & 0 \\ 0 & -1 & 0 \\ 0 & 0 & 1 \end{bmatrix}} = \overset{E}{\begin{bmatrix} 1 & 0 & 0 \\ 0 & 1 & 0 \\ 0 & 0 & 1 \end{bmatrix}}$$

$$\overset{i}{\begin{bmatrix} -1 & 0 & 0 \\ 0 & -1 & 0 \\ 0 & 0 & -1 \end{bmatrix}} \cdot \overset{i}{\begin{bmatrix} -1 & 0 & 0 \\ 0 & -1 & 0 \\ 0 & 0 & -1 \end{bmatrix}} = \overset{E}{\begin{bmatrix} 1 & 0 & 0 \\ 0 & 1 & 0 \\ 0 & 0 & 1 \end{bmatrix}}$$

$$\overset{\sigma_h}{\begin{bmatrix} 1 & 0 & 0 \\ 0 & 1 & 0 \\ 0 & 0 & -1 \end{bmatrix}} \cdot \overset{\sigma_h}{\begin{bmatrix} 1 & 0 & 0 \\ 0 & 1 & 0 \\ 0 & 0 & -1 \end{bmatrix}} = \overset{E}{\begin{bmatrix} 1 & 0 & 0 \\ 0 & 1 & 0 \\ 0 & 0 & 1 \end{bmatrix}}$$

Thus, the collection of the above matrices does represent the group C_{2h}.

4.9 CHARACTER OF MATRIX

The character of a matrix is often called trace of the matrix. Thus,

$$\Gamma = \begin{bmatrix} a_{11} & 0 & 0 & 0 & 0 \\ 0 & a_{22} & 0 & 0 & 0 \\ 0 & 0 & a_{33} & 0 & 0 \\ 0 & 0 & 0 & a_{44} & 0 \\ 0 & 0 & 0 & 0 & a_{nn} \end{bmatrix}$$

It is obvious that its character is:

$$\chi = a_{11} + a_{22} + a_{33} + a_{44} + \ldots. + a_{nn} \tag{4.14}$$

or

$$\chi = \sum_{i=1}^{n} a_{ii} \tag{4.15}$$

It means character is simply the sum of its diagonal elements, which run from upper left to lower right.

4.9.1 WATER MOLECULE

To illustrate the term character, we shall consider the transformation matrices of H_2O molecule, which belongs to the C_{2v} point group. The molecule has 3 atoms (N = 3). In this case, the 3N Cartesian coordinate vectors attached to the atom can be represented as:

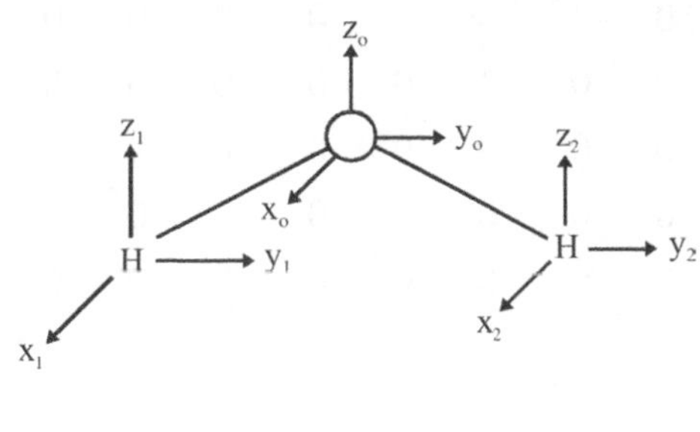

The matrices describing the symmetry transformations of identity (E) can be written as:

$$\begin{bmatrix} x_1' \\ y_1' \\ z_1' \\ x_2' \\ y_2' \\ z_2' \\ x_o \\ y_o \\ z_o \end{bmatrix} = \overset{E}{\begin{bmatrix} 1 & 0 & 0 & 0 & 0 & 0 & 0 & 0 & 0 \\ 0 & 1 & 0 & 0 & 0 & 0 & 0 & 0 & 0 \\ 0 & 0 & 1 & 0 & 0 & 0 & 0 & 0 & 0 \\ 0 & 0 & 0 & 1 & 0 & 0 & 0 & 0 & 0 \\ 0 & 0 & 0 & 0 & 1 & 0 & 0 & 0 & 0 \\ 0 & 0 & 0 & 0 & 0 & 1 & 0 & 0 & 0 \\ 0 & 0 & 0 & 0 & 0 & 0 & 1 & 0 & 0 \\ 0 & 0 & 0 & 0 & 0 & 0 & 0 & 1 & 0 \\ 0 & 0 & 0 & 0 & 0 & 0 & 0 & 0 & 1 \end{bmatrix}} \begin{bmatrix} x_1 \\ y_1 \\ z_1 \\ x_2 \\ y_2 \\ z_2 \\ x_o \\ y_o \\ z_o \end{bmatrix}$$

$\chi = 1 + 1 + 1 + 1 + 1 + 1 + 1 + 1 + 1 = 9$
$\chi = 9$

In this case, the sum of the diagonal elements is 9 and it is called the trace (spur) of the matrix and actual numerical value of the trace is called the character of the representation. Hence, character χ (E) is 9. Similarly, the symmetry transformation matrix for C_2 axis of symmetry, σ_v (xz) and σ_v (yz) planes are:

$$\begin{bmatrix} x_1' \\ y_1' \\ z_1' \\ x_2' \\ y_2' \\ z_2' \\ x_o \\ y_o \\ z_o \end{bmatrix} = \overset{C_2}{\begin{bmatrix} 0 & 0 & 0 & -1 & 0 & 0 & 0 & 0 & 0 \\ 0 & 0 & 0 & 0 & -1 & 0 & 0 & 0 & 0 \\ 0 & 0 & 0 & 0 & 0 & 1 & 0 & 0 & 0 \\ -1 & 0 & 0 & 0 & 0 & 0 & 0 & 0 & 0 \\ 0 & -1 & 0 & 0 & 0 & 0 & 0 & 0 & 0 \\ 0 & 0 & 1 & 0 & 0 & 0 & 0 & 0 & 0 \\ 0 & 0 & 0 & 0 & 0 & 0 & -1 & 0 & 0 \\ 0 & 0 & 0 & 0 & 0 & 0 & 0 & -1 & 0 \\ 0 & 0 & 0 & 0 & 0 & 0 & 0 & 0 & 1 \end{bmatrix}} \begin{bmatrix} x_1 \\ y_1 \\ z_1 \\ x_2 \\ y_2 \\ z_2 \\ x_o \\ y_o \\ z_o \end{bmatrix}$$

$\chi = 0 + 0 + 0 + 0 + 0 + 0 - 1 - 1 + 1$
$\chi = -1$

$$\sigma_v(xz)$$

$$\begin{bmatrix} x_1' \\ y_1' \\ z_1' \\ x_2' \\ y_2' \\ z_2' \\ x_o \\ y_o \\ z_o \end{bmatrix} = \begin{bmatrix} 0 & 0 & 0 & 1 & 0 & 0 & 0 & 0 & 0 \\ 0 & 0 & 0 & 0 & -1 & 0 & 0 & 0 & 0 \\ 0 & 0 & 0 & 0 & 0 & 1 & 0 & 0 & 0 \\ 1 & 0 & 0 & 0 & 0 & 0 & 0 & 0 & 0 \\ 0 & -1 & 0 & 0 & 0 & 0 & 0 & 0 & 0 \\ 0 & 0 & 1 & 0 & 0 & 0 & 0 & 0 & 0 \\ 0 & 0 & 0 & 0 & 0 & 0 & 1 & 0 & 0 \\ 0 & 0 & 0 & 0 & 0 & 0 & 0 & -1 & 0 \\ 0 & 0 & 0 & 0 & 0 & 0 & 0 & 0 & 1 \end{bmatrix} \begin{bmatrix} x_1 \\ y_1 \\ z_1 \\ x_2 \\ y_2 \\ z_2 \\ x_o \\ y_o \\ z_o \end{bmatrix}$$

$\chi = 0 + 0 + 0 + 0 + 0 + 0 + 1 - 1 + 1$
$\chi = 1$

$$\sigma_v(yz)$$

$$\begin{bmatrix} x_1' \\ y_1' \\ z_1' \\ x_2' \\ y_2' \\ z_2' \\ x_o \\ y_o \\ z_o \end{bmatrix} = \begin{bmatrix} -1 & 0 & 0 & 0 & 0 & 0 & 0 & 0 & 0 \\ 0 & 1 & 0 & 0 & 0 & 0 & 0 & 0 & 0 \\ 0 & 0 & 1 & 0 & 0 & 0 & 0 & 0 & 0 \\ 0 & 0 & 0 & -1 & 0 & 0 & 0 & 0 & 0 \\ 0 & 0 & 0 & 0 & 1 & 0 & 0 & 0 & 0 \\ 0 & 0 & 0 & 0 & 0 & 1 & 0 & 0 & 0 \\ 0 & 0 & 0 & 0 & 0 & 0 & -1 & 0 & 0 \\ 0 & 0 & 0 & 0 & 0 & 0 & 0 & 1 & 0 \\ 0 & 0 & 0 & 0 & 0 & 0 & 0 & 0 & 1 \end{bmatrix} \begin{bmatrix} x_1 \\ y_1 \\ z_1 \\ x_2 \\ y_2 \\ z_2 \\ x_o \\ y_o \\ z_o \end{bmatrix}$$

$\chi = -1 + 1 + 1 - 1 + 1 + 1 - 1 + 1 + 1$
$\chi = 3$
$\chi\ (\sigma_v(yz)) = (-1) + 1 + 1 + (-1) + 1 + 1 + (-1) + 1 + 1 = 3$
The character of the C_2 matrix will be
$\chi\ (C_2) = 0 + 0 + 0 + 0 + 0 + 0 + (-1) + (-1) + 1 = -1$
$\chi\ (\sigma_v(xz)) = 0 + 0 + 0 + 0 + 0 + 0 + 1 + (-1) + 1 = 1$

A little observation of the above transformations indicates that the diagonal elements occur only, when a given symmetry operation leaves the position of an atom unchanged. Hence, the character of representation matrix can be obtained by taking into consideration the displacement vectors of the atoms, whose position remain unchanged as a result of the symmetry operation.

Thus in case of H_2O, the vector, which remains unchanged contribute +1 while the vectors, which are reversed contribute -1 along the diagonal. Consequently, the character for operations E is 9 and C_2 is -1. Similarly, it can be confirmed that χ for $\sigma_v(xz)$ is 1 and for $\sigma_v(yz)$, it is 3.

Conjugated matrices have identical characters. The conjugated matrices also follow the same rule, which is obeyed by the conjugated elements of a group, i.e., when matrix R is conjugate to matrix P, then their must be another matrix A, so that:

$$R = A^{-1}. P.A$$

4.10 REPRESENTATION OF GROUPS

Each operation in a point group is represented by set of numbers (matrices). When any two matrices, representing two operations, are multiplied, then it results into another matrix. It represents an operation of the group. Thus, a representation is a set of matrices, which represent the operation of a point group. A representation is donated by T.

C_{2v} point group has four symmetry element $E + C_2 + \sigma_v\,(xz) + \sigma_v\,(yz)$. Thus,

$$\Gamma = \overset{E}{\begin{bmatrix} 1 & 0 & 0 \\ 0 & 1 & 0 \\ 0 & 0 & 1 \end{bmatrix}} \overset{C_2}{\begin{bmatrix} -1 & 0 & 0 \\ 0 & -1 & 0 \\ 0 & 0 & 1 \end{bmatrix}} \overset{\sigma_v(xz)}{\begin{bmatrix} 1 & 0 & 0 \\ 0 & -1 & 0 \\ 0 & 0 & 1 \end{bmatrix}} \overset{\sigma_v(yz)}{\begin{bmatrix} -1 & 0 & 0 \\ 0 & 1 & 0 \\ 0 & 0 & 1 \end{bmatrix}}$$

If two symmetry operations in point group, say, $\sigma_v\,(xz).C_2$ gives $\sigma_v\,(yz)$, then matrix corresponding to C_2 and $\sigma_v\,(xz)$ must multiply to give a result the matrix corresponding $\sigma_v\,(yz)$.

$$\sigma_v(xz).C_2 = \sigma_v\,(yz)$$

$$\begin{bmatrix} 1 & 0 & 0 \\ 0 & -1 & 0 \\ 0 & 0 & 1 \end{bmatrix} . \begin{bmatrix} -1 & 0 & 0 \\ 0 & -1 & 0 \\ 0 & 0 & 1 \end{bmatrix} = \begin{bmatrix} -1 & 0 & 0 \\ 0 & 1 & 0 \\ 0 & 0 & 1 \end{bmatrix}$$

Similarly, if we consider symmetry element σ_v (xz).σ_v(xz), then the product is equal to identity element (E). The corresponding matrices multiply together in same way will give a matrix corresponding to E.

$$\begin{bmatrix} 1 & 0 & 0 \\ 0 & -1 & 0 \\ 0 & 0 & 1 \end{bmatrix} . \begin{bmatrix} 1 & 0 & 0 \\ 0 & -1 & 0 \\ 0 & 0 & 1 \end{bmatrix} = \begin{bmatrix} 1 & 0 & 0 \\ 0 & 1 & 0 \\ 0 & 0 & 1 \end{bmatrix}$$

$$\sigma_v(xz).\ \sigma_v(xz) = E$$

The large size diagonal matrices of element of group C_{2v} can also be constructed with ± 1 as the diagonal number as:

	E	C_2	σ_v(xz)	σ_v (yz)
Γ_1	[1]	[–1]	[1]	[–1]
Γ_1	[1]	[–1]	[–1]	[1]
Γ_1	[1]	[1]	[1]	[1]

Thus, representation for *x*, *y* and *z* coordinates can also be shown by this table.

Let us take another point group C_{2h}, which consists of four symmetry elements, i.e., E + C_2 + σ_h + i. Matrix for each element is as follows:

C_{2h} = E, C_2, σ_h, i

$$\Gamma = \overset{E}{\begin{bmatrix} 1 & 0 & 0 \\ 0 & 1 & 0 \\ 0 & 0 & 1 \end{bmatrix}} \overset{C_2}{\begin{bmatrix} -1 & 0 & 0 \\ 0 & -1 & 0 \\ 0 & 0 & 1 \end{bmatrix}} \overset{\sigma_h}{\begin{bmatrix} 1 & 0 & 0 \\ 0 & 1 & 0 \\ 0 & 0 & -1 \end{bmatrix}} \overset{i}{\begin{bmatrix} -1 & 0 & 0 \\ 0 & -1 & 0 \\ 0 & 0 & -1 \end{bmatrix}}$$

or Γ = [3] [–1] [1] [–3]

Now binary combination of two symmetry elements $C_2.\sigma_h = i$ can be represented as:

$C_2.\sigma_h = i$

So $-1 \times 1 = -1$

Hence,

$$\Gamma = \begin{bmatrix} 1 & 0 \\ 0 & 1 \end{bmatrix} . \begin{bmatrix} 1 & 0 \\ 0 & 1 \end{bmatrix} = \begin{bmatrix} 1 & 0 \\ 0 & 1 \end{bmatrix}$$

or $[1] \times [1] = [1]$

4.11 REDUCIBLE AND IRREDUCIBLE REPRESENTATIONS

4.11.1 REDUCIBLE REPRESENTATION

Suppose, we have a set of matrices A, B, C, D and X, which is a representation of group. If we perform similarity transformation on each matrix (big matrices), we get new set of matrices (smaller matrices).

For example,

$X^{-1}. A. X = A'$

$X^{-1}. B. X = B'$

$X^{-1}. C. X = C'$

$X^{-1}. D. X = D'$ and so on

Thus, on making same similarity transformation on matrices (A, B, C, D, ...), we obtain new set of matrices (A′, B′, C′, D′, ...). The new sets of matrices are also representation of the group. This can be proved mathematically as:

$A . B = C$ then $A' . B' = C'$

Therefore, $= (X^{-1}. A. X) (X^{-1}. B. X)$

$= X^{-1}. A. (X. X^{-1}) B. X$

$= X^{-1} (A. E). B. X$

$= X^{-1} (A. B) X$

$= X^{-1}. C. X$

$= C'$

It means all products is the set of matrices A′, B′, C′, ..., will run parallel to those in the representation A, B. C, ... Thus, the prime set also constitutes a representation.

When the matrix of any element in a group is transformed to new element using some other matrix, then we can find a new matrix in a block-factored matrix as follows:

$$A' = X^{-1}.\, A.\, X = \begin{bmatrix} A_1' & & & \\ & A_2' & & \\ & & A_3' & \\ & & & A_4' \end{bmatrix}$$

Now, a combination (product) of element matrices gives the following form:

$$A_1.\, B_1 = C_1$$

$$\begin{bmatrix} A_1 & & \\ & A_2 & \\ & & A_3 \end{bmatrix} \cdot \begin{bmatrix} B_1 & & \\ & B_2 & \\ & & B_3 \end{bmatrix} = \begin{bmatrix} C_1 & & \\ & C_2 & \\ & & C_3 \end{bmatrix}$$

Thus, C′ is a small dimension matrix and must be the product of A′ and B′. Similarly, each of the matrices block out in the same manner, corresponding block of each matrix can be multiplied together (binary combination) separately. Thus, equation can be written as:

$$C_2 = A_2.\, B_2$$

$$C_3 = A_3.\, B_3$$

In essence, representations of higher dimension, which can be reduced to representation of lower dimension by similarity transformation process are block diagonal matrices. The similarity transformation matrix, which cannot be reduced further to representation of lower dimension is called irreducible representation.

Let us take general example to explain representation. Suppose A, B and X are three symmetry element in which A, B, X are of same dimension whereas B matrix may be a block diagonal matrix, then

$$A = \begin{bmatrix} a_{11} & a_{12} & a_{13} \\ a_{21} & a_{22} & a_{23} \\ a_{31} & a_{32} & a_{33} \end{bmatrix}$$

Reducible representation

On similarity transformation, $X^{-1} A X$, A becomes:

$$A = \left[\begin{array}{cc|c} b_{11} & b_{12} & 0 \\ b_{21} & b_{22} & 0 \\ \hline 0 & 0 & b_{33} \end{array}\right] \quad \text{or} \quad \left[\begin{array}{c|c} B_1 & \\ \hline & B_2 \end{array}\right]$$

[B1] [B2]

Irreducible representations

This is known as block diagonal matrix. Here, similarity transformation of matrix A of 3×3 dimension gives two matrices B_1 and B_2. B_1 is of 2×2 dimension and B_2 is of 1×1 dimension. Thus, these representations are reducible because similarity transformation has block diagonalized the original matrix to matrices of reduced order in block form. If this reducible matrix does not reduce further, then the representation is called irreducible representation. It is irreducible representation of a group that is of fundamental importance for application of group theory to various chemical problems.

Here Γ_1 = 2 x 2 Dimension representation ┐

Γ_2 = 1 x 1 Dimension representation ┘ — Irreducible representation

Matrix, which is broken down by using similarity transformation, also represents the group.

4.11.2 IRREDUCIBLE REPRESENTATION

When dimension of any of the representation may not be reduced by using similarity transformation, then the final representation, which has set of matrix of 1×1 dimension is said to be irreducible representation. The number of irreducible representations for a point group will depend on the classes of symmetry operations of the group, i.e., irreducible representation will be equal to number of classes. The irreducible representations are the characteristics of the group, and are also known as symmetry species. Thus, it can be concluded that an irreducible representation is one, which cannot be broken down into matrices with smaller dimensions. In other words, an irreducible representation cannot be further reduced to a simpler representation.

Point group C_{2v} contains four elements, which indicates that the order of group is 4. This order cannot be reduced by similarity transformation and hence, there are four classes. Therefore, four irreducible representations are there in point group C_{2v} whereas in C_{3v} point group, the order of group is 6, which can be further reduced into three classes, and hence, there are three irreducible representations of point group C_{3v}.

4.11.3 RELATIONSHIP BETWEEN REDUCIBLE AND IRRCDUCIBLE REPRESENTATIONS

We can reduce any reducible representation without knowing the transformation matrix. The relationship between the reducible and irreducible representations is expressed as:

$$a_i = 1/h\ [n_R n_R \chi_i\ (R)\ \chi\ (R)] \tag{4.16}$$

where a_i is the number of times ith reducible representation occurred in a reducible representation, h is the order of group. χ_i (R) is the character of the ith irreducible representation corresponding to operation (R), and χ (R) is the character of reducible representation corresponding to the same operation.

Thus, we can determine the number of times the ith irreducible representation will occur in a reducible representation, if only the characters of the representation are known. Let us consider examples of group C_{3v} and C_{2v} to make this idea more clear.

4.11.4 C_{3V} GROUP

In the group C_{3v}, the irreducible representations are Γ_1, Γ_2 and Γ_3 and two reducible representations are Γ_a and Γ_b.

C_{3v}	E	$2C_3$	$3\sigma_v$
Γ_1	1	1	1
Γ_2	1	1	–1
Γ_3	2	–1	0
Γ_a	5	2	–1
Γ_b	7	1	–3

Using Eq. (4.16), for Γ_a

$$a_{\Gamma_1} = 1/6\,[1(1)(5) + 2(1)(2) + 3(1)(-1)]$$
$$= 1/6\,[5 + 4 - 3]$$
$$= 1/6\,(6) = 1$$

So $a_{\Gamma_1} = 1$

$$a_{\Gamma_2} = 1/6\,[1(1)(5) + 2(1)(2) + 3(-1)(-1)]$$
$$= 1/6\,[5 + 4 + 3]$$
$$= 1/6\,(12) = 2$$

So $a_{\Gamma_2} = 2$

$$a_{\Gamma_3} = 1/6\,[1(2)(5) + 2(-1)(2) + 3(0)(-1)]$$
$$= 1/6\,(10 - 4 + 0) \text{ or } = 1/6 = 1$$

So $a_{\Gamma_3} = 1$

For Γ_b

$$a_{\Gamma_{1'}} = 1/6\,[1(1)(7) + 2(1)(1) + 3(1)(-3)]$$
$$= 1/6\,(7 + 2 - 9)$$
$$= 1/6\,(0) = 0$$

So $a_{\Gamma_{1'}} = 0$

$$a_{\Gamma_{2'}} = 1/6\,[1(1)(7) + 2(1)(1) + 3(-1)(-3)]$$
$$= 1/6\,[7 + 2 + 9] = 1/6–18 = 3$$

So $a_{\Gamma_{2'}} = 3$

$$a_{\Gamma_{3'}} = 1/6\,[1(2)(7) + 2(-1)(1) + 3(0)(-3)]$$
$$= 1/6\,(14 - 2 - 0)$$
$$= 1/6\,(12) = 2$$

So $a_{\Gamma_{3'}} = 2$

The results obtained above will be found to satisfy Eq. (4.14).

For Γ_a

As a_{Γ_1}, a_{Γ_2} and a_{Γ_3} are 1, 2 and 1, respectively; so Γ_1 and Γ_3 are written once only but Γ_2 is 2 and therefore, it is to be written twice.

	E	$2C_3$	$3\sigma_v$
Γ_1	1	1	1
Γ_2	1	1	–1
Γ_2	1	1	–1
Γ_3	2	–1	0
Γ_a	5	2	–1

For Γ_b

Similarly for Γ_b, a_{Γ_1}, a_{Γ_2} and a_{Γ_3} are 0, 3 and 2, respectively and hence, Γ_1 is not written, but Γ_2 and Γ_3 are written thrice and twice, respectively.

	E	$2C_3$	$3\sigma_v$
Γ_2	1	1	–1
Γ_2	1	1	–1
Γ_2	1	1	–1
Γ_3	2	–1	0
Γ_a	2	–1	0
Γ_b	7	1	–3

4.11.5 C_{2V} GROUP

We can determine the irreducible components of Γ_R, i.e., Γ_1, Γ_2, Γ_3 and Γ_4. In this case, the irreducible representation for E will be:

C_{2v}	E	C_2	$\sigma_v(xz)$	$\sigma_v(yz)$
Γ_1	1	1	1	1
Γ_2	1	1	–1	–1
Γ_3	1	–1	1	–1
Γ_4	1	–1	–1	1
Γ_R	9	–1	1	3

$$a_{\Gamma_1} = 1/4\,[1(1)(9) + 1(1)(-1) + 1\,(1)(1) + 1(1)\,(3)]$$
$$= 1/4\,[9 - 1 + 1 + 3]$$
$$= 1/4\,(12) = 3$$

So $a_{\Gamma_1} = 3$

$$a_{\Gamma_2} = 1/4\,[1(1)(9) + 1(1)(-1) + 1\,(-1)(1) + 1(-1)\,(3)]$$
$$= 1/4\,[9 - 1 - 1 - 3] = 1/4\,(4) = 1$$

So $a_{\Gamma_2} = 1$

$$a_{\Gamma_3} = 1/4\,[1(1)(9) + 1(-1)(-1) + 1\,(1)(1) + 1(-1)\,(3)]$$
$$= 1/4\,[9 + 1 + 1 - 3] = 1/4\,(8) = 2$$

So $a_{\Gamma 3} = 2$

$$a_{\Gamma_4} = 1/4\,[1(1)(9) + 1(-1)(-1) + 1\,(-1)(1) + 1(1)\,(3)]$$
$$= 1/4\,[9 + 1 - 1 + 3] = 1/4\,(12) = 3$$

So $a_{\Gamma_4} = 3$

Similarly, $\Gamma_R\,(C_2) = 3(1) + 1 + 2(-1) + 3(-1) = -1$

$$\Gamma_R\,(\sigma_v) = 3(1) - 1 + 2(1) + 3(-1) = 1$$

$$\Gamma_R\,(\sigma_v') = 3(1) - 1 + 2(-1) + 3(1) = 3$$

Hence, we can write that

$$\Gamma_R = 3\,\Gamma_1 + \Gamma_2 + 2\,\Gamma_3 + 3\,\Gamma_4$$

Thus, $\Gamma_R\,(E) = 3(1) + 1 + 2(1) + 3\,(1) = 9$

KEYWORDS

- **Character**
- **Irreducible**
- **Matrix**
- **Reducible**
- **Representation**

CHAPTER 5

CHARACTER TABLES

CONTENTS

There are two theorems of fundamental importance. These are known as Schur's lemmas. These lemmas are useful for the study of irreducible representation (IR).

Schur's lemma 1: If Γ_i is an irreducible representation of a group and if a matrix P commutes with all the matrices of this irreducible representation, then this matrix P must be a constant matrix, i.e., $P = C \times E$, where C is scalar quantity. It means that if a nonconstant-commuting matrix exists, then the representation is reducible. On the other hand, if none exists, then the representation is irreducible.

Schur's lemma 2: If Γ_i and Γ_j are two irreducible representations of a group $G = \{A_i, i = 1, 2, 3, \ldots, n\}$ with l_i and l_j dimensions, respectively and a matrix M (of the order l_i and l_j) satisfy the following relation.

$$\Gamma_i (A_i) (M) = M \Gamma_j (A_i) A_i \in G$$

Then there are two possibilities.
Either (i) M = 0; the none matrix;
or (ii) $M \neq 0$; in this case Γ_i and Γ_j are equivalent representations.
It is clear that two representations Γ_i and Γ_j can be equivalent only in the condition, if their dimensions are equal. In this case, if $l_i \neq l_j$, then case (ii) is not valid and hence, case (i) is applicable.

5.1 THE GREAT ORTHOGONALITY THEOREM

For a particular point group, there could be number of reducible representations (R), but the irreducible representations are finite in number. The number of irreducible representation remains same for the molecules of a specific point group, i.e., irreducible representations are different in two different point groups. In order to derive all the properties of group representation and their character, "Great Orthogonality Theorem" (GOT) was introduced. Basically, this theorem is related to the element of matrices, which constitute the irreducible representations of a group. Therefore, it can be said that GOT is used to derive the properties of irreducible representations.

Mathematically, the great orthogonality theorem may be stated as:

$$\sum_R [\Gamma_i (R)_{mn}] [\Gamma_j (R)_{m'n'}]^* = \frac{h}{\sqrt{l_i l_j}} \delta_{ij} \delta_{mm'} \delta_{nn'} \qquad (5.1)$$

It means that if in a set of matrices (constituting any irreducible representation), any set of corresponding matrix element (one from each matrix) behaves as the component of vector in h-dimensional space in such a way that all these vectors are orthogonal to each other and each one of them is normalized so that the square of its length equals to h/l_i.

Here h denotes the order of a group, Γ_i and Γ_j are the ith and j^{th} irreducible representations of a point group of the order h with dimension l_i and l_j, respectively. The matrix element of the mth row and n^{th} column corresponding to an operation R, belonging to ith irreducible representation is represented as $\Gamma_i (R)_{mn}$ and $[\Gamma_j (R)_{m'n'}]^+$ is complex conjugate of the mth and n^{th} matrix element. It is very important to consider complex conjugate of one factor on the left hand side, in case when complex or imaginary numbers are involved. δ is Kronecker delta function.

The Kronecker deltas (δ) may have values, 0 or 1 depending on i^{th}, j^{th}, m^{th}, n^{th}, m'^{th}, and n'^{th}. Thus, δ have following properties.

(i) $\delta_{ij} = \begin{cases} 1 & \text{(When } i = j\text{)} \\ 0 & \text{(When } i \neq j\text{)} \end{cases}$

(ii) $\delta_{mm'} = \begin{cases} 1 & \text{(When } m = m'\text{)} \\ 0 & \text{(When } m \neq m'\text{)} \end{cases}$

(iii) $\delta_{nn'} = \begin{cases} 1 & \text{(When } n = n'\text{)} \\ 0 & \text{(When } n \neq n'\text{)} \end{cases}$

If matrix elements are real, then Eq. (5.1) can be represented as:

$$\Gamma_i(R)_{m'n'} = \Gamma_j(R)_{m'n'} \tag{5.2}$$

On assuming that the matrix element are real, Eq. (5.1) can be represented in the form of three simpler equations:

(i) For two different irreducible representations.

$$i \neq j,\ m = m' \text{ and } n = n'$$

Then,

$$\sum_R \Gamma_i(R)_{mn}\, \Gamma_j(R)_{m'n'} = 0 \tag{5.3}$$

(ii) For the same irreducible representations.

$$i = j,\ m \neq m' \text{ and } n \neq n'$$

Then,

$$\sum_R \Gamma_i(R)_{mn}\, \Gamma_j(R)_{m'n'} = 0 \tag{5.4}$$

(iii) For irreducible representations.

$$i = j,\ m = m' \text{ and } n = n'$$

$$\sum_{R} [\Gamma_i (R)_{mn}]^2 = h/l_i \qquad (5.5)$$

The vectors are orthogonal, if they are selected from matrices of different representation (Eq. 5.3; $i \neq j$) or if they belong to the same representation but different sets of elements in the matrices representations (Eq. 5.4; $m \neq m'$ and/or $n \neq n'$). While according to Eq. (5.5), the square of the length of such vector is equals to h/l_i.

Five important rules (properties) about the irreducible representation and their character can be derived from this theorem, which are as follows:

(i) The number of irreducible representations of a group is equal to the number of classes in that group.

(ii) The sum of the squares of the dimensions of the irreducible representations of a group is equal to the order of the group, i.e.,

$$l_1^2 + l_2^2 + l_3^2 + \ldots + l_n^2 = \Sigma\, l_i^2 = h \qquad (5.6)$$

As the dimension of an irreducible representation is equal to the character of its E operation. Therefore, the sum of the square of character of identity operation of the irreducible representation is equal to the order of the representation, we can say that

$$\sum_i [\chi_i (E)]^2 = h \qquad (5.7)$$

χ_i (E) is the character of the representation of E in the ith irreducible representation.

That is, the sum of squares of character of identity operation of the irreducible representation of group is equal to the order of the group.

(iii) The sum of the square of the characters in any irreducible representation is equal to order of the group h, i.e.,

$$\sum_{R} [\chi_i (R)]^2 = h \qquad (5.8)$$

(iv) The vectors, whose components are the characters of two different irreducible representations, are orthogonal, i.e., the character of the symmetry operation in two different irreducible representations satisfy the following relation.

$$\sum_R \Gamma_i (R)_{mn} \Gamma_j (R)_{m'n'} = 0 \quad (5.9)$$

when $i \neq j$.

Here n is the order of the class and χ_i (R) and χ_j (R) is the character of the same symmetry operations in the ith and j^{th} irreducible representation. It can be said that character of the irreducible representations of the same group or two different irreducible representations of a group are orthogonal.

(v) The character of a matrix does not change by the process of similarity transformation on it. Therefore in a given representation (R or IR), the characters of the element of all matrices belonging to operation in the same class are identical, i.e.,

$$\sum \chi \text{ (Irreducible)} = \chi \text{ (Reducible)}$$

i.e., the sum of the characters of the irreducible representation matrices and reducible representation matrix are equal.

5.2 CONSTRUCTION OF CHARACTER TABLE

Let us consider the irreducible representations of some typical point groups and construct their character table by using five rules derived from great orthogonality theorem.

5.2.1 THE C_{2V} GROUP

Step 1: According to rule (i), number of irreducible representation of a point group is equal to number of classes of the group. As C_{2v} consists of four elements of symmetry (E, C_2, σ_v (xz), σ_v (yz)) and each of them is a separate class also, i.e., E, C_2, σ_v (xz), σ_v (yz). Hence, according to this rule, number of irreducible representation will also be four and may represented as Γ_1, Γ_2, Γ_3 and Γ_4. Therefore, initially character table can be represented as:

C_{2v}	E	C_2	$\sigma_v(xz)$	$\sigma_v(yz)$
Γ_1				
Γ_2				
Γ_3				
Γ_4				

Step 2: As per rule (ii), the sum of squares of the dimensions of these representations should be equal to order of group (h). Order of group C_{2v} is four, and therefore,

$$\sum_{i=1}^{4} l_i^2 = 4$$

The sum of l_1^2, l_2^2, l_3^2 and l_4^2 will be equal to 4 (four) only, if $l_1 = l_2 = l_3 = l_4 = +1$, i.e., $1^2 + 1^2 + 1^2 + 1^2 = 4$.

The dimension of a representation is equal to the character of the identity operation of irreducible representation. Thus, +1 should be written as the character of E operation for all the four irreducible representations, i.e.,

C_{2v}	E	C_2	$\sigma_v(xz)$	$\sigma_v(yz)$
Γ_1	1			
Γ_2	1			
Γ_3	1			
Γ_4	1			

Step 3: Applying rule (iii), i.e., the sum of squares of the character must be equal to order of the group (h). Hence,

$$\sum_R [\chi_i(R)]^2 = h = 4$$

This is valid only, if

$$[\chi(E)]^2 + [\chi(C_2)]^2 + [\chi(\sigma_v(xz))]^2 + [\chi(\sigma_v(yz))]^2 = 4$$

$$\text{or } 1^2 + 1^2 + 1^2 + 1^2 = 4$$

It means that among all the irreducible representations of a group, one irreducible representation will have all the character equal to +1. Let us say, that Γ_1 will have all the character equal to +1, then

C_{2v}	E	C_2	$\sigma_v(xz)$	$\sigma_v(yz)$
Γ_1	1	1	1	1
Γ_2	1			
Γ_3	1			
Γ_4	1			

Step 4: Now before applying rule (iv), let us assume

Γ_2 has characters A, B, C

Γ_3 has characters P, Q, R, and

Γ_4 has characters *X, Y, Z*.

C_{2v}	E	C_2	$\sigma_v(xz)$	$\sigma_v(yz)$
Γ_1	1	1	1	1
Γ_2	1	A	B	C
Γ_3	1	P	Q	R
Γ_4	1	X	Y	Z

Then, according to rule (iv), the orthogonality condition is applied.

$$\sum n.\chi_i(R), \chi_j(R) = 0$$

It means sum of the product of the characters under two irreducible representations along with order of that class should be equals to zero.

Considering orthogonality of Γ_1 and Γ_2, we have:

$$\text{For } \Gamma_1.\Gamma_2 = n_E.\chi_1(E).\chi_2(E) + n.\chi_1(C_2).\chi_2(C_2) + n.\chi_1((\sigma_v(xz)).\chi_2((\sigma_v(xz)) + n.\chi_1((\sigma_v(yz)).\chi_2((\sigma_v(yz)) = 0$$

In this case, n for each class is equal to one, and therefore,

$$1.1.1 + 1.1.A + 1.1.B + 1.1.C = 0$$

$$1 + A + B + C = 0$$

It is possible only, when out of remaining three characters one character is +1 and the other two are −1.
Let A = 1 and B = C = −1, then
1 + 1 + (−1) + (−1) = 0
In the same manner,

$$\text{For } \Gamma_1.\Gamma_3 = n.\chi_1(E).\chi_3(E) + n.\chi_1(C_2).\chi_3(C_2) + n.\chi_1(\sigma_v(xz)).\chi_3(\sigma_v(xz)) + n.\chi_1(\sigma_v(yz)).\chi_3(\sigma_v(yz)) = 0$$

$$1.1.1 + 1.1.P + 1.1.Q + 1.1.R = 0$$

$$1 + P + Q + R = 0$$

Now this is only possible, when out of remaining three characters, one character is +1 and the other two are −1.
Let P = R = −1 and Q = +1
Then, 1 + (−1) + 1 + (−1) = 0
Similarly,

$$\Gamma_1.\Gamma_4 = n.\chi_1(E).\chi_4(E) + n.\chi_1(C_2).\chi_4(C_2) + n.\chi_1(\sigma_v(xz)).\chi_4(\sigma_v(xz)) + n.\chi_1(\sigma_v(yz).\chi_4(\sigma_v(yz)) = 0$$

$$1.1.1 + 1.1.X + 1.1.Y + 1.1.Z = 0$$

$$1 + X + Y + Z = 0$$

Again, it is possible, if

$$X = Y = -1 \text{ and } Z = +1$$

So that 1 + (−1) + (−1) + 1 = 0
Finally, character table for C_{2v} point group can be written as:

C_{2v}	E	C_2	$\sigma_v(xz)$	$\sigma_v(yz)$
Γ_1	1	1	1	1
Γ_2	1	1	−1	−1
Γ_3	1	−1	1	−1
Γ_4	1	−1	−1	1

5.2.2 C_{3V} GROUP

Step 1: It consists of following six elements, i.e., E, C_3^+, C_3^-, σ_v (a), σ_v (*b*), σ_v (*c*) and the classes are – E, 2 C_3, and 3 σ_v. Hence, according to rule (i), there are three irreducible representations for this group as there are three classes.

C_{3v}	E	$2C_2$	$3\sigma_v$
Γ_1			
Γ_2			
Γ_3			

Step 2: The order of this group is equal to number of elements in this group, i.e., six and hence, h = 6. If we denote their dimensions by l_1, l_2 and l_3; then, according to rule (ii), we have

$$l_1^2 + l_2^2 + l_3^2 = 6$$

The only values of l_i, which satisfy these requirements are 1, 1 and 2.

i.e., $$1^2 + 1^2 + 2^2 = 6$$

As the dimension of representation is equal to the character of the identity operation of irreducible representation. Therefore,

$$E_1 - l_1 = 1$$

$$E_2 = l_2 = 1$$

$$E_3 = l_3 = 2$$

	E	$2C_3$	$3\sigma_v$
Γ_1	1		
Γ_2	1		
Γ_3	2		

It mean l_1 and l_2 are one dimensional and l_3 is two dimensional.

Thus, these values will also satisfy rule (ii), i.e., $\sum [\chi_i (E)]^2 = h$

Step 3: In any group, always there will be a one-dimensional representation, whose all characters are equal to 1. Thus, we have $\chi_1 = \chi_2 = \chi_3 = 1$.

C_{3v}	E	$2C_3$	$3\sigma_v$
Γ_1	1	1	1
Γ_2	1		
Γ_3	2		

For Γ_1

$$(1)^2 + 2\,(1)^2 + 3\,(1)^2 = 6$$

It will satisfy rule (iii), i.e.,

$$\sum_R [\chi_i\,(R)]^2 = h$$

Step 4: According to rule (iv), the Γ_2 should be orthogonal to Γ_1. It means that, one out of both the characters $\chi\,(C_3)$ and $\chi\,(\sigma_v)$ must have negative value so that the condition of orthogonality can be satisfied. The only possibility seems to satisfy this condition is $\chi\,(C_3) = +1$ and $\chi\,(\sigma_v) = -1$

C_{3v}	E	$2C_3$	$3\sigma_v$
Γ_1	1	1	1
Γ_2	1	1	–1
Γ_3	2		

This will satisfy the rule (iv), i.e.,

$$\sum_R n\,\chi_i\,(R)\,\chi_j\,(R) = 0$$

$$= 1\,(1)\,(1) + 2\,(1)\,(1) + 3\,(1)\,(-1)$$

$$= 1 + 2 - 3 = 0$$

Now our third representation Γ_3 is of dimension 2, i.e., χ_3 (E) = 2. In order to find out the values of χ_3 (C_3) and χ_3 (σ_v), we make use of orthogonality relationship according to rule (iv).

$$\sum_R \chi_1(R)\chi_3(R) = 1\,(1)\,(2) + 2\,(1)\,[\chi_3\,(C_3)] + 3\,(1)\,[\chi_3\,(\sigma_v)] = 0$$

$$\sum_R \chi_2(R)\chi_3(R) = 1\ (1)\ (2) + 2\ (1)\ [\chi_3\ (C_3)] + 3\ (-1)\ [\chi_3\ (\sigma_v)] = 0$$

Solving these

$$2\ \chi_3\ (C_3) + 3\ \chi_3\ (\sigma_v) = -2 \quad (5.10)$$

$$2\ \chi_3\ (C_3) - 3\ \chi_3\ (\sigma_v) = -2 \quad (5.11)$$

Addition of Eqs. (5.10) and (5.11) gives:

$$4\ \chi_3\ (C_3) = -4$$

$$\chi_3\ (C_3) = -1$$

Putting this value of $\chi_3\ (C_3)$ in Eq. (5.10), we get

$$2\ (-1) + 3\ \chi_3\ (\sigma_v) = -2$$

$$3\ \chi_3\ (\sigma_v) = 2\text{–}2 = 0$$

or $$\chi_3\ (\sigma_v) = 0$$

Thus, the complete set of characters of the irreducible representations of group C_{3v} is:

C_{2v}	E	$2C_3$	$3\sigma_v$
Γ_1	1	1	1
Γ_2	1	1	–1
Γ_3	2	–1	0

We may note here that there is still a check on the correctness of Γ_3. According to rule (ii), the expression $[\chi_i\ (R)]^2 = h$ must be satisfied. Putting the values, we see that this is true.

$$1\ (2)^2 + 2\ (-1)^2 + 3\ (0)^2 = 6$$

5.3 PRESENTATION OF CHARACTER TABLES

When the motion of a molecule belonging to a particular point group is represented in the form of transformation matrices, then it can be arranged in a tabular form. Such a table is called the character table of a particular point group.

Character table is formed by mathematical technique, which is based on the properties of irreducible representation using several special basis sets. One can derive many useful information about different properties of a molecule from the character tables, like hybridization, crystal field theory, modes of fundamental vibrations, IR and Raman activity, bond order, delocalization energy, free valency, bond length, etc.

In general, the character table can be divided in six different areas as:

I	**II**		
III	**IV**	**V**	**VI**

Different areas of the character table have been assigned Roman numerals for differentiation:

Area I: It gives Schoenflies notation for the point group.

Area II: It lists the distinct elements of the group in the form of classes.

Area III: It lists the special symbols, used to designate the irreducible representation, which are known as Mulliken symbols. They provide information in an extremely concise form about the nature of irreducible representation. Their meaning are as follows:

(i) All one dimensional irreducible representations are designated with A and B, all two dimensional representations as E, three dimensional representations as T or F and so on.

(ii) One dimensional irreducible representation is labeled as A, if it is symmetric with respects to rotation about the proper principal axis C_n (Symmetric means $\chi(C_n) = +1$), but if it is antisymmetric with respect to rotation about C_n (i.e., $\chi(C_n) = -1$), then it is designated is B.

(iii) If the irreducible representation is symmetric with respect to rotation about a C_2 axis perpendicular to C_n or symmetric with respect to reflection in a σ_v plane, a subscript one (1) is attached to A and B to give A_1 and B_1. If it is antisymmetric in this respect, a subscript two (2) is attached to A and B to give A_2 and B_2.

(iv) Primes (') and double primes (") are attached to those representations, which are symmetric and antisymmetric, respectively with respect to reflection in a σ_h plane.

(v) Finally, subscript g (German gerade = even) and u (ungerade = uneven) are given to representations, which are symmetric and antisymmetric, respectively with respect to center of symmetry (i).

(vi) The use of numerical subscript for E's and T's also follows certain rules, but these cannot be that easily stated.

Area IV: It consists of character for irreducible representation corresponding to class.

Area V: It shows the irreducible representation for which the coordinates *x*, *y* and *z* as well as the rotations about the axis specified in the subscripts, i.e., R_x, R_y and R_z provide the bases.

Area VI: It lists how the functions corresponding to the binary combinations of *x*, *y* and *z* provide bases for certain irreducible representations?

5.3.1 CHARACTER TABLE FOR C_{2v} POINT GROUP

Area I: Its Schoenflies notation is C_{2v}.

Area II: It lists the distinct classes of group for C_{2v} point group. These are

$$C_{2v} = \{E, C_2, \sigma_v (xz), \sigma_v (yz)\}$$

Area III: It contains Mulliken symbols. All the irreducible representations of C_{2v} are one dimensional. They are designated as A or B. In order to differentiate these representations, we proceed to the next operation, i.e., C_2, in this point group. It may be seen that Γ_1 and Γ_2 are symmetric with respect to rotational axis C_2. Since $\chi (C_2) = +1$. Hence, these are designated as A. The other two representations Γ_3 and Γ_4 are antisymmetric with this respect to C_2, i.e., $\chi (C_2) = -1$, and hence, both of these are labeled as B.

Γ_1 and Γ_2 can be further differentiated with respect to the reflection plane $\sigma_v (xz)$. Γ_1 is symmetric with respect to reflection plane; $\sigma_v (xz) = +1$ and hence, it is designated as A_1 whereas Γ_2 is designated as A_2 as it is antisymmetric with respect to reflection plane $\sigma_v (xz)$. Similarly, Γ_3 and Γ_4 can be differentiated by $\sigma_v (xz)$. Γ_3 is symmetric and it is designated as B_1 whereas Γ_4 is antisymmetric and hence, it is designated as B_2.

Area IV: It gives the characters for irreducible representation of each class, which have already been calculated with the help of great orthogonality theorem earlier.

C_{2v}	E	C_2	$\sigma_v(xz)$	$\sigma_v(xz)$
$A_1 \leftarrow A \leftarrow \Gamma_1$	1	1	1	1
$A_2 \leftarrow A \leftarrow \Gamma_2$	1	1	–1	–1
$B_1 \leftarrow A \leftarrow \Gamma_3$	1	–1	1	–1
$B_2 \leftarrow A \leftarrow \Gamma_4$	1	–1	–1	1

Area V: It represents translational coordinates *x*, *y*, *z* as well as rotation axes R_x, R_y and R_z. For assigning Cartesian coordinates, all symmetry operations are performed with each irreducible representation. Then characters are written, if unchanged then +1 and inverted, then –1.

Consider a vector along Z-axis. The E, C_2, σ_v (xz) and σ_v (yz) do not change the direction of head of vector. Hence, character of *Z* vector is 1, 1, 1, 1. Character of *Z* vector matches with the irreducible presentation A_1. Similar operations along *X* and *Y* vectors belong to irreducible representation B_1 and B_2, respectively.

	Vector	E	C_2	σ_v (xz)	σ_v (yz)	Symbol
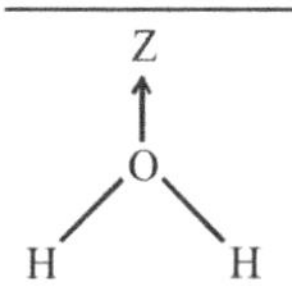	*z*	1	1	1	1	A_1
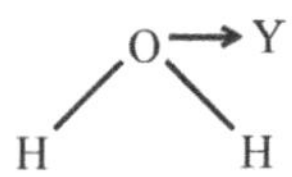	*y*	1	–1	–1	1	B_2
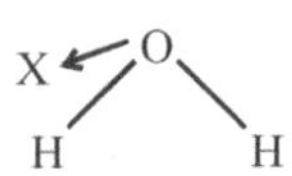	*x*	1	–1	1	–1	B_1

For assignment of R_x, R_y and R_z, a curved arrow should be considered around the axis and all symmetry operations are to be performed along this arrow. If direction of head of arrow does not change after operation, the character is +1 and if it becomes opposite, then the character is –1. Character of R_x, R_y and R_z matches the irreducible representation of B_2, B_1 and A_2, respectively.

	E	C_2	σ_v (xz)	σ_v (yz)	Symbol
	1	1	–1	–1	A_2

	E	C_2	σ_v (xz)	σ_v (yz)	Symbol
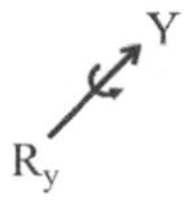	1	–1	1	–1	B_1
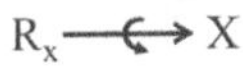	1	–1	–1	1	B_2

Area VI: It shows binary coordinates. It can be obtained by product of two Cartesian coordinates.

	E	C_2	σ_v (xz)	σ_v (yz)
x	1	–1	1	–1
y	1	–1	–1	1
z	1	1	1	1

	E	C_2	σ_v (xz)	σ_v (yz)	Symbol
x.x	1.1 = 1	–1. –1 = 1	1. 1 = 1	–1. –1 = 1	A_1
x.y	1.1 = 1	–1. –1 = 1	1. –1 = –1	–1. 1 = –1	A_2
x.z	1.1 = 1	–1. 1 = –1	1. 1 = 1	–1. 1 = –1	B_1
y.z	1.1 = 1	–1. 1 = –1	–1. 1 = –1	1. 1 = 1	B_2

Now character table of C_{2v} point group is given by –

C_{2v}	E	C_2	σ_v (xz)	σ_v (yz)		
A_1	1	1	1	1	*z*	x^2, y^2, z^2
A_2	1	1	–1	–1	R_z	xy
B_1	1	–1	1	–1	*x*, R_y	xz
B_2	1	–1	–1	1	*y*, R_x	yz

5.3.2 CHARACTER TABLE OF C_{3V} POINT GROUP

Area I: Its Schoenflies notation is C_{3v}.

Area II: It lists the distinct classes of point group C_{3v}. These are C_{3v} = {E, 2 C_3, 3 σ_v}.

Area III: It contains Mulliken's symbols. The characters of identity operation having one dimension for Γ_1 and Γ_2 whereas Γ_3 represents two dimensional representation. Hence, Γ_1 and Γ_2 both are represented either by symbol A or B while Γ_3 is represented by E. Both Γ_1 and Γ_2 representation are symmetric with respect to proper axis C_3 and, hence, designated as A. Now, we should differentiate Γ_1 and Γ_2 representation. For this, we proceed to next symmetry operation, σ_v. Γ_1 is symmetric with respect to σ_v plane and hence, designated as A_1 whereas Γ_2 is represented as A_2 because it is antisymmetric with respect to σ_v.

Area IV: It gives the character for irreducible representation of each class.

Area V: It represents translational coordinates *x*. *y* and *z* as well as rotational axes R_x, R_y and R_z.

The vector along Z-axis remains unchanged with respect to E, C_3 and σ_v operation. Hence, its characters are 1, 1, and 1, which belong to irreducible representation of A_1.

Vectors along *x* and *y* coordinates are not independent of each other while performing irreducible representation of two-dimension, i.e., E.

Similarly, rotational axis R_z can be designated as a curve arrow around Z-axis. During operation *E* and C_3, the direction of head of arrow remained same, while its direction changes after operation σ_v. Hence, the characters are 1, 1, and –1, which are similar to the irreducible representation of A_2. R_y and R_z form a two-dimensional representation and, hence, these belong to irreducible representation E.

Area VI: It shows binary coordinates. It can be obtained, by product of Cartesian coordinates, which one can calculate as –

	E	2 C_3	3 σ_v	Symbol
x.y	2	–1	0	E
z	1	1	1	
x.z	2	–1	0	E
y.z	2	–1	0	E
z^2	1	1	1	A_1

The character table of C_{3v} is:

C_{3v}	E	2 C_2	3 σ_v		
A_1	1	1	1	*z*	$x^2 + y^2, z^2$
A_2	1	1	–1	R_z	
E	2	–1	0	(*x*, *y*) (R_x, R_y)	($x^2 - y^2$, xy) (xz, yz)

5.3.3 CHARACTER TABLE FOR C_{2h} POINT GROUP

Area I: Its Schoenflies notation is C_{2h}.

Area II: It lists the distinct classes of C_{2h} point group. These are C_{2h} = $\{E, C_2, i, \sigma_h\}$.

Area III: It represents Mulliken symbols. All the irreducible representations of C_{2h} group are one dimensional. These are designated as A or B. In order to differentiate between these representations, we proceed to the next operation, i.e., C_2, in the present case. It is shown that Γ_1 and Γ_3 are symmetric with respect to rotational axis C_2, since $\chi(C_2) = +1$. Hence, they are designated as A, while other two representations Γ_2 and Γ_4 are antisymmetric with this respect to C_2, i.e., $\chi(C_2) = -1$, and thus, these can be labeled as B.

Γ_1 and Γ_2 can be further differentiated using symmetry with respect to inversion center (i). Γ_1 is symmetric with respect to inversion center and therefore, it is represented as A_g (g = gerade) whereas Γ_3 is designated as A_u (ungerade), due to its antisymmetric character with respect to inversion center. Similarly, Γ_2 and Γ_4 can also be differentiated using symmetry with respect to inversion center and these are designated as B_g and B_u, respectively.

Area IV: It gives the characters for irreducible representation of each class operation, which has been already calculated with the help of great orthogonality theorem.

Area V: It represents translational coordinates (x, y, z) as well as rotational axes R_x, R_y, and R_z. For assigning Cartesian coordinates, all symmetry operations are performed with each irreducible representation. Then characters are written, if it remain unchanged, i.e., $\chi(R) = +1$ and if it inverts, then $\chi(R) = -1$.

Considering vector along Z axis, E and C_2, operations on this vector do not change the direction of its head. But operations i and σ_h inverts the vector z. Hence, character of z vector is 1, 1, –1, –1 and it matches the irreducible presentation of A_u. Similar operations along x and y vectors also belong to irreducible representation of B_u as the operations E and σ_h do not change the direction of vectors while operations C_2 and i give the opposite vectors.

For assignment of R_x, R_y and R_z, a curved arrow should be considered around the axis and all symmetry operations are performed along this. If the direction of head of arrow does not change after operation, the character is +1 and if it becomes opposite after that operation, the character is –1. Characters of R_x and R_y match the irreducible representation of B_g, while R_z matches to A_g representations.

Area VI: It shows binary coordinates. These can be obtained by products of two Cartesian coordinates.

Now complete character table of C_{2h} is given by.

C_{2h}	E	C_2	i	σ_h		
A_g	1	1	1	1	R_z	x^2, y^2, z^2, xy
B_g	1	–1	1	–1	R_x, R_y	xz, yz
A_u	1	1	–1	–1	z	
B_u	1	–1	–1	1	x, y	

KEYWORDS

- **Character table**
- **Great orthogonality theorem**
- **Mulliken symbol**

CHAPTER 6

HYBRIDIZATION

CONTENTS

The concept of hybridization of orbitals, i.e., mixing the atomic orbitals (AOs) having relatively closer energy and their redistribution into equivalent hybrid orbitals, was introduced to explain the tetracovalency of carbon in methane. Hybridization in the case of methane (CH_4) is sp^3. Other kinds of hybridization are also known, i.e., sp, sp^2, sp^3d, sp^3d^2, sp^3d^3, dsp^2, etc. This concept of hybridization can be given a mathematical background on the basis of group theory and one can prove that there is a measure contribution of sp^3 hybridization in methane.

6.1 HYBRIDIZATION

Construction of hybrid orbitals can be divided into two parts:

- Determination of appropriate AO's of the central atom, which on combination will give the hybrid orbitals of the desired symmetry of molecule/ion and
- Actual determination of the hybrid orbitals as a function of linear combination (LCAO) of AOs including the values of the coefficients in it.

The subscript of an orbital symbol indicates that how an atomic orbital is transformed? These are Cartesian coordinates or their binary products, which are given in the character table of the point group to which the molecule/ion belongs. Hence, AOs can be selected on the basis of the representation properties of the subscripts, such as *x*, *y*, *z*, representing p_x, p_y, p_z orbitals, respectively.

6.2 CONSTRUCTION OF SIGMA HYBRID ORBITALS

The hybrid orbitals, which form sigma bonds, can be treated as vectors. These form the basis of representation.

(i) Point group of the molecule/ion is decided.
(ii) All symmetry operations of the class of the point group are performed.
(iii) Matrices are constructed to represent these operations.
(iv) Character of each symmetry operation of the class is written under its column. This is reducible representation of molecule/ion.
(v) Now, reducible representations (R) are reduced to irreducible representation (IRs) using reduction formula:

$$a_i = \Sigma\, n_R\, \chi(R)\, \chi_i(R)$$

Each IR so obtained, points to the transformation properties of the Cartesian coordinate mentioned against each IR in the extreme right column of the character table. This gives the AOs to form the hybrid orbitals.

6.2.1 *TRIGONAL PLANAR MOLECULE AB_3 (BF_3)*

It has three vectors r_1, r_2, and r_3 corresponding to each B-F bond. This molecule belongs to D_{3h} point group.

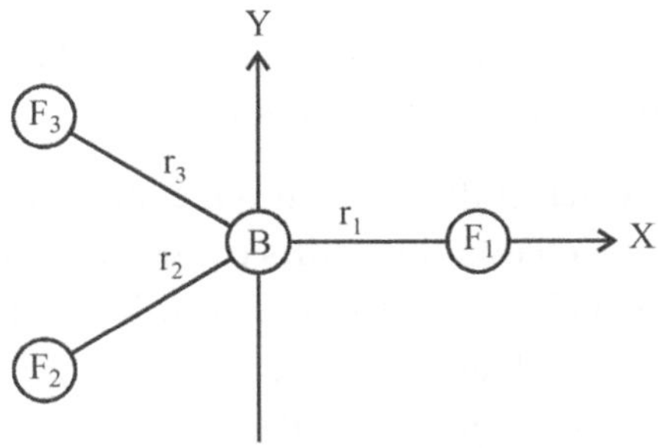

Its molecular formula and the geometry suggest that it is necessary to construct three equivalent orbitals of boron (central atom) to form three σ-bonds pointing along the directions of B-F bonds. These three equivalent orbitals form a basis of representation. The character χ (R) is equal to sum of the diagonal characters of the matrix.

Operation		**Matrix notation**	**Characters χ (R)**
E	$r_1 = 1.r_1 + 0.r_2 + 0.r_3$ $r_2 = 0.r_1 + 1.r_2 + 0.r_3$ $r_3 = 0.r_1 + 0.r_2 + 1.r_3$	$\begin{bmatrix} 1 & 0 & 0 \\ 0 & 1 & 0 \\ 0 & 0 & 1 \end{bmatrix}$	3
$C_3(z)$	$r_1 = 0.r_1 + 1.r_2 + 0.r_3$ $r_2 = 0.r_1 + 0.r_2 + 1.r_3$ $r_3 = 1.r_1 + 0.r_2 + 0.r_3$	$\begin{bmatrix} 0 & 1 & 0 \\ 0 & 0 & 1 \\ 1 & 0 & 0 \end{bmatrix}$	0
$C_2(x)$ or $\sigma_v(x)$	$r_1 = 1.r_1 + 0.r_2 + 0.r_3$ $r_2 = 0.r_1 + 0.r_2 + 1.r_3$ $r_3 = 0.r_1 + 1.r_2 + 0.r_3$	$\begin{bmatrix} 1 & 0 & 0 \\ 0 & 0 & 1 \\ 0 & 1 & 0 \end{bmatrix}$	1
σ_h	$r_1 = 1.r_1 + 0.r_2 + 0.r_3$ $r_2 = 0.r_1 + 1.r_2 + 1.r_3$ $r_3 = 0.r_1 + 0.r_2 + 1.r_3$	$\begin{bmatrix} 1 & 0 & 0 \\ 0 & 1 & 0 \\ 0 & 0 & 1 \end{bmatrix}$	3

$$S_3(z) = \sigma_h.C_3(z); \text{ since } \chi\, C_3(z) = 0 \text{ and hence, } S_3(z) = 0.$$

Therefore, reducible representation for this case is given by:

D_{3h}	E	$2C_3(z)$	$3C_2$	σ_h	$2S_3$	$3\sigma_v$
$\Gamma_\sigma(R)$	3	0	1	3	0	1

Here, some general rules are used to determine the character of matrices corresponding to a given symmetry operation.

(i) The vector, which shifts its position, contributes zero (0).
(ii) The vector, which does not shifts contributes one (1).
(iii) The vector, which rotates from its position contributes cos θ, where θ is the angle of rotation.

Character table for point group D_{3h} is:

D_{3h}	E	$2C_3$	$3C_2$	σ_h	$2S_3$	$3\sigma_v$		
A_1'	1	1	1	1	1	1		x^2+y^2, z^2
A_2'	1	1	−1	1	1	−1	R_z	
E'	2	−1	0	2	−1	0	(x, y)	(x^2-y^2, xy)
A_1''	1	1	1	−1	−1	−1		
A_2''	1	1	−1	−1	−1	1	z	
E''	2	1	0	−2	1	0	(R_x, R_y)	(xy, yz)
$\Gamma_\sigma(R)$	3	0	1	3	0	1		

Using reduction formula:

$$a_{A_1}' = \frac{1}{12}(1 \times 1 \times 3 + 2 \times 1 \times 0 + 3 \times 1 \times 1 + 1 \times 1 \times 3 + 2 \times 1 \times 0 + 3 \times 1 \times 1) = 1$$

$$a_{A_2}' = \frac{1}{12}(1 \times 1 \times 3 + 2 \times 1 \times 0 + 3 \times -1 \times 1 + 1 \times 1 \times 3 + 2 \times 1 \times 0 + 3 \times -1 \times 1) = 0$$

$$a_E' = \frac{1}{12}(1 \times 2 \times 3 + 2 \times -1 \times 0 + 3 \times 0 \times 1 + 1 \times 2 \times 3 + 2 \times -1 \times 0 + 3 \times 0 \times 1) = 1$$

$$a_{A_1}'' = \frac{1}{12}(1 \times 1 \times 3 + 2 \times 1 \times 0 + 3 \times 1 \times 1 + 1 \times -1 \times 3 + 2 \times -1 \times 0 + 3 \times -1 \times 1) = 0$$

$$a_{A2}" = \frac{1}{12}(1 \times 1 \times 3 + 2 \times 1 \times 0 + 3 \times -1 \times 1 + 1 \times -1 \times 3 + 2 \times -1 \times 0 + 3 \times 1 \times 1) = 0$$

$$a_{E}" = \frac{1}{12}(1 \times 2 \times 3 + 2 \times 1 \times 0 + 3 \times 0 \times 1 + 1 \times -2 \times 3 + 2 \times 1 \times 0 + 3 \times 0 \times 1) = 0$$

Thus, reducible representation can be reduced as:

$$\Gamma_{\sigma}(R) = 1.A_1' + 0.A_2' + 1.E' + 0.A_1" + 0.A_2" + 0.E"$$

or $$\Gamma_{\sigma}(R) = A_1' + E'$$

These representations represent orbitals. Cartesian coordinates *x*, *y* and *z* in character table represents p_x, p_y and p_z, respectively. Similarly $(x^2 + y^2)$, z^2, $(x^2 - y^2)$, xy, yz, xz represent s, d_{z^2}, $d_{x^2-y^2}$, d_{xy}, d_{yz} and d_{xz} orbital, respectively.

A_1'	E'
s	p_x, p_y
d_{z^2}	$d_{x^2-y^2}$, d_{xy}

Thus, there are four possible combination; $s.p_x.p_y$; $s.d_{x^2-y^2}.d_{xy}$; $d_{z^2}.p_x.p_y$ and $d_{z^2}.d_{x^2-y^2}.d_{xy}$.

Total wave function is the linear combination of all these possibilities.

$$\psi_{\sigma} = a\,(sp^2) + b\,(sd^2) + c\,(dp^2) + e\,(d^3)$$

where *a*, *b*, *c* and e are coefficients of these hybrial orbitals, indicating their contribution in formation of hybrid orbitals. Values of coefficients can be determined by using variation method. In this case, the values of *b*, *c* and *e* coefficients are much less and these are considered negligible. Secondly, the last three terms involve the participation of *d* orbitals, which has no significance in boron atom as there are no *d* orbitals and therefore, their contribution is almost zero. Thus, resultant hybridization of BF_3 molecule can be represented as:

$$\psi_{BF_3} = sp^2$$

6.2.2 SQUARE PLANAR ION AB_4 $[PTCL_4]^{2-}$

It has four vectors r_1, r_2, r_3 and r_4 corresponding to each Pt – Cl bond. This ion belongs to D_{4h} point group.

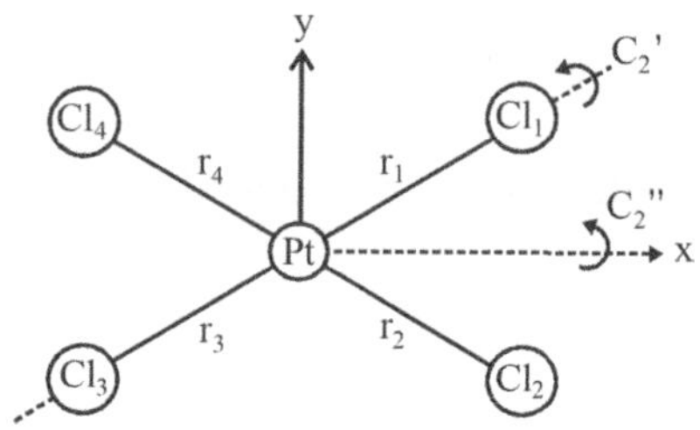

Molecular formula and the geometry of this ion suggest that it is necessary to construct four equivalent orbitals of *Pt* (central) atom to form four σ-bonds pointing along the direction of Pt – Cl bonds. These four equivalent orbitals form a basis of representation. The character χ (R) is equal to sum of the diagonal characters of the matrix.

Operation		Matrix notation	Characters χ (R)
E	$r_1 = 1.r_1 + 0.r_2 + 0.r_3 + 0.r_4$ $r_2 = 0.r_1 + 1.r_2 + 0.r_3 + 0.r_4$ $r_3 = 0.r_1 + 0.r_2 + 1.r_3 + 0.r_4$ $r_4 = 0.r_1 + 0.r_2 + 0.r_3 + 1.r_4$	$\begin{bmatrix} 1 & 0 & 0 & 0 \\ 0 & 1 & 0 & 0 \\ 0 & 0 & 1 & 0 \\ 0 & 0 & 0 & 1 \end{bmatrix}$	4
$C_4(z)$	$r_1 = 0.r_1 + 1.r_2 + 0.r_3 + 0.r_4$ $r_2 = 0.r_1 + 0.r_2 + 1.r_3 + 0.r_4$ $r_3 = 0.r_1 + 0.r_2 + 0.r_3 + 1.r_4$ $r_4 = 1.r_1 + 0.r_2 + 0.r_3 + 0.r_4$	$\begin{bmatrix} 0 & 1 & 0 & 0 \\ 0 & 0 & 1 & 0 \\ 0 & 0 & 0 & 1 \\ 1 & 0 & 0 & 0 \end{bmatrix}$	0
$C_2(z)$	$r_1 = 0.r_1 + 0.r_2 + 1.r_3 + 0.r_4$ $r_2 = 0.r_1 + 0.r_2 + 0.r_3 + 1.r_4$ $r_3 = 1.r_1 + 0.r_2 + 0.r_3 + 0.r_4$ $r_4 = 0.r_1 + 1.r_2 + 0.r_3 + 0.r_4$	$\begin{bmatrix} 0 & 0 & 1 & 0 \\ 0 & 0 & 0 & 1 \\ 1 & 0 & 0 & 0 \\ 0 & 1 & 0 & 0 \end{bmatrix}$	0
$C_{2'}$	$r_1 = 1.r_1 + 0.r_2 + 0.r_3 + 0.r_4$ $r_2 = 0.r_1 + 0.r_2 + 0.r_3 + 1.r_4$ $r_3 = 0.r_1 + 0.r_2 + 1.r_3 + 0.r_4$ $r_4 = 0.r_1 + 1.r_2 + 0.r_3 + 0.r_4$	$\begin{bmatrix} 1 & 0 & 0 & 0 \\ 0 & 0 & 0 & 1 \\ 0 & 0 & 1 & 0 \\ 0 & 1 & 0 & 0 \end{bmatrix}$	2

Operation		Matrix notation	Characters χ (R)
C_2''	$r_1 = 0.r_1 + 1.r_2 + 0.r_3 + 0.r_4$ $r_2 = 1.r_1 + 0.r_2 + 0.r_3 + 0.r_4$ $r_3 = 0.r_1 + 0.r_2 + 0.r_3 + 1.r_4$ $r_4 = 0.r_1 + 0.r_2 + 1.r_3 + 0.r_4$	$\begin{bmatrix} 0 & 1 & 0 & 0 \\ 1 & 0 & 0 & 0 \\ 0 & 0 & 0 & 1 \\ 0 & 0 & 1 & 0 \end{bmatrix}$	0
i	$r_1 = 0.r_1 + 0.r_2 + 1.r_3 + 0.r_4$ $r_2 = 0.r_1 + 0.r_2 + 0.r_3 + 1.r_4$ $r_3 = 1.r_1 + 0.r_2 + 0.r_3 + 0.r_4$ $r_4 = 0.r_1 + 1.r_2 + 0.r_3 + 0.r_4$	$\begin{bmatrix} 0 & 0 & 1 & 0 \\ 0 & 0 & 0 & 1 \\ 1 & 0 & 0 & 0 \\ 0 & 1 & 0 & 0 \end{bmatrix}$	0
σ_h	$r_1 = 1.r_1 + 0.r_2 + 0.r_3 + 0.r_4$ $r_2 = 0.r_1 + 1.r_2 + 0.r_3 + 0.r_4$ $r_3 = 0.r_1 + 0.r_2 + 1.r_3 + 0.r_4$ $r_4 = 0.r_1 + 0.r_2 + 0.r_3 + 1.r_4$ $S_4\,(z) = \sigma_h.C_4\,(z) = \sigma_h.0 = 0$	$\begin{bmatrix} 1 & 0 & 0 & 0 \\ 0 & 1 & 0 & 0 \\ 0 & 0 & 1 & 0 \\ 0 & 0 & 0 & 1 \end{bmatrix}$	4

$$\sigma_v \text{ are along } C_{2'} \text{ and hence, } \chi\,(R) = 2$$

$$\sigma_d \text{ are along } C_2'' \text{ and hence, } \chi\,(R) = 0$$

Therefore, reducible representations for this ion are:

D_{4h}	E	$2C_4$	C_2	$2C_2'$	$2C_2''$	i	$2S_4$	σ_h	$2\sigma_v$	$2\sigma_d$
$\Gamma_{\tilde{A}}(R)$	4	0	0	2	0	0	0	4	2	0

The character table for point group D_{4h} is:

D_{4h}	E	$2C_4$	C_2	$2C_2'$	$2C_2''$	i	$2S_4$	σ_h	$2\sigma_V$	$2\sigma_d$		
A_{1g}	1	1	1	1	1	1	1	1	1	1		x^2, y^2, z^2
A_{2g}	1	1	1	−1	−1	1	1	1	−1	−1	R_z	
B_{1g}	1	−1	1	1	−1	1	−1	1	1	−1		$x^2 - y^2$
B_{2g}	1	−1	1	−1	1	1	−1	1	−1	1		xy
E_g	2	0	−2	0	0	2	0	−2	0	0	(R_x, R_y)	(xz, yz)
A_{1u}	1	1	1	1	1	−1	−1	−1	−1	−1		
A_{2u}	1	1	1	−1	−1	−1	−1	−1	1	1	z	
B_{1u}	1	−1	1	1	−1	−1	1	−1	−1	1		
B_{2u}	1	−1	1	−1	1	−1	1	−1	1	−1		
E_u	2	0	−2	0	0	−2	0	2	0	0	(x, y)	

Using reduction formula

$$a_{A_{1g}} = \frac{1}{16}[1 \times 1 \times 4 + 2 \times 1 \times 0 + 1 \times 1 \times 0 + 2 \times 1 \times 2 + 2 \times 1 \times 0 + 1 \times 1 \times 0 + 2 \times 1 \times 0 + 1 \times 1 \times 4 + 2 \times 1 \times 2 + 2 \times 1 \times 0] = \frac{1}{16} \times = 1$$

$$a_{A_{2g}} = \frac{1}{16}[1 \times 1 \times 4 + 2 \times 1 \times 0 + 1 \times 1 \times 0 + 2 \times -1 \times 2 + 2 \times -1 \times 0 + 1 \times 1 \times 0 + 2 \times 1 \times 0 + 1 \times 1 \times 4 + 2 \times -1 \times 2 + 2 \times -1 \times 0] = 0$$

$$a_{B_{1g}} = \frac{1}{16}[1 \times 1 \times 4 + 2 \times -1 \times 0 + 1 \times 1 \times 0 + 2 \times 1 \times 2 + 2 \times -1 \times 0 + 1 \times 1 \times 0 + 2 \times -1 \times 0 + 1 \times 1 \times 4 + 2 \times 1 \times 2 + 2 \times -1 \times 0] = 1$$

$$a_{E_g} = \frac{1}{16}[1 \times 2 \times 4 + 2 \times 0 \times 0 + 1 \times -2 \times 0 + 2 \times 0 \times 2 + 2 \times 0 \times 0 + 1 \times 2 \times 0 + 2 \times 0 \times 0 + 1 \times -2 \times 4 + 2 \times 0 \times -2 + 2 \times 0 \times 0] = 0$$

$$a_{E_u} = \frac{1}{16}[1 \times 2 \times 4 + 2 \times 0 \times 0 + 1 \times -1 \times 0 + 2 \times 0 \times 2 + 2 \times 0 \times 0 + 1 \times 2 \times 0 + 2 \times 0 \times 0 + 1 \times 2 \times 4 + 2 \times 0 \times 2 + 2 \times 0 \times 0] = 1$$

Similarly, $B_{2g} = A_{1u} = A_{2u} = B_{1u} = B_{2u} = 0$

Thus, reducible representation for this case can be reduced as:

$$\Gamma_\sigma(R) = 1.A_{1g} + 0.A_{2g} + 1.B_{1g} + 0.B_{2g} + 0.E_g + 0.A_{1u} + 0.A_{2u} + 0.B_{1u} + 0.B_{2u} + 1.E_u$$

or

$$\Gamma_\sigma(R) = A_{1g} + B_{1g} + E_u$$

These representations represent orbitals. Cartesian coordinates *x*, *y*, and *z* in character table respresent p_x, p_y and p_z, respectively. Similarly $x^2 + y^2$, z^2 and $x^2 - y^2$ represent s, d_{z^2} and $d_{x^2-y^2}$ orbitals, respectively.

A_{1g}	B_{1g}	E_u
s	$d_{x^2-y^2}$	p_x, p_y
d_{z^2}		

Thus, there are two possible combinations. There are:

$$s, d_{x^2-y^2}.p_x.p_y \text{ and } d_{z^2}.d_{x^2-y^2}.p_x.p_y$$

Total wave function is the linear combination of all these possibilities.

$$\psi_\sigma = a\,(dsp^2) + b\,(d^2p^2)$$

where a and b are coefficient of these hybrid orbitals and values of these coefficients can be determined by variation method. The value of b was found negligible and therefore, resultant hybridization of $[PtCl_4]^{2-}$ is:

$$\psi_{PtCl_4^{2-}} = dsp^2$$

6.2.3 TETRAHEDRAL MOLECULE AB_4 (CH_4)

CH_4 molecule has 4 vectors r_1, r_2, r_3 and r_4 corresponding to each C-H bond. This molecule belongs to T_d point group.

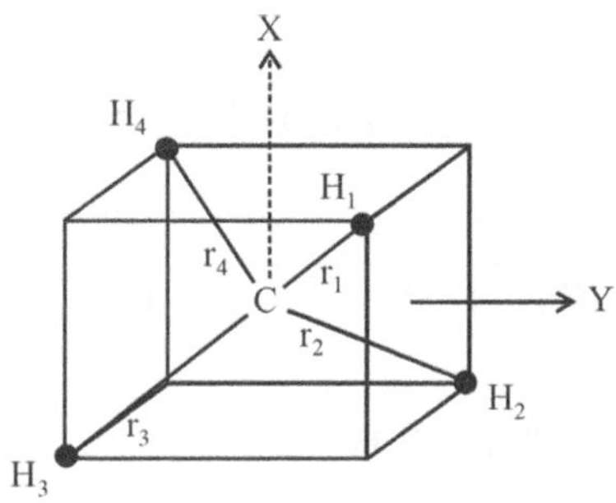

Its molecular formula and geometry suggest that it is necessary to construct four equivalent orbitals of carbon atom (central atom) to form four σ-bonds pointing along the directions of C-H bonds. These four equivalent orbitals form a basis of representation. The character χ (R) is equal to the sum of the diagonal characters of the matrix.

Operation		Matrix notation	Characters χ (R)
E	$r_1 = 1.r_1 + 0.r_2 + 0.r_3 + 0.r_4$ $r_2 = 0.r_1 + 1.r_2 + 0.r_3 + 0.r_4$ $r_3 = 0.r_1 + 0.r_2 + 1.r_3 + 0.r_4$ $r_4 = 0.r_1 + 0.r_2 + 0.r_3 + 1.r_4$	$\begin{bmatrix}1&0&0&0\\0&1&0&0\\0&0&1&0\\0&0&0&1\end{bmatrix}$	4
C_2	$r_1 = 1.r_1 + 0.r_2 + 0.r_3 + 0.r_4$ $r_2 = 0.r_1 + 0.r_2 + 1.r_3 + 0.r_4$ $r_3 = 0.r_1 + 0.r_2 + 0.r_3 + 1.r_4$ $r_4 = 0.r_1 + 1.r_2 + 0.r_3 + 0.r_4$	$\begin{bmatrix}1&0&0&0\\0&0&1&0\\0&0&0&1\\0&1&0&0\end{bmatrix}$	1

Operation		Matrix notation	Characters χ (R)
$C_2(x)$	$r_1 = 0.r_1 + 0.r_2 + 0.r_3 + 1.r_4$ $r_2 = 0.r_1 + 0.r_2 + 1.r_3 + 0.r_4$ $r_3 = 0.r_1 + 1.r_2 + 0.r_3 + 0.r_4$ $r_4 = 1.r_1 + 0.r_2 + 0.r_3 + 0.r_4$	$\begin{bmatrix} 0 & 0 & 0 & 1 \\ 0 & 0 & 1 & 0 \\ 0 & 1 & 0 & 0 \\ 1 & 0 & 0 & 0 \end{bmatrix}$	0
σ_d (along CH_1H_2)	$r_1 = 1.r_1 + 0.r_2 + 0.r_3 + 0.r_4$ $r_2 = 0.r_1 + 1.r_2 + 0.r_3 + 0.r_4$ $r_3 = 0.r_1 + 0.r_2 + 0.r_3 + 1.r_4$ $r_4 = 0.r_1 + 0.r_2 + 1.r_3 + 0.r_4$	$\begin{bmatrix} 1 & 0 & 0 & 0 \\ 0 & 1 & 0 & 0 \\ 0 & 0 & 0 & 1 \\ 0 & 0 & 1 & 0 \end{bmatrix}$	2
$S_4(x)$	$r_1 = 0.r_1 + 0.r_2 + 1.r_3 + 0.r_4$ $r_2 = 0.r_1 + 0.r_2 + 0.r_3 + 1.r_4$ $r_3 = 1.r_1 + 0.r_2 + 0.r_3 + 0.r_4$ $r_4 = 0.r_1 + 1.r_2 + 0.r_3 + 0.r_4$	$\begin{bmatrix} 0 & 0 & 1 & 0 \\ 0 & 0 & 0 & 1 \\ 1 & 0 & 0 & 0 \\ 0 & 1 & 0 & 0 \end{bmatrix}$	0

Therefore, reducible representations for this case are:

T_d	E	$8C_3$	$3C_2$	$6S_4$	$6\sigma_d$
$\Gamma_{\tilde{A}}(R)$	4	1	0	0	2

Character table of T_d group point is:

T_d	E	$8C_3$	$3C_2$	$6S_4$	$6\sigma_d$		
A_1	1	1	1	1	1		$x^2 + y^2 + z^2$
A_2	1	1	1	−1	−1		
E	1	−1	1	1	−1		$(2z^2 - x^2 - y^2, x^2 - y^2)$
T_1	1	−1	1	−1	1	(R_x, R_y, R_z)	
T_2	2	0	−2	0	0	(x, y, z)	(xy, xz, yz)

Using reduction formula:

$$a_{A_1} = \frac{1}{24}[1 \times 1 \times 4 + 8 \times 1 \times 1 + 3 \times 1 \times 0 + 6 \times 1 \times 0 + 6 \times 1 \times 2] = 1$$

$$a_{A_2} = \frac{1}{24}[1 \times 1 \times 4 + 8 \times 1 \times 1 + 3 \times 1 \times 0 + 6 \times -1 \times 0 + 6 \times -1 \times 2] = 0$$

$$a_E = \frac{1}{24}[1 \times 2 \times 4 + 8 \times -1 \times 1 + 3 \times 2 \times 0 + 6 \times 0 \times 0 + 6 \times 0 \times 2] = 0$$

$$a_{T_1} = \frac{1}{24}[1 \times 3 \times 4 + 8 \times 0 \times 1 + 3 \times -1 \times 0 + 6 \times 1 \times 0 + 6 \times -1 \times 2 = 0$$

$$a_{T_2} = \frac{1}{24}[1 \times 3 \times 4 + 8 \times 0 \times 1 + 3 \times -1 \times 0 + 6 \times -1 \times 0 + 6 \times 1 \times 2] = 1$$

Thus, reducible representation for this case can be reduced as:

$$\Gamma_\sigma (R) = 1.A_1 + 0.A_2 + 0.E + 0.T_1 + 1.T_2$$

or $$\Gamma_\sigma (R) = A_1 + T_2$$

These representations represent orbitals. Cartesian coordinates *x*, *y*, and *z* represent p_x, p_y and p_z, respectively. Similarly, $x^2 + y^2 + z^2$, xy, yz and xz represent s, d_{xy}, d_{yz} and d_{xz}, respectively.

A_1	T_2
s	(p_x, p_y, p_z)
	(d_{xy}, d_{yz}, d_{xz})

Hence, there are only two possibilities of combinations. These are s, p_x, p_y, p_z and s, d_{xy}, d_{yz}, d_{xz}. Total wave function is a linear combination of these two possibilities.

$$\psi_\sigma - a\,(sp^3) + b\,(sd^3)$$

where *a* and *b* are coefficients and values of these coefficients can be calculated by variation method. The value of coefficient *b* is negligable as there is no significant contribution of *d* orbitals in carbon atoms (*d* orbitals are not there in the ground state of carbon) and therefore, sd^3 will not contribute in hybridization and resultant hybridization of CH_4 molecule is represented as:

$$\psi_{CH_4} = sp^3$$

6.2.4 *TRIGONAL BIPYRAMIDAL MOLECULE AB_5 (PCl_5)*

It has five vectors r_1, r_2, r_3, r_4 and r_5 corresponding to each P-Cl bond. This molecule also belongs to D_{3h} point group.

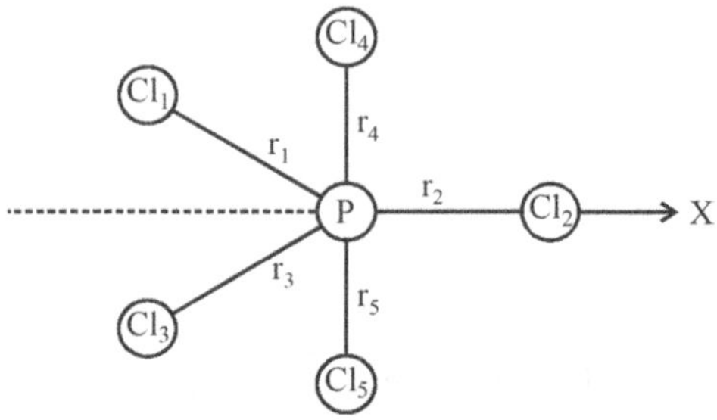

Phosphorous atom forms five equivalent hybrid orbitals to form five σ-bonds pointing along bond directions of P-Cl. These five equivalent hybrid orbitals form a basis of representation. The character χ (R) is equal to the sum of the diagonal characters of the matrix.

Operation		Matrix notation	Characters χ (R)
E	$r_1 = 1.r_1 + 0.r_2 + 0.r_3 + 0.r_4 + 0.r_5$ $r_2 = 0.r_1 + 1.r_2 + 0.r_3 + 0.r_4 + 0.r_5$ $r_3 = 0.r_1 + 0.r_2 + 1.r_3 + 0.r_4 + 0.r_5$ $r_4 = 0.r_1 + 0.r_2 + 0.r_3 + 1.r_4 + 0.r_5$ $r_5 = 0.r_1 + 0.r_2 + 0.r_3 + 0.r_4 + 1.r_5$	$\begin{bmatrix} 1 & 0 & 0 & 0 & 0 \\ 0 & 1 & 0 & 0 & 0 \\ 0 & 0 & 1 & 0 & 0 \\ 0 & 0 & 0 & 1 & 0 \\ 0 & 0 & 0 & 0 & 1 \end{bmatrix}$	5
C_3	$r_1 = 0.r_1 + 1.r_2 + 0.r_3 + 0.r_4 + 0.r_5$ $r_2 = 0.r_1 + 0.r_2 + 1.r_3 + 0.r_4 + 0.r_5$ $r_3 = 1.r_1 + 0.r_2 + 0.r_3 + 0.r_4 + 0.r_5$ $r_4 = 0.r_1 + 0.r_2 + 0.r_3 + 1.r_4 + 0.r_5$ $r_5 = 0.r_1 + 0.r_2 + 0.r_3 + 0.r_4 + 1.r_5$	$\begin{bmatrix} 0 & 1 & 0 & 0 & 0 \\ 0 & 0 & 1 & 0 & 0 \\ 1 & 0 & 0 & 0 & 0 \\ 0 & 0 & 0 & 1 & 0 \\ 0 & 0 & 0 & 0 & 1 \end{bmatrix}$	2
C_2	$r_1 = 1.r_1 + 0.r_2 + 0.r_3 + 0.r_4 + 0.r_5$ $r_2 = 0.r_1 + 0.r_2 + 1.r_3 + 0.r_4 + 0.r_5$ $r_3 = 0.r_1 + 1.r_2 + 0.r_3 + 0.r_4 + 0.r_5$ $r_4 = 0.r_1 + 0.r_2 + 0.r_3 + 0.r_4 + 1.r_5$ $r_5 = 0.r_1 + 0.r_2 + 0.r_3 + 1.r_4 + 0.r_5$	$\begin{bmatrix} 1 & 1 & 0 & 0 & 0 \\ 0 & 0 & 1 & 0 & 0 \\ 0 & 1 & 0 & 0 & 0 \\ 0 & 0 & 0 & 0 & 1 \\ 0 & 0 & 0 & 1 & 0 \end{bmatrix}$	1

Operation		Matrix notation	Characters χ (R)
σ_h	$r_1 = 1.r_1 + 0.r_2 + 0.r_3 + 0.r_4 + 0.r_5$ $r_2 = 0.r_1 + 1.r_2 + 0.r_3 + 0.r_4 + 0.r_5$ $r_3 = 0.r_1 + 0.r_2 + 1.r_3 + 0.r_4 + 0.r_5$ $r_4 = 0.r_1 + 0.r_2 + 0.r_3 + 0.r_4 + 1.r_5$ $r_5 = 0.r_1 + 0.r_2 + 0.r_3 + 1.r_4 + 0.r_5$	$\begin{bmatrix}1&0&0&0&0\\0&1&0&0&0\\0&0&1&0&0\\0&0&0&0&1\\0&0&0&1&0\end{bmatrix}$	3
$C_2(x)$	$r_1 = 0.r_1 + 0.r_2 + 1.r_3 + 0.r_4 + 0.r_5$ $r_2 = 0.r_1 + 1.r_2 + 0.r_3 + 0.r_4 + 0.r_5$ $r_3 = 1.r_1 + 0.r_2 + 0.r_3 + 0.r_4 + 0.r_5$ $r_4 = 0.r_1 + 0.r_2 + 0.r_3 + 0.r_4 + 1.r_5$ $r_5 = 0.r_1 + 0.r_2 + 0.r_3 + 1.r_4 + 0.r_5$	$\begin{bmatrix}0&0&1&0&0\\0&1&0&0&0\\1&0&0&0&0\\0&0&0&0&1\\0&0&0&1&0\end{bmatrix}$	1
σ_d	$r_1 = 1.r_1 + 0.r_2 + 0.r_3 + 0.r_4 + 0.r_5$ $r_2 = 0.r_1 + 0.r_2 + 1.r_3 + 0.r_4 + 0.r_5$ $r_3 = 0.r_1 + 1.r_2 + 0.r_3 + 0.r_4 + 0.r_5$ $r_4 = 0.r_1 + 0.r_2 + 0.r_3 + 1.r_4 + 0.r_5$ $r_5 = 0.r_1 + 0.r_2 + 0.r_3 + 0.r_4 + 1.r_5$	$\begin{bmatrix}1&0&0&0&0\\0&0&1&0&0\\0&1&0&0&0\\0&0&0&1&0\\0&0&0&0&1\end{bmatrix}$	3

The characters of $S_3 = \sigma_h.C_3 = 0$ because all the vectors shift from their original positions. Therefore, reducible representation for this case is:

D_{3h}	E	$2C_3$	$3C_2$	σ_h	$2S_3$	$3\sigma_d$
Γ_σ(R)	5	2	1	3	0	3

Character table of D_{4h} point group is:

D_{3h}	E	$2C_3$	$2C_2$	σ_h	$2S_3$	$3\sigma_v$		
A_1'	1	1	1	1	1	1		x^2+y^2, z^2
A_2'	1	1	−1	1	1	−1	R_z	
E'	2	−1	0	2	−1	0	(x, y)	(x^2-y^2, xy)
A_1''	1	1	1	−1	−1	−1		
A_2''	1	1	−1	−1	−1	1	z	
E''	2	−1	0	−2	1	0	(R_x, R_y)	(xz, yz)

Using reduction formula

$$a_{A_1}' = [1 \times 1 \times 5 + 2 \times 1 \times 2 + 3 \times 1 \times 1 + 1 \times 1 \times 3 + 2 \times 1 \times 0 + 3 \times 1 \times 3] = 2$$

$$a_{A_2}' = [1 \times 1 \times 5 + 2 \times 1 \times 2 + 3 \times -1 \times 1 + 1 \times 1 \times 3 + 2 \times 1 \times 0 + 3 \times -1 \times 3] = 0$$

$$a_{E}' = [1 \times 2 \times 5 + 2 \times -1 \times 2 + 3 \times 0 \times 1 + 1 \times 2 \times 3 + 2 \times -1 \times 0 + 3 \times 0 \times 3] = 1$$

$$a_{A_1}'' = [1 \times 1 \times 5 + 2 \times 1 \times 2 + 3 \times 1 \times 1 + 1 \times -1 \times 3 + 2 \times -1 \times 0 + 3 \times -1 \times 3] = 0$$

$$a_{A_2}'' = [1 \times 1 \times 5 + 2 \times 1 \times 2 + 3 \times -1 \times 1 + 1 \times -1 \times 3 + 2 \times -1 \times 0 + 3 \times 1 \times 3] = 1$$

$$a_{E}'' = [1 \times 2 \times 5 + 2 \times -1 \times 2 + 3 \times 0 \times 1 + 1 \times -2 \times 3 + 2 \times 1 \times 0 + 3 \times 0 \times 3] = 0$$

Thus, reducible representation for this case can be reduced as:

$$\Gamma_\sigma (R) = 2.A_1' + 0.A_2' + 1.E' + 0.A_1'' + 1.A_2'' + 0.E''$$

$$\Gamma_\sigma (R) = 2\,A_1' + E' + A_2''$$

These representations represent orbitals, Cartesian coordinates *x*, *y* and *z* represent p_x, p_y and p_z respectively. Similarly,

A_1'	A_2''	E'
s	p_z	(p_x, p_y)
d_{z^2}		$(d_{xy}, d_{x^2-y^2})$

A_1' occur twice in Γ_σ (R) and hence, following six combinations are possible.

(i) s, s, p_z, p_x, p_y
(ii) s, d_{z^2}, p_z, p_x, p_y
(iii) d_{z^2}, d_{z^2}, p_z, p_x, p_y
(iv) s, s, p_z, d_{xy}, $d_{x^2-y^2}$
(v) s, d_{z^2}, p_z, d_{xy}, $d_{x^2-y^2}$
(vi) d_{z^2}, d_{z^2}, p_z, d_{xy}, $d_{x^2-y^2}$

Total wave function is a linear combination of all these possibilities:

$$\psi_\sigma = a\,(s^2p^3) + b\,(sp^3d) + c\,(d^2p^3) + e\,(s^2pd^2) + f(spd^3) + g\,(pd^4)$$

where *a*, *b*, *c*, *e*, *f* and *g* are coefficients of these hybrid orbitals and values of these coefficients can be determined by variation method. The values of coefficients of *a*, *b*, *c*, *e*, *f* and *g* are negligibly small and therefore, resultant hybridization of PCl_5 is

$$\psi_{PCl_5} = sp^3d$$

6.2.5 OCTAHEDRAL MOLECULE AB_6 (SF_6)

It has six vectors r_1 to r_6 corresponding to each S-F bond. The molecule belongs to O_h point group.

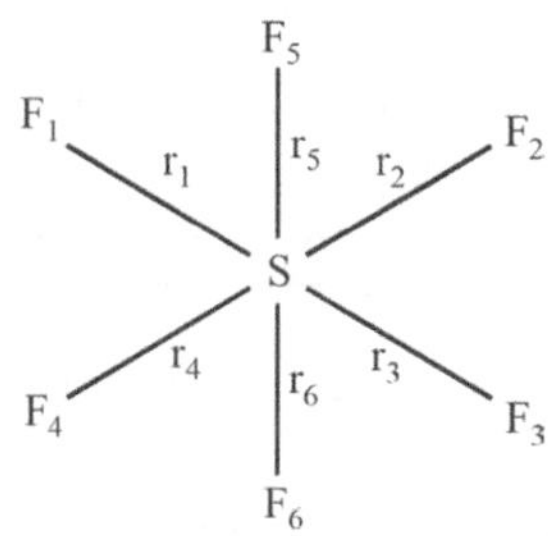

Its molecular formula and geometry suggest that it is necessary to construct six equivalent hybrid orbitals of S atom to form six σ bonds along the direction of S-F bonds. These six equivalent orbitals form basis of representation. The character χ (R) is equal to the sum of diagonal characters of the matrix.

Operation		Matrix notation	Characters χ (R)
E	$r_1 = 1.r_1 + 0.r_2 + 0.r_3 + 0.r_4 + 0.r_5 + 0.r_6$ $r_2 = 0.r_1 + 1.r_2 + 0.r_3 + 0.r_4 + 0.r_5 + 0.r_6$ $r_3 = 0.r_1 + 0.r_2 + 1.r_3 + 0.r_4 + 0.r_5 + 0.r_6$ $r_4 = 0.r_1 + 0.r_2 + 0.r_3 + 1.r_4 + 0.r_5 + 0.r_6$ $r_5 = 0.r_1 + 0.r_2 + 0.r_3 + 0.r_4 + 1.r_5 + 0.r_6$ $r_6 = 0.r_1 + 0.r_2 + 0.r_3 + 0.r_4 + 0.r_5 + 1.r_6$	$\begin{bmatrix} 1 & 0 & 0 & 0 & 0 & 0 \\ 0 & 1 & 0 & 0 & 0 & 0 \\ 0 & 0 & 1 & 0 & 0 & 0 \\ 0 & 0 & 0 & 1 & 0 & 0 \\ 0 & 0 & 0 & 0 & 1 & 0 \\ 0 & 0 & 0 & 0 & 0 & 1 \end{bmatrix}$	6

Operation		Matrix notation	Characters χ (R)
C_3	$r_1 = 0.r_1 + 0.r_2 + 0.r_3 + 1.r_4 + 0.r_5 + 0.r_6$ $r_2 = 0.r_1 + 0.r_2 + 0.r_3 + 0.r_4 + 1.r_5 + 0.r_6$ $r_3 = 0.r_1 + 1.r_2 + 0.r_3 + 0.r_4 + 0.r_5 + 0.r_6$ $r_4 = 0.r_1 + 0.r_2 + 0.r_3 + 0.r_4 + 0.r_5 + 1.r_6$ $r_5 = 0.r_1 + 0.r_2 + 1.r_3 + 0.r_4 + 0.r_5 + 0.r_6$ $r_6 = 1.r_1 + 0.r_2 + 0.r_3 + 0.r_4 + 0.r_5 + 0.r_6$	$\begin{bmatrix} 0 & 0 & 0 & 1 & 0 & 0 \\ 0 & 0 & 0 & 0 & 1 & 0 \\ 0 & 1 & 0 & 0 & 0 & 0 \\ 0 & 0 & 0 & 0 & 0 & 1 \\ 0 & 0 & 1 & 0 & 0 & 0 \\ 1 & 0 & 0 & 0 & 0 & 0 \end{bmatrix}$	0

When C_2 is in between bonds, in that case, all vectors shift from their positions and hence, the character of $C_2 = 0$

C_4	$r_1 = 0.r_1 + 1.r_2 + 0.r_3 + 0.r_4 + 0.r_5 + 0.r_6$ $r_2 = 0.r_1 + 0.r_2 + 1.r_3 + 0.r_4 + 0.r_5 + 0.r_6$ $r_3 = 0.r_1 + 0.r_2 + 0.r_3 + 1.r_4 + 0.r_5 + 0.r_6$ $r_4 = 1.r_1 + 0.r_2 + 0.r_3 + 0.r_4 + 0.r_5 + 0.r_6$ $r_5 = 0.r_1 + 0.r_2 + 0.r_3 + 0.r_4 + 1.r_5 + 0.r_6$ $r_6 = 0.r_1 + 0.r_2 + 0.r_3 + 0.r_4 + 0.r_5 + 1.r_6$	$\begin{bmatrix} 0 & 1 & 0 & 0 & 0 & 0 \\ 0 & 0 & 1 & 0 & 0 & 0 \\ 0 & 0 & 0 & 1 & 0 & 0 \\ 1 & 0 & 0 & 0 & 0 & 0 \\ 0 & 0 & 0 & 0 & 1 & 0 \\ 0 & 0 & 0 & 0 & 0 & 1 \end{bmatrix}$	2

As C_2 is coinciding with C_4, its character will be 2

i	$r_1 = 0.r_1 + 0.r_2 + 1.r_3 + 0.r_4 + 0.r_5 + 0.r_6$ $r_2 = 0.r_1 + 0.r_2 + 0.r_3 + 1.r_4 + 0.r_5 + 0.r_6$ $r_3 = 1.r_1 + 0.r_2 + 0.r_3 + 0.r_4 + 0.r_5 + 0.r_6$ $r_4 = 0.r_1 + 1.r_2 + 0.r_3 + 0.r_4 + 0.r_5 + 0.r_6$ $r_5 = 0.r_1 + 0.r_2 + 0.r_3 + 0.r_4 + 0.r_5 + 1.r_6$ $r_6 = 0.r_1 + 0.r_2 + 0.r_3 + 0.r_4 + 1.r_5 + 0.r_6$	$\begin{bmatrix} 0 & 0 & 1 & 0 & 0 & 0 \\ 0 & 0 & 0 & 1 & 0 & 0 \\ 1 & 0 & 0 & 0 & 0 & 0 \\ 0 & 1 & 0 & 0 & 0 & 0 \\ 0 & 0 & 0 & 0 & 0 & 1 \\ 0 & 0 & 0 & 0 & 1 & 0 \end{bmatrix}$	0

But $i = S_4 = S_6$ and hence, character of S_4 and S_6 will also be 0, as the character of i is zero. As σ_d also concides with C_2, it character will also be 2

σ_h	$r_1 = 1.r_1 + 0.r_2 + 0.r_3 + 0.r_4 + 0.r_5 + 0.r_6$ $r_2 = 0.r_1 + 1.r_2 + 0.r_3 + 0.r_4 + 0.r_5 + 0.r_6$ $r_3 = 1.r_1 + 0.r_2 + 1.r_3 + 0.r_4 + 0.r_5 + 0.r_6$ $r_4 = 0.r_1 + 0.r_2 + 0.r_3 + 1.r_4 + 0.r_5 + 0.r_6$ $r_5 = 0.r_1 + 0.r_2 + 0.r_3 + 0.r_4 + 0.r_5 + 1.r_6$ $r_6 = 0.r_1 + 0.r_2 + 0.r_3 + 0.r_4 + 1.r_5 + 0.r_6$	$\begin{bmatrix} 1 & 0 & 0 & 0 & 0 & 0 \\ 0 & 1 & 0 & 0 & 0 & 0 \\ 0 & 0 & 1 & 0 & 0 & 0 \\ 0 & 0 & 0 & 1 & 0 & 0 \\ 0 & 0 & 0 & 0 & 0 & 1 \\ 0 & 0 & 0 & 0 & 1 & 0 \end{bmatrix}$	4

Therefore, reducible representation for this case is:

O_h	E	$8C_3$	$6C_2$	$6C_4$	$3C_2$	i	$6S_4$	$8S_6$	$3\sigma_h$	$6\sigma_d$
$\Gamma_\sigma(R)$	6	0	0	2	2	0	0	0	4	2

Character Table of O_h point group is:

O_h	E	$8C_3$	$6C_2$	$6C_4$	$3C_2(=C_4^2)$	i	$6S_4$	$8S_6$	$3\sigma_h$	$6\sigma_d$		
A_{1g}	1	1	1	1	1	1	1	1	1	1		$x^2+y^2+z^2$
A_{2g}	1	1	−1	−1	1	1	−1	1	1	−1		
E_g	2	−1	0	0	2	2	0	−1	2	0		$(2z^2-x^2-y^2, x^2-y^2)$
T_{1g}	3	0	−1	1	−1	3	1	0	−1	−1	(R_x, R_y, R_z)	
T_{2g}	3	0	1	−1	−1	3	−1	0	−1	1		(xz, yz, xy)
A_{1u}	1	1	1	1	1	−1	−1	−1	−1	−1		
A_{2u}	1	1	−1	−1	1	−1	1	−1	−1	1		
E_u	2	−1	0	0	2	−2	0	1	−2	0		
T_{1u}	3	0	−1	1	−1	−3	−1	0	1	1	(x, y, z)	
T_{2u}	3	0	1	−1	−1	−3	1	0	1	−1		

Using reduction formula:

$$a_{A_{1g}} = \frac{1}{48}[1\times1\times6 + 8\times1\times0 + 6\times1\times0 + 6\times1\times2 + 3\times1\times2 + 1\times1\times0$$
$$+ 6\times1\times0 + 8\times1\times0 + 3\times1\times4 + 6\times1\times2] = 1$$

$$a_{E_g} = \frac{1}{48}[1\times2\times6 + 8\times-1\times0 + 6\times0\times0 + 6\times0\times2 + 3\times2\times2 + 1\times2\times0$$
$$+ 6\times0\times0 + 8\times-1\times0 + 3\times2\times4 + 6\times0\times2] = 1$$

$$a_{T_{1u}} = \frac{1}{48}[1\times3\times6 + 8\times0\times0 + 6\times-1\times0 + 6\times1\times2 + 3\times-1\times2$$
$$+ 1\times-3\times0 + 6\times-1\times0 + 8\times0\times0 + 3\times1\times4 + 6\times1\times2] = 1$$

$$a_{T_{1g}} = \frac{1}{48}[1\times1\times6 + 8\times1\times0 + 6\times-1\times0 + 6\times-1\times2 + 3\times1\times2 +$$
$$1\times 1\times1 + 6\times-1\times0 + 8\times1\times0 + 3\times1\times4 + 6\times-1\times2] = 0$$

Similarly $a_{T_{2g}} = a_{A_{1u}} = a_{A_{2u}} = a_{E_u} = a_{T_{2u}} = 0$

Thus, reducible representation for this case can be reduced as:

$$\Gamma_\sigma(R) = 1.A_{1g} + 0.A_{2g} + 1.E_g + 0.T_{1g} + 0.T_{2g} + 0.A_{1u} + 0.A_{2u}$$
$$+ 0.E_u + 1.T_{1u} + 0.T_{2u}$$

$$\Gamma_\sigma(R) = A_{1g} + E_g + T_{1u}$$

These representations represent orbitals. Cartesian coordinates *x*, *y* and *z* in the character table represents p_x, p_y and p_z orbitals, respectively. Similarly $x^2 + y^2 + z^2$, $x^2–y^2$, $2z^2–x^2–y^2$, represents *s*, $d_{x^2-y^2}$ and d_{z^2} respectively

A_{1g}	E_g	T_{1u}
s	$(d_{z^2}, d_{x^2-y^2})$	(p_x, p_y, p_z)

By this, only one combination is possible for hybridization, i.e., sp^3d^2 and hence, hybridization in SF_6 is sp^3d^2.

6.3 HYBRIDIZATION SCHEMES FOR π-BONDING

A similar principle is involved in the construction of π-hybrid orbitals as σ-hybrid orbitals. The basic difference between σ- and π-orbitals, from symmetry point of view, is that, the π-orbitals have nodal plane containing the bond axis, whereas σ-orbitals do not have nodal planes containing the bond axis.

6.3.1 PLANAR MOLECULE ION AB_3 (BF_3 OR NO_3^-)

BF_3 molecule belongs to D_{3h} point group. The orbitals on the F-atoms suitable for π-BF bonds are p_x and p_y orbitals, since p_z orbitals of the F-atoms are used for σ-bonding with sp^2-hybrid orbitals of boron. It is known that p_x and p_y orbitals of the F-atoms are either perpendicular or parallel to the molecular plane. Hence, a maximum of two π AOs are permitted on each F-atom. We are now interested to know, which boron orbitals can overlap with those orbitals of fluorine atom to form π-perpendicular [π (⊥)] and π-parallel [π (||)] bonds. The same procedure is used for π-bonding as followed for σ-bonding. The vectors representing the p-orbitals of the fluorine atoms, which are perpendicular and parallel to the molecular plane, are used to determine the characters of reducible representation of the system. The orientations of the six vectors attached to the fluorine atoms are:

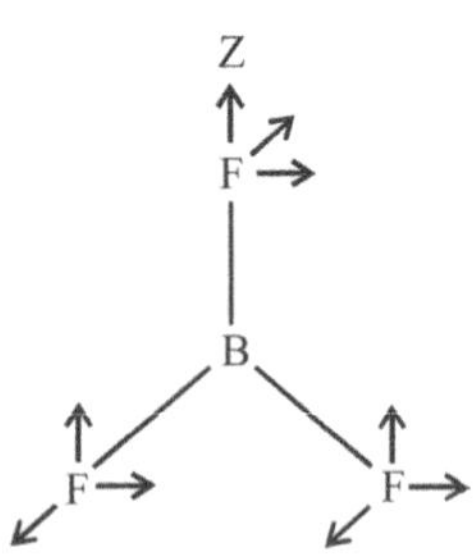

The rules used previously for the σ-systems are also applicable here with only one change. Here, one may find that if a vector inverts (may changes into its own negative vector) due to symmetry operation, it contributes –1 to the character of the class.

In D_{3h} system, no symmetry operation interchanges an out-of-plane vector with an in-plane vector. Therefore, these two sets may be considered independently.

In case of Γ(π) in the plane for E, the three vectors in the plane remain unshifted and hence, contributes +3 to its character, while C_3 shifts one vector into the other and therefore, $\chi(C_3) = 0$. C_2 shifts each Γ(π) vector into its own negative and hence, $\chi(C_2) = -1$, while other two (π) vectors mutually exchange with each other. In case of σ_h, all the three (π) vectors remain unshifted and hence, $\chi(\sigma_h) = +3$. Similarly, $\chi(S_3) = 0$ as $\chi(C_3) = 0$. As σ_v is similar to C_2; hence, $\chi(\sigma_v) = -1$.

Thus, Γ(π) is:

D_{3h}	E	$2C_3$	$3C_2$	σ_h	$2S_3$	$3\sigma_v$
$\Gamma_{\parallel}$	3	0	–1	3	0	–1
$\Gamma_{\perp}$	3	0	–1	–3	0	1

Similar pattern is followed with Γ(⊥), which differs in planes only. In case of σ_h, all the Γ(⊥) vectors are converted into their own negative vectors and thus, contributes – 1. Hence, $\chi(\sigma_h) = -3$. σ_v causes two vectors to exchange mutually with each other and contributes zero, while one vector remains unshifted, i.e., $\chi(\sigma_v) = 1$.

Thus, we get the total π orbitals representation by adding both representations, i.e., $\Gamma_\pi(\perp) + \Gamma_\pi(\parallel)$ and the results are as follows:

D_{3h}	E	$2C_3$	$3C_2$	σ_h	$2S_3$	$3\sigma_{1/2}$
$\Gamma_\pi(R)$	6	0	–2	0	0	0

On reducing, $\Gamma_\perp$ and $\Gamma_\parallel$ separately using character table of D_{3h} point group and the reduction formula to obtain irreducible representations (IRs), one obtains:

$$\Gamma_\perp = A_2'' + E''$$

Here $A_2'' = p_z$; $E'' = (d_{xz}, d_{yz})$, $d_{x^2-y^2}$

$$\Gamma_{||} = A_2' + E'$$

Here $A_2' = \text{None}; E' = (p_x, p_y) \text{ and } (d_{xy}, d_{x^2-y^2})$

It is found from the character table of D_{3h}, that atomic orbitals of A_2'' symmetry, i.e., p_z and E'' symmetry, i.e., d_{xz}, d_{yz} are appropriate for $\Gamma_\pi(\perp)$ representation (out-of-plane vectors) whereas atomic orbitals A_2' symmetry, i.e., none and E' symmetry, i.e., p_x, p_y and $d_{x^2-y^2}$, d_{xy} are appropriate for $\Gamma_\pi(||)$ representation (in-plane vectors).

Thus, in order to form π bonds ⊥ to the plane of molecule, the central atom B must use three hybrid orbitals; one transforming as A_2" and two as E". The orbitals having these transformation properties are A_2" = p_z and E" = (d_{xz}, d_{yz}). Therefore, only one combination is possible, i.e., $p_z + (d_{xy}, d_{yz})$, which leads to pd^2 hybrid. Thus, this gives a set of three equivalent hybrid orbitals for forming π (⊥) bonds.

The $\Gamma_{||}$ set is found to have $A_{2'}$ = none and E' = (p_x, p_y) and $(d_{x^2-y^2}; d_{xy})$. Hence, two sets (combination) are possible.

$$0 + p_x, p_y = p^2$$

or $$0 + (d_{xy}, d_{xz}) = d^2$$

Since there is no *d*-orbital on B atoms having suitable energy, and hence, in-plane π-bond cannot be formed. This result further suggests that only $2p_z$ orbitals of boron will be available for one π (⊥) bond formation, because 2 s, $2p_x$ and $2p_y$ orbitals are used to form a set of three σ-hybrid orbitals (sp^2). This indicates that $2p_z$ orbital of boron can be used to form a π-bond with 2p orbital of any one of fluorine atom. This gives rise to three equivalent resonance structures for BF_3.

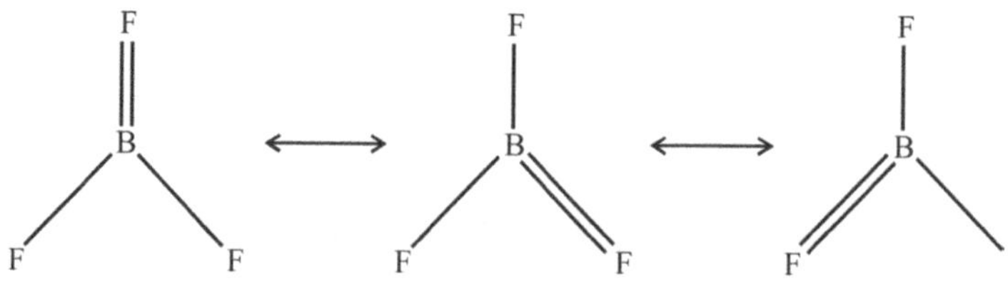

The nonavailability of the A_2' orbitals does not mean that no π bonds can be formed, nor does it mean that only two bonds of the B-atom can be π bonded further. It only means that there can be only two π bonds shared equally among the three B atoms. This general situation, i.e., lack of a complete set of AOs to form a complete set of π-bonds arises very frequently in other systems also.

Thus, it can be concluded that, in AB_3 type molecules, there are two possibilities by which central atom can form π-bond using its *d*-orbital having appropriate energy. These possibilities are:

- A set of $3\pi - (\perp)$ hybrid orbitals; and
- A set of $2\pi - (\|)$ hydrid orbitals.

6.3.2 *TETRAHEDRAL MOLECULE AB_4*

In order to have eight possible AB-π bonds, which from a basis of representation, we attach two vectors to each B atoms, along the direction ⊥ to the bonds and also ⊥ to each other as:

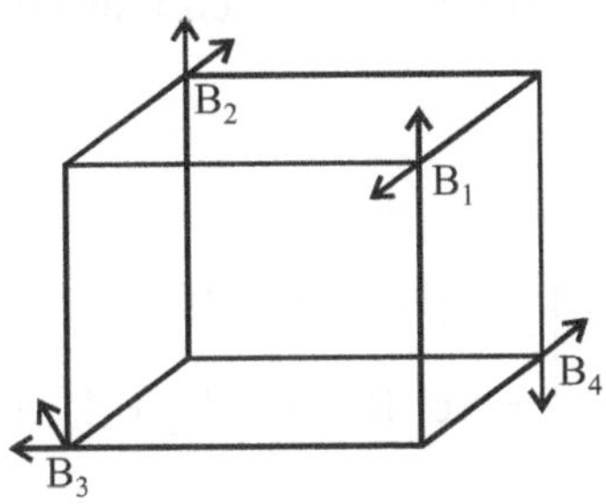

It can be noted that all the eight vectors are equivalent because they can be interchanged by symmetry operation (σ_d) of the tetrahedral molecule. The characters of the reducible representation for this system are obtained by carrying out all the symmetry operations. These vectors are:

T_d	E	$8C_3$	$3C_2$	$6S_4$	$6\sigma_d$
$\Gamma_\pi(R)$	8	−1	0	0	0

C_3 operations have zero character because it shift all the six vectors at the corner of triangular face. When *x* and *y* vectors along the principal axis are rotated by 120°, then

$$\begin{bmatrix} x' \\ y' \end{bmatrix} = \begin{bmatrix} -\frac{1}{2} & -\sqrt{\frac{3}{2}} \\ \sqrt{\frac{3}{2}} & -\frac{1}{2} \end{bmatrix} \begin{bmatrix} x \\ y \end{bmatrix}$$

$$\chi(C_3) = -1$$

The reducible representation of T_d can be reduced to sum of irreducible representations using reduction formula, we get:

$$\Gamma_\pi = E + T_1 + T_2$$

Following *s, p,* and *d* orbitals belonging of the irreducible representations constituting Γ_π are:

$$E = d_{x^2-y^2}, d_{z^2}; T_1 = \text{None}; T_2 = p_x, p_y, p_z \text{ and } d_{xy}, d_{yz}, d_{zx}$$

There AOs belonging to the same representations are required for both; σ- and π-bonding, σ hybridization required AOs of A_1 and T_2 symmetry. In situation like this, it is usually assumed that the σ-bond formation takes precedence over π-bond.

σ-hybrid

$$A_1 + T_2 \begin{cases} \text{Set I} = s + p_x + p_y + p_z\ (= sp^3) \\ \text{Set II} = s + d_{xy} + d_{xz} + d_{yz}\ (= sd^3) \end{cases}$$

π-hybrid

$$E + T_1 + T_2 \begin{cases} \text{Set I} = d_{x^2-y^2} + d_{z^2} + p_x + p_y + p_z\ (= d^2p^3) \\ \text{Set II} = d_{x^2-y^2} + d_{z^2} + d_{xy} + d_{xz} + d_{yz}\ (= d^5) \end{cases}$$

There are no AOs belonging to T_1 symmetry and, therefore, if p_x, p_y and p_z are used for σ-hybrid, i.e., sp^3, then, a set of only five hybrid orbitals may be formed of the central atom A possessing s, p and *d* valence orbitals.

If A requires sp^3 hybrids for σ-bonding, then the pure set of d^5 is constructed from two orbitals of E and 3 orbitals of T_2 representation. If A requires sd^3 hybrids for bonding, then only p^3d^2 set is available for π-hybrids.

Finally, there is a whole range of intermediate cases, in which σ-orbitals are a mixture of the sp^3 and sd^3 limiting cases and then π-orbitals are a complementary mixture of the d^5 and p^3d^2 limiting cases. Only group theory alone can suggest such various possibilities.

6.3.3 SQUARE PLANAR MOLECULE AB_4

In AB_4 type molecules having square planar structure, the central atom (A) either uses $d_{x^2-y^2}$, s, p_x and p_y atomic orbitals, i.e., dsp^2 hybrids or d_{z^2}, $d_{x^2-y^2}$, p_x and p_y, i.e., d^2p^2 hybrid to form σ-hybrid. But among these, two dsp^2 σ-hybrids are most appropriate for σ-hybrid formation.

These molecules belong to point group D_{4h}. We divide the eight possible bonds into two subsets, four of them are perpendicular to molecular plane and the other four are lying in the molecular plane.

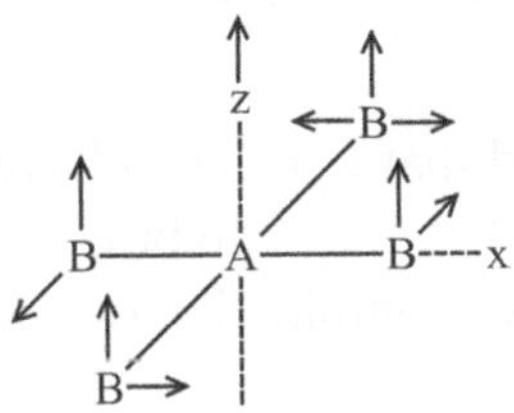

Here, z axis is perpendicular (⊥) to the plane of paper.

On rotating along x and y axis, only the vectors in one direction will interchange. But as x and y are not operations of the group and therefore, four ⊥ and four || vectors (two set of vectors) may be operated independently. It means eight possible π-bonds can be divided into above two sets of vectors. The total character of the two representations, of which two sets of vectors form the basis, can be worked out as:

D_{4h}	E	$2C_4$	C_2	$2C_2'$	$2C_2''$	i	$2S_4$	σ_h	$2\sigma_v$	$2\sigma_d$
$\Gamma_\pi(\perp)$	4	0	0	−2	0	0	0	−4	2	0
$\Gamma_\pi(\parallel)$	4	0	0	−2	0	0	0	+4	−2	0

The combination of the two representations gives total π-representation as:

$$\Gamma_\pi(R) = \Gamma_\pi(\perp) + \Gamma_\pi(\parallel)$$

The total π-representations can be reduced to following irreducible representations using standard reduction formula as:

$$\Gamma_\pi(\perp) = A_{2u} + B_{2u} + E_g$$

$$\Gamma_\pi(\parallel) = A_{2g} + B_{2g} + E_u$$

The atomic orbitals corresponding to these irreducible representations are:

$\Gamma_\pi(\perp)$ —
- Atomic orbital symmetry $A_{2u} \rightarrow p_z$
- Atomic orbital symmetry $B_{2u} \rightarrow$ None
- Atomic orbital symmetry $E_g \rightarrow d_{xz}, d_{yz}$

$\Gamma_\pi(\parallel)$ —
- Atomic orbital symmetry $A_{2g} \rightarrow$ None
- Atomic orbital symmetry $B_{2g} \rightarrow d_{xz}$
- Atomic orbital symmetry $E_u \rightarrow p_x, p_y$

From this, it can be noticed that $\Gamma_\pi(\perp)$ can have combination of one *p* and two *d* orbitals, i.e., $p_z + d_{xz} + d_{yz} = pd^2$ hybrid.

Similarly, $\Gamma_\pi(\parallel)$ can have combination of one *d* and two p orbitals, i.e.,

$$d_{xz} + p_x + p_y = dp^2 \text{ hybrid}$$

But both these sets are not complete, as these sets do not possess B_{2u} and A_{2g} orbital in atom A. Therefore, in this case, there may be three perpendicular π-bond, which may be shared among all the four A-B bonds. p_z, d_{xz} and d_{yz} orbitals are not used for σ-bond formation because p_z is perpendicular to molecular plane and $d_{x^2-y^2}$ orbital is required. So, p_z, d_{xz} and d_{yz} orbital may be used for formation of π-bond (⊥) between A-B.

Therefore, π-bonds in-plane of σ-bond may be formed by utilization of the d_{xy}, p_x and p_y orbitals. As p_x and p_y orbitals have already been used to form σ-hybrid, only d_{xy} orbital of the central atom can be used to form one π-bond. This π-bond will be shared equally by all four A-B pairs of the molecules.

6.3.4 OCTAHEDRAL MOLECULE AB_6

AB_6 type octahedral molecule belongs to O_h point group, in which 12 possible A-B π-bonds form a basis. Two vectors can be attached to each B atom. In such molecules, one vector can be exchanged with rest of the 11 vectors by one symmetry operation (C_4) or by any other operations. It shows that possible 12 π-bonds belong to same set (all the 12 vectors are equivalent).

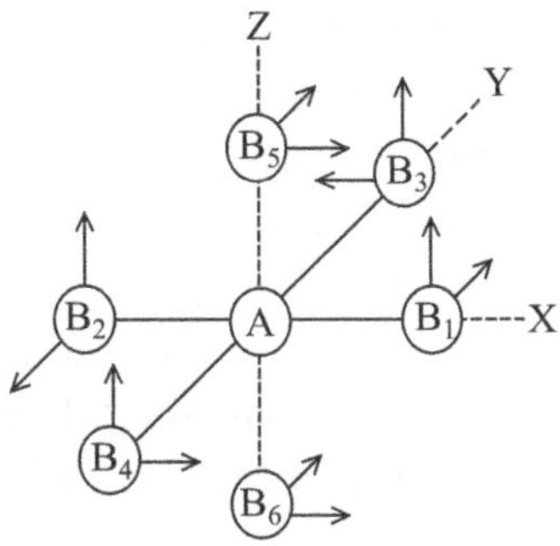

The total character is obtained by performing the operation of each class. If vectors shift, then character becomes zero, if it remains unshifted, then character is +1 and, if there is a change in the direction (opposite to original), then character becomes –1.

O_h	E	$8C_3$	$6C_2$	$6C_4$	$3C_2(=C_4^2)$	i	$6S_4$	$8S_6$	$3\sigma_h$	$6\sigma_d$
$\Gamma_\pi(R)$	12	0	0	0	–4	0	0	0	0	0

The total π-representations will be the addition of reducible representations, then the irreducible representation becomes:

$$\Gamma_\pi(R) = T_{1g} + T_{2g} + T_{1u} + T_{2u}$$

The atomic orbitals corresponding to these irreducible representations are:

Γ_π
- Atomic orbital symmetry $T_{1g} \rightarrow$ None
- Atomic orbital symmetry $T_{2g} \rightarrow d_{xy}, d_{xz}, d_{yz}$
- Atomic orbital symmetry $T_{1u} \rightarrow p_x, p_y, p_z$
- Atomic orbital symmetry $T_{2u} \rightarrow$ None

It gives a conclusion that 12 π-bonds cannot be formed because the atomic orbital T_{1g} and T_{2u} are not available on atom A. As T_{1u} orbitals (p_x, p_y and p_z) have already been used for σ-bond formation between A-B; thus, these orbitals cannot be of any use for π-bonding. Now, only one atomic orbital, T_{2g} having d_{xy}, d_{xz} and d_{yz} orbitals are present for π-bonding.

Therefore, there is a possibility that three π-bonds can be formed between A-B, which is shared equally among the 6 A-B pairs of AB_6 molecules. In T_{2g} atomic orbital, d_{xy} orbital can form π-bond with 4 B atoms (B_1, B_2, B_3 and B_4) equally. In the same manner d_{xz} orbital can form π-bond with four B atoms (B_1, B_2, B_5 and B_6) equally, and d_{yz} orbitals of central atom also form π-bonds equally well with B_3, B_4, B_5 and B_6 atoms. It means, that there is actual sharing of π-bond equally among six A-B bonds. This will result in the one-half of a π-bond per A-B pair in AB_6 octahedral molecules.

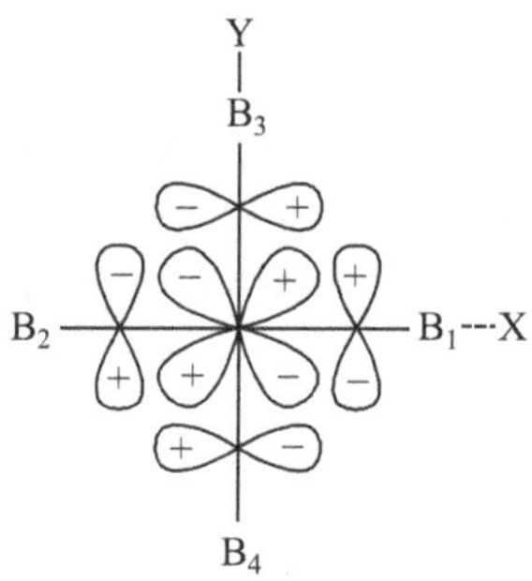

d_{xy}

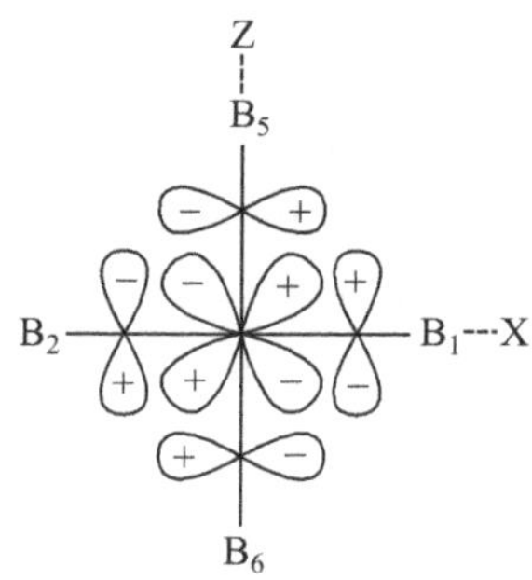

d_{xz}

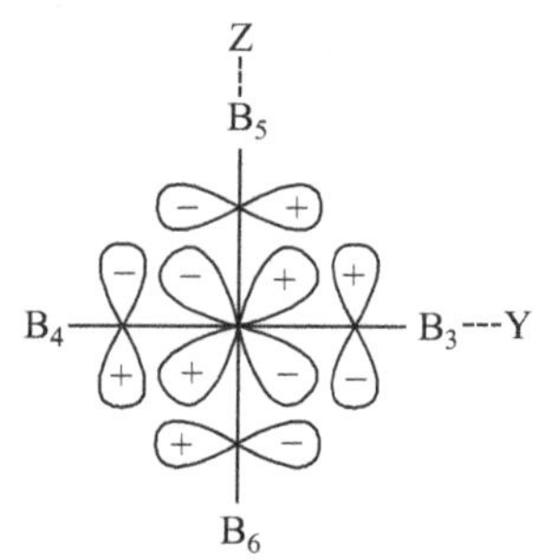

d_{yz}

6.4 MATHEMATICAL FORM OF THE EQUIVALENT HYBRID ORBITALS

It has been explained that how one can obtain the symmetry orbitals required to construct a set of equivalent orbitals suitable for σ and π bonds in a given system using group theory. However, it has not been explained that what proportions of each AOs is used regarding the bond strengths, energies and overlap integrals, which requires the use of equivalent hybrid orbitals? It is often necessary to make use of the algebraic expression known as mathematical forms of the combination by considering the set of equivalent orbitals as a linear combination of AOs. Now, one can find out linear coefficients.

Let us consider the ith hybrid ψ_i, which is obtained by a linear combination of ϕ_i AOs. Thus:

$$\psi_i = C_{i1}\,\phi_1 + C_{i2}\,\phi_2 + C_{i3}\,\phi_3 + \ldots + C_{ij}\,\phi_j$$

$$\psi_i = \sum^{N} C_{ij}\phi_j C_{ij}\phi_j$$

where C_{ij} are the linear coefficients of AOs used to construct the equivalent hybrid orbital ψ_i and φ_i representing AOs of ith atom.

As we are interested in evaluating these coefficients, two principles are used here:

(a) The set of equivalent orbitals forms an orthonormal set.
   (i) Each equivalent orbital of the set may be normalized, i.e.,

$$\int \psi_i\,\psi_k^*\,d\tau = 1 \quad \text{if } i = k \text{ or } \int \psi^2\,d\tau = 1$$

   (ii) Each orbitals of the equivalent set must be orthogonal to all the other hybrid orbitals of the set.

$$\int \psi_i\,\psi_k^*\,d\tau = 0 \quad \text{if } i \neq k$$

(b) Each hybrid orbital is equivalent to the other hybrid orbitals in the set under the appropriate symmetry operations of the group. The coefficients must be so adjusted that when a symmetry operation (R) is carried out on one hybrid orbital of the set, then it is transformed into its equivalent member, i.e.,

$$R.\,\psi_j = \psi_j,\ \text{i.e., } R\sum_j C_{jk}\,\phi_j = \sum_j C_{ki}\,\phi_i$$

6.4.1 LINEAR MOLECULE AB_2 (sp HYBRID ORBITALS)

Linear molecules belong to special category of the C_{nv} and D_{nh} point groups.

- AB_2 type molecules with center of symmetry belong to $D_{\infty h}$, e.g., $BeCl_2$, BeH_2.

- BAC type molecules without center of symmetry belong to $C_{\infty v}$ point group, i.e., COS, HCN.

Generally, composition of σ hybrids is known form reduction of the reducible representation to the irreducible representation components. But in AB_2 type linear molecule, it is very simple to identify σ-hybrid. Here, σ-hybrid is formed from s and p_z orbitals with equal contribution form each orbital.

In $C_{\infty v}$ and $D_{\infty h}$ infinite order groups, C_{2v} subgroup and D_{2h} subgroup, respectively are selected, which have low order, where *x* and *y* coordinates do not intermix by rotation operation of the subgroup.

Correlation between infinite group with subgroup are:

$C_{\infty v}$	C_{2v}
$A_1 = \Sigma^+$	$\rightarrow A_1$
$A_1 = \Sigma^-$	$\rightarrow A_2$
$E_1 = \pi$	$\rightarrow B_1 + B_2$
$E_2 = \Delta$	$\rightarrow A_1 + A_2$

$D_{\infty h}$	D_{2h}
Σ_g^+	$\rightarrow A_g$
Σ_g^-	$\rightarrow B_{1g}$
π_g	$\rightarrow B_{2g} + B_{3g}$
Δ_g	$\rightarrow A_g + B_{1g}$
Σ_u^+	$\rightarrow B_{1u}$
Σ_u^-	$\rightarrow A_u$
π_u	$\rightarrow B_{2u} + B_{3u}$
Δ_u	$\rightarrow A_u + B_{1u}$

The total character for the two σ hybrid representations have σ hybrid along the X-axis (internuclear axis), which form basis of group D_{2h}.

Total σ-representations are:

	E	$C_2(z)$	$C_2(y)$	$C_2(x)$	i	σ_{xy}	σ_{xz}	σ_{yz}
$\Gamma_\sigma(R)$	2	0	0	2	0	2	2	0

In D_{2h} point group, A_g becomes Σ_g^+ and B_{1u} becomes Σ_u^+. The atomic orbitals corresponding to irreducible representations are:

- Atomic orbital with A_g symmetry is s.
- Atomic orbital with B_{1u} symmetry is p_z.

Therefore, there can be one combination possible, i.e., s + p_z (= sp hybrid), but there is one more possibility of combination, i.e., d_{z^2} + p_z (= dp hybrid) as d_{z^2} also belongs to A_g. If we take example of $BeCl_2$, then Be does not have *d* orbitals, so set of dp hybrids cannot be considered.

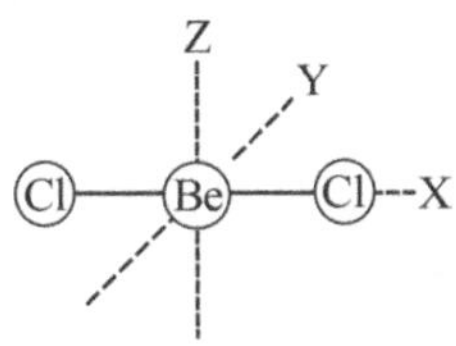

If ϕ_1 and ϕ_2 are the two combining atomic orbitals, the resulting hybridized orbitals are ψ_{h_1} and ψ_{h_2}:

Hence, σ hybrid in $BeCl_2$ is formed of:

$$\psi_{h_1} = \frac{1}{\sqrt{2}}\psi_s + b_1\ \psi p_z$$

and

$$\psi_{h_2} = \frac{1}{\sqrt{2}}\psi_s - b_2\ \psi p_z$$

$$\psi_{h1} = a_1\ \phi_1 + b_1\ \phi_{pz} \tag{6.1}$$

$$\psi_{h2} = a_2\ \phi_1 + b_2\ \phi_{pz} \tag{6.2}$$

Since s orbital is spherically symmetric, it contributes equally to the making of two hybrid orbitals.

Therefore,
$$a_1 = a_2 = \frac{1}{\sqrt{2}} \tag{6.3}$$

Now $\psi_1 = \frac{1}{\sqrt{2}}\phi_s + b_1\ \phi_{p_z}$ since is normalized.

Therefore, $a_1^2 + b_1^2 = 1$

$$\frac{1}{2} + b_1^2 = 1$$

or $$b_1^2 = \frac{1}{2}$$

or $$b_1 = \frac{1}{\sqrt{2}} \qquad (6.4)$$

Now considering that ψ_1 and ψ_2 are orthogonal to each other.

$$a_1 a_2 + b_1 b_2 = 0$$

or $$\frac{1}{2} + \frac{1}{\sqrt{2}} b_2 = 0$$

$$b_2 = -\frac{1}{\sqrt{2}} \qquad (6.5)$$

Hence, $$\psi_{h2} = \frac{1}{\sqrt{2}} \phi_s - \frac{1}{\sqrt{2}} \phi_{pz} = \frac{1}{\sqrt{2}} (\phi_s - \phi_{pz}) \qquad (6.6)$$

6.4.2 TRIGONAL PLANAR MOLECULE AB_3-sp^2 HYBRID ORBITALS

The composition of the three hybrid orbitals can be shown as:

$$\psi_1 = a_1\,\phi_s + b_1\,\phi_{px} + C_1\,\phi_{py} \qquad (6.7)$$

$$\psi_2 = a_2\,\phi_s + b_2\,\phi_{px} + C_2\,\phi_{py} \qquad (6.8)$$

$$\psi_3 = a_3\,\phi + b_3\,\phi_{px} + C_3\,\phi_{py} \qquad (6.9)$$

Since s orbital is spherically symmetric, it contributes equally to the making of the three hybrid orbitals.

Therefore, $$a_1 = a_2 = a_3 = 1/\sqrt{3}$$

ψ_{h1} is formed along X-axis and hence, it cannot has any contribution from p_y, i.e., $C_1 = 0$

Therefore,

$$\psi_1 = \frac{1}{\sqrt{3}}\,\varphi_s + b_1\,\varphi_{px} \quad \text{since } \psi_1 \text{ is normalized}$$

Therefore, $$a_1^{\,2} + b_1^{\,2} = 1$$

$$\frac{1}{3} + b_1^{\,2} = 1$$

or $$b_1^{\,2} = \frac{2}{3}$$

or $$b_1 = \sqrt{\frac{2}{3}}$$

So $$\psi_1 = \frac{1}{\sqrt{3}}\,\phi_s + \sqrt{\frac{2}{3}}\,\phi_{px} \tag{6.10}$$

Now considering that ψ_1 and ψ_2 are orthogonal to each other, we have:

$$a_1\,a_2 + b_1\,b_2 = 0$$

or $$\frac{1}{3} + \sqrt{\frac{2}{3}}\;b_2 = 0$$

or $$b_2 = -\frac{1}{\sqrt{6}} \tag{6.11}$$

Further, the normalization condition requires that $a_2^{\,2} + b_2^{\,2} + c_2^{\,2} = 1$

$$\frac{1}{3} + \frac{1}{6} + c_2^{\,2} - 1$$

or $$c_2^{\,2} = \frac{1}{2}$$

or $$c_2 = \frac{1}{\sqrt{2}} \tag{6.12}$$

Hence, $$\psi_2 = \frac{1}{3}\,\phi_s - \frac{1}{\sqrt{6}}\,\phi_{px} + \frac{1}{\sqrt{2}}\,\phi_{py}$$

Considering orthogonality of ψ_1 and ψ_3, we have,

$$a_1 a_3 + b_2 b_3 + c_2 c_3 = 0$$

Hence,
$$\frac{1}{3} + \sqrt{\frac{2}{3}}\, b_3 = 0$$

or
$$b_3 = -\frac{1}{\sqrt{6}} \qquad (6.13)$$

Again ψ_2 and ψ_3 are also orthogonal and hence,

$$a_2 a_3 + b_2 b_3 + c_2 c_3 = 0$$

or
$$\frac{1}{3} + \frac{1}{6} + \frac{1}{\sqrt{2}} c_3 = 0$$

or
$$\frac{1}{\sqrt{2}} c_3 = \frac{1}{2}$$

or
$$c_3 = -\frac{1}{\sqrt{2}} \qquad (6.14)$$

Hence,
$$\psi_3 = \frac{1}{\sqrt{3}} \phi_s - \frac{1}{\sqrt{6}} \phi_{px} - \frac{1}{\sqrt{2}} \phi_{py} \qquad (6.15)$$

6.4.3 TETRAHEDRAL MOLECULE AB_4 (sp^3 HYBRID ORBITALS)

The composition of the four hybrid orbitals can be shown as:

$$\psi_1 = a_1 \phi_s + b_1 \phi_{px} + c_1 \phi_{py} + d_1 \phi_{pz} \qquad (6.16)$$

$$\psi_2 = a_2 \phi_s + b_2 \phi_{px} + c_2 \phi_{py} + d_2 \phi_{pz} \qquad (6.17)$$

$$\psi_3 = a_3 \phi_s + b_3 \phi_{px} + c_3 \phi_{py} + d_3 \phi_{pz} \qquad (6.18)$$

$$\psi_4 = a_4 \phi_s + b_4 \phi_{px} + c_4 \phi_{py} + d_4 \phi_{pz} \qquad (6.19)$$

Considering four hybrid wave functions and that ψ_1 is in direction of Z-axis (i.e., $\theta = 0^0$) and hence, ϕ_{px} and ϕ_{py} do not contribute, i.e., $c_1 = 0$ and $d_1 = 0$

$$\psi_1 = a_1\, \phi_s + b_1\, \phi_{pz} \tag{6.20}$$

Square of Eq. (6.20) on integration gives –

$$\int \psi_1^{\,2}\, d\tau = a_1^{\,2} \int \phi_s^{\,2}\, d\tau + b_1^{\,2} \int \phi_{pz}^{\,2}\, d\tau + 2a_1, b_1 \int \phi_s\, \phi_{pz}\, d\tau$$

ψ_1, ϕ_s and ϕ_{pz} are normalized and mutually orthogonal and hence,

$$a_1^{\,2} + b_1^{\,2} = 1 \tag{6.21}$$

Since s-orbital is spherically symmetric and therefore, it contributes equally for making the four hybrid orbitals,

Therefore, $$a_1^{\,2} = a_2^{\,2} = a_3^{\,2} = a_4^{\,2} = \frac{1}{4}$$

$$a_1 = a_2 = a_3 = a_4 = \frac{1}{2} \tag{6.22}$$

Placing value of a_1 in Eq. (6.21)

$$\frac{1}{4} + b_1^{\,2} = 1$$

or $$b_1^{\,2} = \frac{3}{4}$$

or $$b_1 = \frac{\sqrt{3}}{2} \tag{6.23}$$

Now putting the values of both; a_1 and b_1 in Eq. (6.20).

$$\psi_1 = \frac{1}{2}\, \phi_s + \frac{\sqrt{3}}{2}\, \phi_{pz} \tag{6.24}$$

Secondly, hybrid orbital is in xz plane and hence,

$$\psi_1 = \frac{1}{2}\, \phi_s + b_2\, \phi_{pz} + c_2\, \phi_{px} \tag{6.25}$$

Square of the Eq. (6.25) on integration gives:

$$\frac{1}{4} + b_2^{\,2} + c_2^{\,2} = 1 \tag{6.26}$$

Since all the hybrid orbitals are orthogonal to each other,

$$\int \psi_1 \psi_2 \, d\tau = 0 \tag{6.27}$$

hence,
$$\frac{1}{4} + \frac{\sqrt{3}}{2} c_2 = 0$$

or
$$c_2 = -\frac{1}{4} \cdot \frac{2}{\sqrt{3}}$$

or
$$c_2 = -\frac{1}{2\sqrt{3}} \tag{6.28}$$

Putting the value of c_2 in Eq. (6.26), we have,

$$\frac{1}{4} + b_2^2 + \frac{1}{12} = 1$$

or
$$b_2^2 = 1 - \frac{1}{4} - \frac{1}{12} = \frac{8}{12} = \frac{2}{3}$$

or
$$b_2 = \pm\sqrt{\frac{2}{3}} \tag{6.29}$$

Hence,
$$\psi_2 = \frac{1}{2}\phi_s + \sqrt{\frac{2}{3}}\,\phi_{px} - \frac{1}{2\sqrt{3}}\,\phi_{pz}$$

Hence, the complete wave equations of hybrid orbitals are given as,

$$\psi_1 = \frac{1}{2}\phi_s + \frac{\sqrt{3}}{2}\phi_{pz} \tag{6.30}$$

$$\psi_2 = \frac{1}{2}\phi_s + \sqrt{\frac{2}{3}}\,\phi_{px} - \frac{1}{2\sqrt{3}}\,\phi_{pz} \tag{6.31}$$

$$\psi_3 = \frac{1}{2}\phi_s - \frac{1}{\sqrt{6}}\phi_{px} + \frac{1}{\sqrt{2}}\phi_{py} - \frac{1}{2\sqrt{3}}\phi_{pz} \tag{6.32}$$

$$\psi_4 = \frac{1}{2}\phi_s - \frac{1}{\sqrt{6}}\phi_{px} - \frac{1}{\sqrt{2}}\phi_{py} - \frac{1}{2\sqrt{3}}\phi_{pz} \tag{6.33}$$

A correlation between irreducible representation and orbitals like s, p_x, p_y, p_z, d_{xy}, d_{z^2}, etc. can be obtained by using mathematically function as bases. We shall use only atomic wave function, which are of importance to a chemist as these atomic wave functions define atomic orbitals. Symmetric properties of atomic orbitals are also important in determining their participation in the formation of hybrid orbitals. Angular wave functions $\psi(\theta, \phi)$ of *s*, *p* and *d* orbitals are only used, as radial part is not affected by any symmetry operation. These wave functions are hydrogenic function (orbitals) and are normalized. Angular wave functions of these atomic orbitals are –

Atomic orbital	**Wave function $\psi(\theta, \varphi)$**
s	$\left(\frac{1}{4\pi}\right)^{1/2}$
p_x	$\left(\frac{3}{4\pi}\right)^{1/2} \sin\theta \cos\phi$
p_y	$\left(\frac{3}{4\pi}\right)^{1/2} \sin\theta \sin\phi$
p_z	$\left(\frac{3}{4\pi}\right)^{1/2} \cos\phi$
d_{z^2}	$\left(\frac{5}{16\pi}\right)^{1/2} (3\cos^2\theta - 1)$
d_{xz}	$\left(\frac{15}{4\pi}\right)^{1/2} \sin\theta \cos\theta \cos\phi$
d_{yz}	$\left(\frac{15}{4\pi}\right)^{1/2} \sin\theta \cos\theta \sin\phi$
$d_{x^2-y^2}$	$\left(\frac{15}{16\pi}\right)^{1/2} \sin\theta \cos 2\phi$
d_{xy}	$\left(\frac{15}{16\pi}\right)^{1/2} \sin^2\theta \sin 2\phi$

Here, s-orbital is highly symmetric as $\psi(\theta, \phi)$ for s-orbital is independent of θ and ϕ and remains unchanged by any operation R like E, C_n, σ_v, σ_h, i, S_n, etc., for example, R $(\psi_s) = +1\ (\psi_s)$.

In case of spherically symmetric environment of a free ion, the three *p* and five *d* orbitals are degenerate, but in case of other symmetry, they may loose their degeneracy.

Let us consider the case of water molecule, which belongs to C_{2v} point group. This point group has four operations, i.e., E, C_2, σ_v (xz), σ_v (yz).

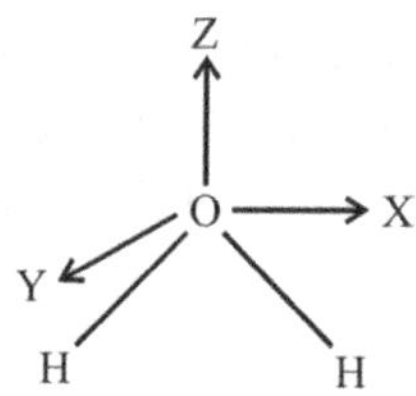

$$E(\psi_s) = (+1)\,\psi_s = \psi_s$$

$$C_2(\psi_s) = (+1)\,\psi_s = \psi_s$$

$$\sigma_v(xz)(\psi_s) = (+1)\,\psi_s = \psi_s$$

$$\sigma_v(yz)(\psi_s) = (+1)\,\psi_s = \psi_s$$

Similarly, for p-orbitals

$$E(\psi_{pz}) = E(\cos\theta) = (+1)\cos\theta = \cos\theta = \psi_{pz}$$

$$C_2(\psi_{pz}) = C_2(\cos\theta) = (+1)\cos\theta = \cos\theta = \psi_{pz}$$

$$\sigma_v(xz)(\psi_{pz}) = \sigma_{xz}(\cos\theta) = (+1)\cos\theta = \cos\theta = \psi_{pz}$$

$$\sigma_v(yz)(\psi_{pz}) = \sigma_{yz}(\cos\theta) = (+1)\cos\theta = \cos\theta = \psi_{pz}$$

$$E(\psi_{px}) = E(\sin\theta\cos\phi) = (+1)\sin\theta\cos\phi = \sin\theta\cos\phi = \psi_{px}$$

$$C_2(\psi_{px}) = C_2(\sin\theta\cos\phi) = \sin\theta\cos(\phi + 180)\sin\theta(-\cos\phi) = (-1)\sin\theta\cos\phi = (-1)\,\psi_{px}$$

$$\sigma_v(yz)(\psi_{px}) = \sigma_v(xz).(\sin\theta\cos\phi) = \sin\theta\cos\phi = (+1)\,\psi_{px}$$

$$\sigma_v(yz)(\psi_{px}) = \sigma_v(yz).(\sin\theta\cos\phi) = \sin\theta(\cos\phi + 180) = \sin\theta(-\cos\phi) = (-1)\sin\theta\cos\phi = (-1)\,\psi_{px}$$

Similarly, it can be proved that:

$$E\,(\psi_{py}) = (+1)\,\psi_{py}$$

$$C_2\,(\psi_{pz}) = (-1)\,\psi_{py}$$

$$\sigma_v\,(xz)\,(\psi_{py}) = (-1)\,\psi_y$$

$$\sigma_v\,(yz)\,(\psi_{py}) = (+1)\,\psi_{py}$$

Summarizing all these observations, we get:

C_{2v}	E	C_2	$\sigma_v(xz)$	$\sigma_v(yz)$	
A_1	+1	+1	+1	+1	ψ_s
A_1	+1	+1	+1	+1	ψ_{pz}
B_1	+1	−1	+1	−1	ψ_{px}
B_2	+1	−1	−1	+1	ψ_{py}

Carrying out same process with *d*-orbitals, one can determine the irreducible representation corresponding to a particular *d*-orbital like d_{xy}, d_{yz}, d_{xz}, d_{z^2} and $d_{x^2-y^2}$.

For d_{xy}

$$E.\,(\Psi_{d_{xy}}) = E.(\sin^2\theta.\,\sin 2\phi) = (+1)\sin^2\theta\,\sin 2\phi = \Psi_{d_{xy}}$$

$$C_2.\,(\Psi_{d_{xy}}) = C_2.\,(\sin^2\theta.\,\sin 2\phi) = \sin^2\theta.\,\sin 2(\phi + 180) = \sin^2\theta.\,\sin(2\phi + 360)\,\sin^2\theta\,\sin 2\phi = (+1)\,\Psi_{d_{xy}}$$

$$\sigma_v(xz).\,(\Psi_{d_{xy}}) = \sigma_v(xz).\,(\sin^2\theta.\,\sin 2\phi) = \sin^2\theta.\,\sin 2(\phi + 90) = \sin^2\theta.\,\sin(2\phi + 180)\,\sin^2\theta.\,2\cos\phi\,(-\sin\phi) = -\sin^2\theta\,\sin 2\phi = (-1)\,\Psi_{d_{xy}}$$

$$\sigma_v(yz):\,(\Psi_{d_{xy}}) = \sigma_v(yz).(\sin^2\theta.\,\sin 2\phi) = \sin^2\theta.\,\sin 2(\phi + 90) = \sin^2\theta.\,\sin(2\phi + 180)\,\sin^2\theta.\,2\cos\phi - (\sin\phi) = -\sin^2\theta\,\sin 2\phi = (-1)\,\Psi_{d_{xy}}$$

Thus,

C_{2v}	E	C_2	$\sigma_v\,(xz)$	$\sigma_v\,(yz)$	
A_2	+1	+1	−1	−1	$\psi_{d_{xy}}$

Similarly, it can be proved that $\Psi_{d_{z^2}}$ and $\Psi_{d_{x^2-y^2}}$ in point group C_{2v} will have A_1 representation and $\Psi_{d_{xz}}$ and $\Psi_{d_{yz}}$ will have B_1 and B_2 representation, respectively.

These correlations between irreducible representation and orbitals varies from one to other character table and therefore, it should be determined individually for each case.

KEYWORDS

- **Hybridization**
- **Linear**
- **Octahedral**
- **Square planar**
- **Tetrahedral**
- **Trigonal bipyramidal**
- **Trigonal planar**

CHAPTER 7

MOLECULAR ORBITAL THEORY

CONTENTS

Molecular orbital theory (MOT) was introduced to explain the abnormal behavior of O_2 molecule in the ground state. Valence bond theory (VBT) suggests the diamagnetic behavior of O_2 molecule with a double bond character, which is not true. If VBT is used to explain the paramagnetic behavior of O_2 molecule corresponding to two unpaired electrons, then it cannot explain the double bond character. On the other hand, if it explains the double bond character than it is unable to explain the paramagnetic behavior of oxygen molecule. This was the first failure of VB theory and the beginning of success story of MOT, which has successfully explained both these facts simultaneously, i.e., double bond character and paramagnetic behavior of O_2 molecule in its ground state.

Group theory can be applied to obtain some of the important properties of a molecule like charge density, electron density, bond order, free valence, delocalization energy, etc. of course, utilizing MO theory. This is being done by constructing M.O. by linear combination of atomic orbitals (LCAO) and solving the corresponding secular determinant.

7.1 SECULAR DETERMINANT

A secular determinant of nth order is reduced to determinants of reduced sizes, perhaps with a 3×3 as the largest one (instead of $N \times N$, where N is the number of carbon atoms of the conjugated molecules). The conjugated molecules may be acyclic or cyclic. These organic molecules are having alternate single and double bonds. Thus:

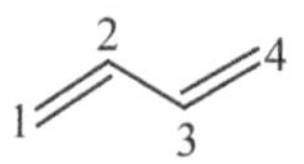

Trans-butadiene

Acyclic

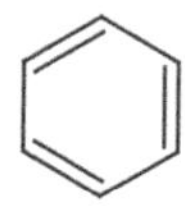

Benzene

Cyclic

The secular determinant is used to calculate the energy of the p-electrons. The Schrodinger equation is used with Huckel approximation to evaluate the energy of molecular orbitals using LCAO (Linear Combination of Atomic Orbitals) concept. The wave equation for the MO is:

$$H\psi = E\psi \tag{7.1}$$

where H is Hamiltonian operator, ψ is the wave function and E is Eigen value (Energy of the system).

The usual method of constructing MO is the linear combination of AOs. For i^{th} MO, it can be generalized as:

$$\psi_i = \sum_{j=1}^{n} C_{ij}\varphi_j \tag{7.2}$$

For i = 1 and n = 2 (Diatomic molecule ethylene)

$$\psi_1 = C_{11}\,\phi_1 + C_{12}\,\phi_2 \tag{7.3}$$

The energy of this MO is obtained with the help of Eq. (7.1). Thus, $H\psi_1 = E\psi_1$ is multiplied by ψ_1 (ψ_1 is a real function) on both sides and integrated with respect to $d\tau$ (dx, dy, dz). Then expression for E is:

$$E = \frac{\int \psi_1 H \psi_1 \, d\tau}{\int \psi_1 \psi_1 \, d\tau} \tag{7.4}$$

Putting the value of ψ_1 from Eq. (7.3) in Eq. (7.4), we get:

$$E = \frac{\int (C_{11}\,\varphi_1 + C_{12}\,\varphi_2)\,H\,(C_{11}\,\varphi_1 + C_{12}\,\varphi_2)\,d\tau}{\int (C_{11}\,\varphi_1 + C_{12}\,\varphi_2)^2\,d\tau}$$

or
$$E = \frac{C_{11}^2 \int \varphi_1\,H\,\varphi_1\,d\tau + 2\,C_{11}\,C_{12}\int \varphi_1\,H\,\varphi_2\,d\text{Ä}\!+ C_{12}^2 \int \varphi_2\,H\,\varphi_2\,d\tau}{C_{11}^2 \int \varphi_1^2\,d\tau + 2\,C_{11}\,C_{12}\int \varphi_1\,\varphi_2\,d\tau + C_{12}^2 \int \varphi_2^2\,d\tau}$$

$$= \frac{C_{11}^2 H_{11} + 2\,C_{11}C_{12}H_{12} + C_{12}^2 H_{22}}{C_{11}^2 S_{11} + 2\,C_{11}C_{12}S_{12} + C_{12}^2 S_{22}} \qquad (7.5)$$

where $H_{ij} = H_{ji} = \int \phi_i\,H\,\phi_j\,d\tau$, i.e., $H_{12} = H_{21} = \int \phi_1\,H\,\phi_2\,d\tau = \int \phi_2\,H\,\phi_1\,d\tau$;

$$H_{11} = \int \phi_1\,H\,\phi_1\,d\tau;\ H_{22} = \int \phi_2\,H\,\phi_2\,d\tau$$

$$S_{ij} = \int \phi_i\,\phi_j\,d\tau$$

$$S_{12} = \int \phi_1\,\phi_2\,d\tau$$

$$S_{11} = \int \phi_1^{\,2}\,d\tau \text{ and } S_{22} = \int \phi_2^{\,2}\,d\tau.$$

H_{ij} is called Coulomb integral (when $i = j$), H_{ij} resonance integral; (when $i \neq j$) and S_{ij} is overlap integral $S_{ij} = 1$, (when $i = j$) and $S_{ij} = 0$ (when $i \neq j$).

To determine coefficients, C_{11} and C_{12}, we have to minimize energy E with respect to each using Eq. (7.5). Thus, differentiating w.r.t. C_{11}, we get:

$$\frac{\partial E}{\partial C_{11}} = \frac{(C_{11}^2 S_{11} + 2\,C_{11}C_{12}S_{12} + C_{12}^2 S_{22})\,(2\,C_{11}H_{11} + 2\,C_{12}H_{12})}{(C_{11}^2 S_{11} + 2\,C_{11}C_{12}S_{12} + C_{12}^2 S_{22})^2}$$

$$\frac{-(C_{11}^2 H_{11} + 2\,C_{11}C_{12}H_{12} + C_{12}^2 H_{22})\,(2\,C_{11}S_{11} + 2\,C_{12}S_{12})}{(C_{11}^2 S_{11} + 2\,C_{11}C_{12}S_{12} + C_{12}^2 S_{22})^2} = 0$$

or
$$(C_{11}^2 S_{11} + 2\,C_{11}C_{12}S_{12} + C_{12}^2 S_{22})\,(2\,C_{11}H_{11} + 2\,C_{12}H_{12}) -$$

$$(C_{11}^2 H_{11} + 2\,C_{11}C_{12}H_{12} + C_{12}^2 H_{22})\,(2\,C_{11}S_{11} + 2\,C_{12}S_{12}) = 0$$

or
$$(C_{11}^2 S_{11} + 2\,C_{11}C_{12} + C_{12}^2 S_{22})\,(2\,C_{11}H_{11} + 2\,C_{12}H_{12})$$

$$= (C_{11}^2 H_{11} + 2\,C_{11}C_{12}H_{12} + C_{12}^2 H_{22})(2\,C_{11}S_{11} + 2\,C_{12}S_{12})$$

or

$$(C_{11}\,H_{11} + C_{12}\,H_{12})$$

$$= \frac{(C_{11}^2 H_{11} + 2\,C_{11}C_{12}H_{12} + C_{12}^2 H_{22})(C_{11}S_{11} + C_{12}S_{12})}{(C_{11}^2 S_{11} + 2\,C_{11}C_{12}S_{12} + C_{12}^2 S_{22})} = 0 \qquad (7.6)$$

From Eqs. (7.5) and 7.6) we get:

$$C_{11}\,H_{11} + C_{12}\,H_{12} = E\,(C_{11}\,S_{11} + C_{12}\,S_{12})$$

or

$$C_{11}\,(H_{11} - E\,S_{11}) + C_{12}\,(H_{12} - E\,S_{12}) = 0 \qquad (7.7)$$

Similarly $\frac{\partial E}{\partial C_{12}} = 0$ gives:

$$C_{11}\,(H_{21} - ES_{21}) + C_{12}\,(H_{22} - ES_{22}) = 0 \qquad (7.8)$$

Equations (7.7) and (7.8) are called Secular equations. These are homogeneous linear equations in C_{11} and C_{12}. The only nontrivial solution of this system requires that the determinants of C_{11} and C_{12} should vanish, i.e.,

$$\begin{vmatrix} H_{11} - ES_{11} & H_{12} - ES_{12} \\ H_{21} - ES_{21} & H_{22} - ES_{22} \end{vmatrix} = 0 \qquad (7.9)$$

This is the secular determinant for ethylene. Generalizing it for n carbon atoms (np electrons) gives:

$$\begin{vmatrix} H_{11} - ES_{11} & H_{12} - ES_{12} & \cdots\cdot & H_{1n} - ES_{1n} \\ H_{21} - ES_{21} & H_{22} - ES_{22} & \cdots\cdot & H_{2n} - ES_{2n} \\ \cdots\cdot & \cdots\cdot & \cdots\cdot & \cdots\cdot \\ H_{n1} - ES_{n1} & H_{n2} - ES_{n2} & \cdots\cdot & H_{nn} - ES_{nn} \end{vmatrix} = 0 \qquad (7.10)$$

The Huckel approximation assumes that:

(i) All S_{ij} are equal to zero if $i \neq j$ and
(ii) $H_{ij} = H_{ji} = 0$ for non-neighbor (not adjacent) i and j atoms, i.e., $H_{13} = H_{14} = H_{24}$ but β for neighboring atoms, i.e., $H_{12} = H_{23} = H_{34}$; $H_{ii} = \alpha$, i.e., $H_{11} = H_{22} = H_{33}$ ….etc. = α. For i = j. We have already seen that $S_{ii} = 1$, i.e., $S_{11} = S_{22}$ …. = 1 and $S_{ij} = 0$, i.e., $S_{12} = S_{23} =$

S_{13} = ... = 0. Putting these values in (7.10) simplifies the secular determinant to give:

$$\begin{vmatrix} \pm - E & {}^{2} & 0 & 0 & \ldots & 0 \\ {}^{2} & \pm - E & {}^{2} & 0 & \ldots & 0 \\ 0 & 0 & \pm - E & {}^{2} & \ldots & 0 \\ \ldots & \ldots & \ldots & \ldots & \ldots & \ldots \\ 0 & 0 & 0 & 0 & \ldots & \pm - E \end{vmatrix} \quad (7.11)$$

For a molecule like naphthalene, the secular determinant is of larger size, i.e., 10 × 10. In such a case, evaluation of E is a laborious take and time consuming process. Since MOs encompass whole of the molecule, and group theory can be applied to the problem in reducing a large secular determinant to smaller determinants of 1 × 1, 2 × 2 and 3 × 3 sizes, their solutions become easy.

Step 1: Each π-orbital is labeled and then point group of the molecule is determined. A lower group involving rotation symmetries can be used and the reducible representation Γ_π is obtained on the basis of the character of the operation (R).

If the orbital changes the sign of its wave function, then the contribution to χ_i is negative, but if it is unaffected, then it is positive.

Step 2: The reducible representation Γ_π is then reduced using the reduction formula or by inspection. A little practice will enable one to reduce the reducible representation.

Step 3: The molecular orbitals (ψ_i) can be obtained by the symmetry-adapted linear combination (SALC) of atomic orbitals. These are then normalized.

Step 4: The secular determinant is then set up using these SALC's in various representations, and solved for determining levels of the system and wave functions coefficients. These energy levels and coefficients are in terms of SALC of AO's, which can be converted back in terms of starting AO's.

Let us apply these four steps in case of the ethylene molecule.

7.2 ETHYLENE

Step 1: Ethylene belong to D_{2h} point group. Γ_π for ethylene molecule can be obtained using a simple point group C_2 considering only rotational axis).

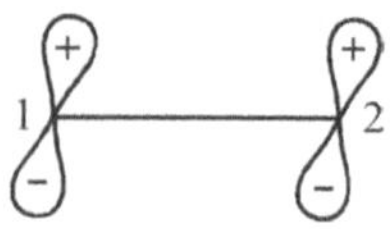

C_2	E	C_2
A	1	1
B	1	−1
“$_{\text{Ã}}$(R)	2	0

Step 2: Using the reduction formula $a_i = \frac{1}{2}\,\Sigma\chi(R)\,n_R\,\chi i\,(R)$

We have:

$$a_A = \frac{1}{2}[1 \times 1 \times 2 + 1 \times 1 \times 0] = \frac{1}{2}[2 + 0] = \frac{1}{2}[2] = 1$$

It means that irreducible A representation occurs once only. Similarly:

$$a_B = \frac{1}{2}[1 \times 1 \times 2 + 1 \times -1 \times 0] = 1 = \frac{1}{2}[2 + 0] = \frac{1}{2}[2] = 1$$

Hence, irreducible representation B also occurs once only.
Therefore,

$$\Gamma_\pi(R) = A + B$$

Step 3: Each irreducible representation can be represented by one wave function. Since, these is a set of carbon atoms (two atoms) with the wave functions ϕ_1 and ϕ_2, therefore:

$$\psi_A = \phi_1 + \phi_2 \quad (7.12)$$

$$\psi_B = \phi_1 - \phi_2 \quad (7.13)$$

Normalized SALC’S are therefore:

$$\psi_A = \frac{1}{\sqrt{2}}(\phi_1 + \phi_2) \quad (7.14)$$

$$\psi_B = \frac{1}{\sqrt{2}} (\phi_1 - \phi_2) \tag{7.15}$$

Step 4: These two wave functions are represented by secular determinant equation.

$$\begin{vmatrix} H_{11} - ES_{11} & H_{12} - ES_{12} \\ H_{21} - ES_{21} & H_{22} - ES_{22} \end{vmatrix} = 0$$

Huckel Approximation in this case is:

$H_{AA} = \alpha$

$H_{AB} = \beta$, if A and B are adjacent C-atoms.

$H_{AB} = 0$, if A and B are non-adjacent C-atoms.

$S_{AA} = 1$

$S_{AB} = 0$ (Zero) Differential overlap integral

Now

$$H_{11} = \frac{1}{\sqrt{2}} \frac{1}{\sqrt{2}} \int (\phi_1 + \phi_2) H (\phi_1 + \phi_2)\, d\tau$$

$$= \frac{1}{2} [\int \phi_1 H \phi_1\, d\tau + + \phi_1 H \phi_2\, d\tau + \int \phi_2 H \phi_1\, d\tau + \int \phi_2 H \phi_2\, d\tau]$$

$$= \frac{1}{2} [H_{11}' + H_{12}' + H_{21}' + H_{22}']$$

Applying the Huckel approximation

$$H_{11} = \frac{1}{2} [\alpha + \beta + \beta + \alpha]$$

$$H_{11} = \alpha + \beta$$

Similarly:

$$H_{12} = \frac{1}{\sqrt{2}} \frac{1}{\sqrt{2}} \int (\phi_1 + \phi_2) H (\phi_1 - \phi_2)\, d\tau$$

$$= \frac{1}{\sqrt{2}} [\int \phi_1 H \phi_1 d\tau - \int \phi_1 H \phi_2 d\tau - \int \phi_2 H \phi_1 d\tau + \int \phi_2 H \phi_2 d\tau]$$

$$= \frac{1}{\sqrt{2}} [H_{11}' - H_{12}' - H_{21}' + H_{22}']$$

$$= \frac{1}{2} (\alpha - \beta - \alpha + \beta)$$

or $$H_{12} = 0$$

Similarly $$H_{21} = 0$$

$$H_{22} = \frac{1}{\sqrt{2}} \frac{1}{\sqrt{2}} \int (\phi_1 - \phi_2) H (\phi_1 - \phi_2) d\tau$$

$$= \frac{1}{\sqrt{2}} [\int \phi_1 H \phi_1 d\tau - \int \phi_1 H \phi_2 d\tau - \int \phi_2 H \phi_1 d\tau + \int \phi_2 H \phi_2 d\tau]$$

$$= \frac{1}{2} [H_{11}' - H_{12}' - H_{21}' + H_{22}']$$

$$= \frac{1}{2} [\alpha - \beta - \beta + \alpha]$$

or $$H_{22} = \alpha - \beta$$

Placing these values in secular determinant, we get:

$$\begin{vmatrix} \alpha + \beta - E & 0 \\ 0 & \alpha - \beta - E \end{vmatrix} = 0$$

The dimension of matrix is 2 × 2 and it can be reduced to 1 × 1 by "block out" method.

$$\alpha + \beta - E = 0 \quad (7.16)$$

or $$E_1 = \alpha + \beta$$

$$\alpha - \beta - E = 0$$

or $$E_2 = \alpha - \beta \quad (7.17)$$

Thus, two MOs in ethylene molecule can be represented as:

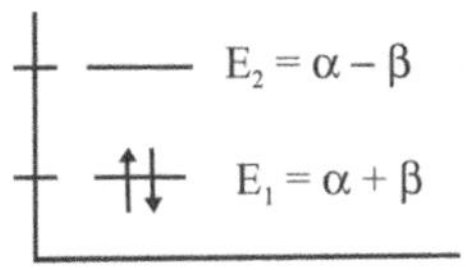

Energy level diagram

Out of these two wave functions, only filled energy level is used further in calculation.

$$\text{MO with lower energy is } \psi_1 = (\phi_1 + \phi_2) \tag{7.18}$$

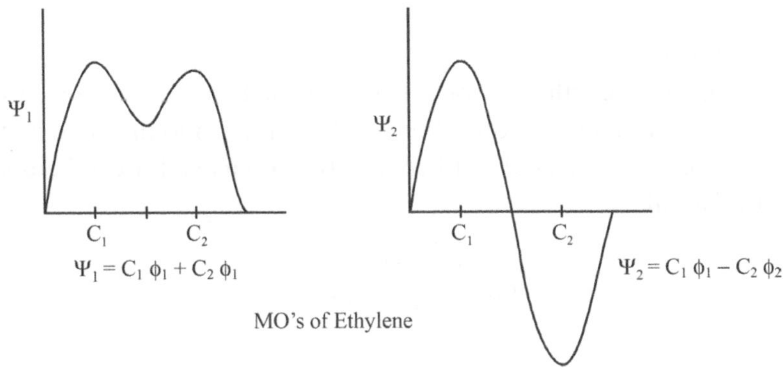

MO's of Ethylene

(i) Electron density

Electron density of a conjugate system is given by the following formula:

$$ED_i = \sum_{J}^{\text{occ MOs}} C_{ij}^2 \tag{7.19}$$

where, n_{ij} = Number of electrons in j[th] energy level; C_{ij} – Coefficient of ith atom in j[th] energy level.

The electron density on two carbon atoms of ethylene can be calculated by putting the value of n_{ij} C_{ij} in the Eq. (7.19).

$$ED_1 = 2 \times \left(\frac{1}{\sqrt{2}}\right)^2 = 1$$

$$ED_2 = 2 \times \frac{1}{\sqrt{2}} = 1$$

(ii) Charge density

Charge density is the charge or electron deficiency of an atom and it can be calculated by following formula:

$$q_i = 1 - E\,D_i \tag{7.20}$$

The charge density on two carbon atoms of ethylene molecule can be determined by putting the value of $E\,D_i$ in Eq. (7.20).

where q_i = Charge of ith atom, and $E\,D_i$ = electron density of ith atom.

$$q_1 = 1-1 = 0$$

$$q_2 = 1-1 = 0$$

(iii) Bond order

It represents the strength of a bond. Higher is the bond order, more stronger will be the bond. In other words, bond order is related to the bond strength. Bond order is always calculated between two atoms and it can be done by using the formula:

$$p^{J}_{kl} = \sum_{J}^{\text{occ MOs}} n_j\, C_{kj}\, C_{ij} \tag{7.21}$$

where n_j = number of e^- in j^{th} energy level; C_{kj} = coefficient of kth atom in j^{th} energy level; C_{ej} = coefficient of ith atom in j^{th} energy level.

$$p_{12} = 2 \times \frac{1}{\sqrt{2}} \times \frac{1}{\sqrt{2}} = 1$$

(iv) Free valence

The concept of free valence at an atom is used as an index to indicate the possibility of attack at that atom. In other words, it represents reactivity of that atom. More is the free valence, more reactive that atom is. Free valence can be calculated using the formula:

$$F_r = N_{max} - N_r \tag{7.22}$$

where N_{max} = maximum possible bonding than an atom is capable to have, i.e., 4.73 ($3 + \sqrt{3}$); N_r = actual σ-bond formed + bond orders for the other bonds formed by that atom.

$$F_1 = 4.73 - (3 + 1) = 0.73$$

$$F_2 = 4.73 - (3 + 1) = 0.73$$

Results show that both the carbon atoms of ethylene are having same free valence and therefore, these are equally reactive.

7.3 BUTADIENE

Step 1: Trans-butadiene belongs to C_{2h} group (cis-butadiene belongs to C_{2v} group). Γ_π can be obtained using C_2 point group as it is a simple form of both, C_{2v} and C_{2h} point groups.

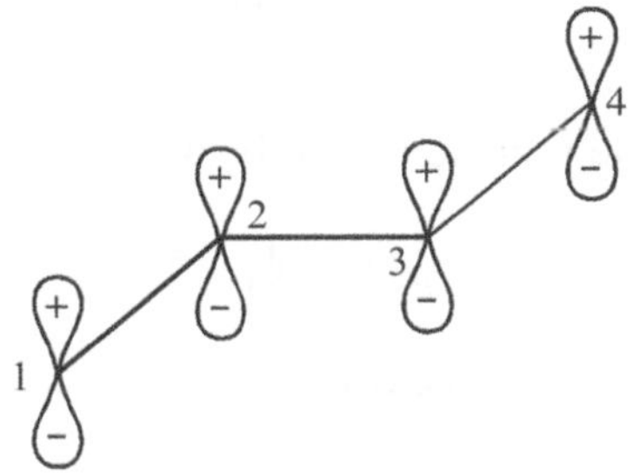

C_2	E	C_2
A	1	1
B	1	−1
$\Gamma_\pi(R)$	4	0

Step 2: Using reduction formula $a_i = \Sigma\chi(R)\, n_R\, \chi_i(R)$
We have

$$a_A = \frac{1}{2}[1 \times 1 \times 4 + 1 \times 1 \times 1 + 0] = [4] = 2$$

That is, A is the irreducible representation, which occurs twice.

Similarly, $a_B = \frac{1}{2}[1 \times 1 \times 4 + 1 \times (-1) \times 0] = [4] = 2$

Irreducible representation B also occurs twice. Therefore, Γ_π (R) = 2 A + 2 B

Step 3: There are also non-equivalent wave functions represented by each irreducible representation. Since, in case of butadiene, there are two sets of carbon, i.e., terminal carbons (1,4) and central carbons (2,3). The wave functions for irreducible representation A are:

$$\psi_A(1) = \phi_1 + \phi_4$$

$$\psi_A(2) = \phi_2 + \phi_3$$

Normalized SALC'S are therefore

$$\psi_A(1) = \frac{1}{\sqrt{2}}(\phi_1 + \phi_4) \tag{7.23}$$

$$\psi_A(2) = \frac{1}{\sqrt{2}}(\phi_2 + \phi_3) \tag{7.24}$$

Similarly for irreducible representation B, we have:

$$\psi_B(1) = \phi_1 - \phi_4$$

$$\psi_B(2) = \phi_2 - \phi_3$$

Normalization of these wave functions gives:

$$\psi_B(1) = \frac{1}{\sqrt{2}}(\phi_1 - \phi_4) \tag{7.25}$$

$$\psi_B(2) = \frac{1}{\sqrt{2}}(\phi_2 - \phi_3) \tag{7.26}$$

Step 4: These four wave functions are represented in the form of following Secular determinant:

According to Huckel approximation:

$$\begin{bmatrix} H_{11} - ES_{11} & H_{12} - ES_{12} & H_{13} - ES_{13} & H_{14} - ES_{14} \\ H_{21} - ES_{21} & H_{22} - ES_{22} & H_{23} - ES_{23} & H_{24} - ES_{24} \\ H_{31} - ES_{31} & H_{32} - ES_{32} & H_{33} - ES_{33} & H_{34} - ES_{34} \\ H_{41} - ES_{41} & H_{42} - ES_{42} & H_{43} - ES_{43} & H_{44} - Es_{44} \end{bmatrix} = 0$$

$H_{AA} = H_{11}, H_{22} \ldots = \alpha$;
$H_{AB} = \beta$, if A and B are adjacent C-atoms;

$H_{AB} = 0$, if A and B are non-adjacent C-atoms;
$S_{AA} = 1$ (Normal sized overlap integral); and
$S_{AB} = 0$ (Zero overlap integral).

Now:

$$H_{11} = \frac{1}{2}\int (\phi_1 + \phi_4)\, H\, (\phi_1 + \phi_4)\, d\tau$$

$$= \frac{1}{2}\int [\varphi_1 H \varphi_1 d\tau + \int \varphi_1 H \varphi_4 d\tau + \int \varphi_4 H \varphi_1 dE + \int \varphi_4 H \varphi_4 d\tau]$$

$$= \frac{1}{2}\ [H_{11}' + H_{14}' + H_{41}' + H_{44}']$$

Putting the values of different columbic integral, we get:

$$H_{11} = \frac{1}{2}\ [\alpha + 0 + 0 + \alpha]$$

or $$H_{22} = H_{33} = H_{44} = \alpha \qquad (7.27)$$

Similarly,

$$H_{12} = H_{21} = \frac{1}{2}\int (\phi_1 + \phi_4)\, H\, (\phi_2 + \phi_3)\, d\tau$$

$$= \frac{1}{2}\ (H_{12}' + H_{13}' + H_{42}' + H_{43}')$$

$$= (\beta + 0 + 0 + \beta)$$

or $$H_{12} - H_{21} - \beta \qquad (7.28)$$

$$H_{13} = H_{31} = 1 \int(\phi_1 + \phi_4)\, H\, (\phi_1 - \phi_4)\, d\tau$$

$$= [H_{11}' - H_{14}' + H_{41}' - H_{44}']$$

$$= \frac{1}{2}\ [\alpha - 0 + 0 - \alpha]$$

or $$H_{13} = H_{31} = 0 \qquad (7.29)$$

Hence,

$$H_{14} = H_{41} = 0$$

$$H_{24} = H_{42} = 0$$

$$H_{22} = \frac{1}{2}[\int (\phi_2 + \phi_3)\, H\, (\phi_2 + \phi_3)\, d\tau$$

$$= \frac{1}{2}\,(H_{22}' + H_{23}' + H_{32}' + H_{33}')$$

$$= \frac{1}{2}\,[\alpha + \beta + \beta + \alpha]$$

or

$$H_{22} = \alpha + \beta \tag{7.30}$$

$$H_{33} = \frac{1}{2}\int (\phi_1 - \phi_4)\, H\, (\phi_1 - \phi_4)\, d\tau$$

$$= \frac{1}{2}(H_{11}' - H_{14}' - H_{41}' + H_{44}')$$

$$= \frac{1}{2}(\alpha - 0 - 0 + \alpha)$$

or

$$H_{33} = \alpha \tag{7.31}$$

$$H_{34} = \frac{1}{2}\int (\phi_1 - \phi_4)\, H\, (\phi_2 - \phi_3)\, d\tau$$

$$= \frac{1}{2}(H_{12}' - H_{13}' - H_{42}' + H_{43}')$$

$$= \frac{1}{2}(\beta - 0 - 0 + \beta)$$

$$H_{34} = \beta \tag{7.32}$$

$$H_{44} = \frac{1}{2}\int (\phi_2 - \phi_3)\, H\, (\phi_2 - \phi_3)\, d\tau$$

$$= \frac{1}{2}(H_{22}' - H_{23}' - H_{32}' + H_{33}')$$

$$= \frac{1}{2}[\alpha - \beta - \beta + \alpha]$$

or

$$H_{44} = \alpha - \beta \tag{7.33}$$

These calculated values are then placed in secular determinant.

$$\left[\begin{array}{cc|cc} (\alpha - E) & \beta & 0 & 0 \\ \beta & \alpha + \beta - E & 0 & 0 \\ \hline 0 & 0 & (\alpha - E) & \beta \\ 0 & 0 & \beta & \alpha - \beta - E \end{array}\right] = 0$$

The above determinant is of the order 4 × 4. It can be reduced to form two smaller blocks by "blocking" method, each with the order 2 × 2. Out of these four 2 × 2 determinants, two are having zero only. Hence, the rest two determinants are:

$$\left|\begin{array}{c|c} \alpha - E & \beta \\ \hline \beta & \alpha + \beta - E \end{array}\right| = 0$$

$$\left|\begin{array}{c|c} \alpha - E & \beta \\ \hline \beta & \alpha - \beta - E \end{array}\right| = 0$$

These can be further solved by dividing by β and substituting x in place of the $\frac{\alpha - E}{\beta}$:

$$\left|\begin{array}{c|c} x & 1 \\ \hline 1 & (x + 1) \end{array}\right| = 0$$

$$\left|\begin{array}{c|c} x & 1 \\ \hline 1 & (x - 1) \end{array}\right| = 0$$

Then these are further solved to find out values of x

$$x\,(x + 1) - 1 = 0$$

$$x^2 + x - 1 = 0$$

or $$x = +\,0.621, \text{ or } -\,1.621$$

$$x\,(x - 1) - 1 = 0 \tag{7.34}$$

$$x^2 - x - 1 = 0 \tag{7.35}$$

or $$x = -0.621, \text{ or } +1.621$$

Putting all these values of x, we have four values of energy, i.e., E_1, E_2, E_3 and E_4

$$x = \frac{\alpha - E}{\beta} = -1.621$$

or $$E_1 = \alpha + 1.621\ \beta \tag{7.36}$$

$$x = \frac{\alpha - E}{\beta} = -0.621$$

or $$E_2 = \alpha + 0.621\ \beta \tag{7.37}$$

$$x = \frac{\alpha - E}{\beta} = +0.621$$

or $$E_3 = \alpha - 1.621\ \beta \tag{7.38}$$

$$x = \frac{\alpha - E}{\beta} = +1.621$$

or $$E_4 = \alpha - 1.621\ \beta \tag{7.39}$$

Thus, four MOs in butadiene molecule can be represented as:

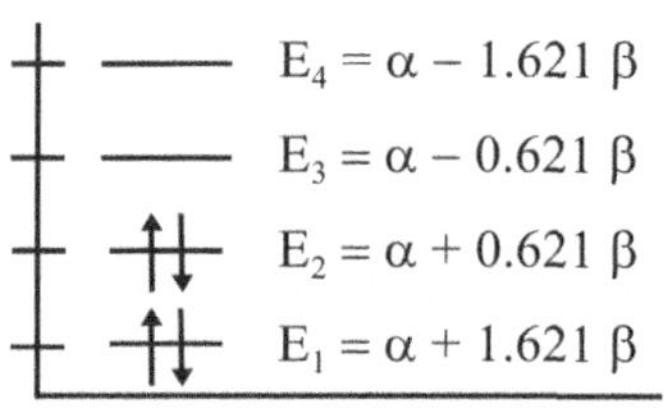

Energy level diagram

As there are four electrons in all, two bower MOs will be filled with two electrons each. Therefore, total energy of the system will be:

$$\text{Total energy} = 2\,(\alpha + 1.621\ \beta) + 2\,(\alpha + 0.621\ \beta)$$

$$= 4\alpha + 4.472\ \beta \tag{7.40}$$

The total energy of butadiene molecule is less than the energy of two ethylene molecules. As the energy of ethylene molecule is $2\alpha + 2\beta$. Therefore, for two ethylene molecule, it will be $4\alpha + 4\beta$.

The value of β is negative and therefore there is a loss of 0.472 β energy in case of butadiene.

This loss in energy in case of butadiene molecule is because of resonance, as two double bonds are in conjugated position. Had there been no resonance in butadiene molecule, then the total energy of butadiene should be number of ethylene components multiplied by number of electrons and in there, the energy of an electron in ψ_1 orbital $= 2 \times 2 \times (\alpha + \beta)$, i.e., $4\alpha + 4\beta$.

It seems that the energy of butadiene molecule ($4\alpha + 4.472\ \beta$) is 0.472 β larger than the energy of two-ethylene molecule ($4\alpha + 4\beta$) but it is not like that, rather these is a loss of energy due to resonance because the value of β is negative.

Resonance Energy = Actual energy – Energy of two ethylene molecules.

$= (4\alpha + 4.472\ \beta) - 2\ (2\alpha + 2\ \beta)$

Resonance energy = 0.472 β

Four equations can be derived from the secular determinant. These are:

$$C_1x + C_2 = 0 \tag{7.41}$$

$$C_1 + C_2x + C_3 = 0 \tag{7.42}$$

$$C_2 + C_3x + C_4 = 0 \tag{7.43}$$

$$C_3 + C_4x = 0 \tag{7.44}$$

Putting the value of $x = +\ 0.618$ in Eq. (7.41), we get:

$$C_1\ (0.618) + C_2 = 0$$

$$C_1 = -\ C_2/0.618 \tag{7.45}$$

Putting the value of C_1 in Eq. (7.42).

$$\frac{-C_2}{0.618} + C_2\ (0.618) + C_3 = 0$$

$$C_2 = C_3 \tag{7.46}$$

Putting the value of C_2 and C_3 in Eq. (7.43).

$$-0.618\ C_1 + C_3\ (0.618) + C_4 + 0$$

$$-0.618\ C_1 - (0.618)\ (0.618)\ C_1 + C_4 + 0$$

$$C_1 = C_4 \tag{7.47}$$

The sum of the squares of the coefficients is equal to unity. Hence,

$$C_1^2 + C_2^2 + C_3^2 + C_4^2 = 1 \tag{7.48}$$

Putting the value of C_2, C_3 and C_4 in Eq. (7.48).

$$C_1^2 + (-0.618\ C_1)^2 + (-0.618\ C_1)^2 + C_1^2 = 1$$

$$C_1^2 + 0.372\ C_1{}^2 + 0.372\ C_1^2 + C_1^2 = 1$$

$$2.744\ C_1^2 = 1$$

$$C_1 = \frac{1}{\sqrt{2.744}} = 0.602 \tag{7.49}$$

Then value of $C_2 = -0.618 \times 0.602 = -0.378$

$$C_3 = C_2 = -0.378$$

$$C_4 = C_1 = 0.602$$

Now using the values of coefficients C_1, C_2, C_3 and C_4, the third function ψ_3 is:

$$\psi_3 = 0.602\ \phi_1 - 0.372\ \phi_2 - 0.372\ \phi_3 - 0.602\ \phi_4 \tag{7.50}$$

Similarly, using the value of $x = -1.621$, the coefficients C_1, C_2, C_3 and C_4 can be determined and putting their values, the wave function ψ_1 is:

$$\psi_1 = 0.372\ \phi_1 + 0.602\ \phi_2 + 0.602\ \phi_3 + 0.372\ \phi_4 \tag{7.51}$$

The wave function ψ_2 can be determined by finding out values of coefficients C_1, C_2, C_3 and C_4 by putting the value of $x = -0.621$ as:

$$\psi_2 = 0.602\ \phi_1 - 0.372\ \phi_2 - 0.372\ \phi_3 - 0.602\ \phi_4 \quad (7.52)$$

The wave function ψ_4 can also be derived by using $x = +\ 1.621$ as:

$$\psi_4 = 0.372\ \phi_1 - 0.602\ \phi_2 + 0.602\ \phi_3 - 0.372\ \phi_4 \quad (7.53)$$

Thus, four wave functions ψ_1, ψ_2, ψ_3 and ψ_4 for butadiene can be written as:

$$\psi_1 = 0.372\ \phi_1 + 0.602\ \phi_2 + 0.602\ \phi_3 + 0.372\ \phi_4$$

$$\psi_2 = 0.602\ \phi_1 - 0.372\ \phi_2 - 0.372\ \phi_3 - 0.602\ \phi_4$$

$$\psi_3 = 0.602\ \phi_1 - 0.372\ \phi_2 - 0.372\ \phi_3 - 0.602\ \phi_4$$

$$\psi_4 = 0.372\ \phi_1 - 0.602\ \phi_2 + 0.602\ \phi_3 - 0.372\ \phi_4$$

Then, these wave functions are diagrammatically presented as:

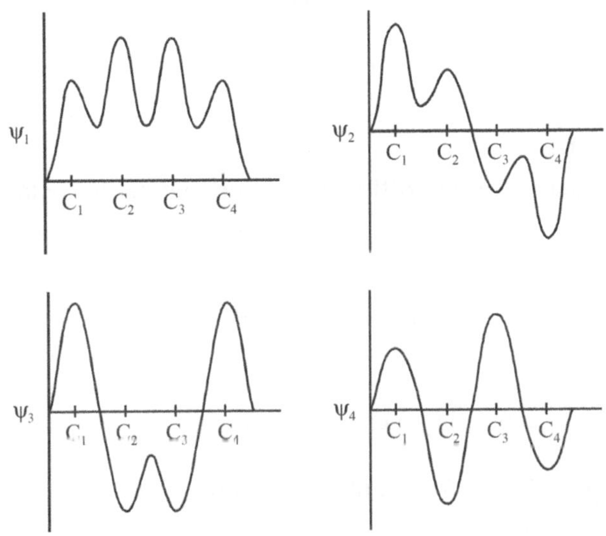

MO's Butadiene

Out of four π-electrons, two electrons are accommodated in each, i.e., ψ_1 and ψ_2, while ψ_3 and ψ_4 remain vacant.

(i) Electron density

Using the equation for electron density

$$ED_i = \sum_{J}^{\text{occ MOs}} n_j\, C_{ij}^2$$

where n_{ij} = number of electrons in j^{th} energy level; C_{ij} = coefficient of ith atom in j^{th} energy level.

$$ED_1 = 2 \times (0.372)^2 + 2 \times (0.602)^2 = 1$$

$$ED_2 = 2 \times (0.602)^2 + 2 \times (0.372)^2 = 1$$

Similarly, $ED_3 = ED_4 = 1$.

(ii) Charge density

It can be calculated on the basis of electron density.

$$q_i = 1 - E\,D_i$$

$$q_1 = 1 - E\,D_1 = 1 - 1 = 0$$

Similarly, $q_2 = q_3 = q_4 = 0.$

(iii) Bond order

The bond order between different carbon atoms of butadiene can be calculated.

$$p_{kl}^{J} = \sum_{J}^{\text{occ MOs}} n_j\, C_{kj}\, C_{ej}$$

n_j = Number of electron in jth energy level.

$$p_{12} = 2 \times 0.372 \times 0.602 + 2 \times 0.602 \times 0.372$$
$$= 0.896$$

$$p_{23} = 2 \times 0.602 \times 0.602 + 2 \times 0.372 \times (-0.372)$$
$$= 0.448$$

$$p_{34} = 2 \times 0.602 \times 0.372 + 2 \times (-0.372) \times (-0.602)$$
$$= 0.896$$

If there is no delocalization of π-electrons in butadiene molecule, then the bond order due to π-electron between $C_2 - C_3$ atom should be zero and the bond order between $C_1 - C_2$ as well as $C_3 - C_4$ atom should be one.

But this is not observed as bond orders between $C_1 - C_2$ and $C_3 - C_4$ has reduced to 0.896. On the other hand, the bond order between $C_2 - C_3$ is 0.448. This clearly reflects that delocalization of π-electrons is taking place in butadiene molecule; thus, reducing the π-bond character between $C_1 - C_2$ and $C_3 - C_4$ atoms and developing some π-bond character between $C_2 - C_3$ atoms.

(iv) Free valence

The concept of free valence at an atom is used as an index to indicate the possibility of attack at that atom. In other words, it represents reactivity of that atom. More is the free valence, more reactive that atom is. Free valence can be calculated by using the formula

$$F_r = N_{max} - N_r \tag{7.54}$$

where N_{max} = Maximum possible bonding than an atom is capable to have, i.e., 4.73 $(3 + \sqrt{3})$.

N_r = Actual σ-bond formed + Bond orders for the other bonds formed by that atom.

Hence for carbon atom = $3 + \sqrt{3} = 3 + 1.73 = 4.73$

$$F_1 = 4.73–3.896 = 0.834$$

$$F_2 = 4.73–3.896 = 0.834$$

For carbon atom 2, the total π-bond order is $p_{12} + p_{23} = 0.896 + 0.448 = 1.344$

$$F_2 = \sqrt{3} - 1.344$$

$$= 0.388$$

$$F_3 = 0.388$$

Results show that first and fourth positions are more reactive.

As the free valence is more at C_1 and C_4; therefore, C_1 and C_4 show more reactivity as compared to C_2 and C_3. It is clearly reflected in the formation of 1, 4-substitution product the major one and 1, 2-substitution product only in minor amount under ordinary conditions.

7.4 CYCLOBUTADIENE

The cyclobutadiene has square planar geometry, in which σ-skeleton is made up of sp^3 hybridized carbon atoms. In ring, two adjacent carbon and hydrogen atom are bonded with σ bond along with all four carbon atoms have p-orbital perpendicular to the plane of the molecules, which combine to form 4 π-MO's. Among these 4 π-MO's, two are non-degenerate and two are doubly degenerate.

Step 1: This molecule belongs to point group D_{4h}.

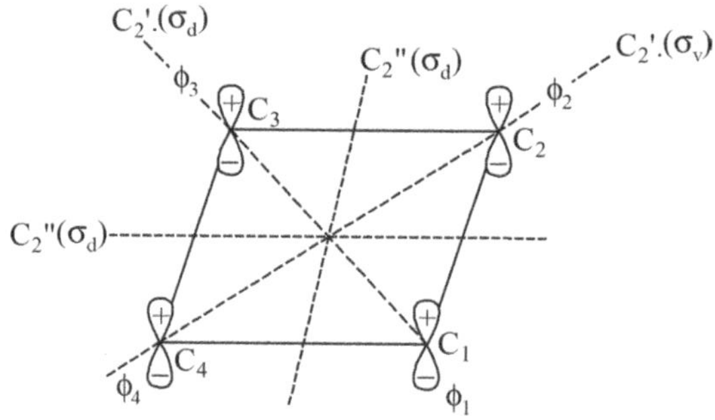

Γ_π can be obtained by C_4 group.

C_4	E	C_4	$C_2 (=C_4^2)$	C_4^3
A	1	1	1	1
B	1	−1	1	−1
E	{1	i	−1	−i}
	{1	−i	−1	i}
$\Gamma_\pi(R)$	4	0	0	0

Step 2: Using reduction formula $a_i = \frac{1}{h}\sum_R n_R\, \chi(R).\chi_i(R)$ we have:

$$\Gamma_\pi(R) = A + B + E$$

Step 3: Here, number of carbon atoms are equal to order C_4 group, hence, wave function is given by:

$$\psi_A = \varphi_1 + \varphi_2 + \varphi_3 + \varphi_4 \qquad (7.55)$$

$$\psi_B = \varphi_1 - \varphi_2 + \varphi_3 - \varphi_4 \tag{7.56}$$

$$\psi_E = \varphi_1 + i\,\varphi_2 - \varphi_3 - i\,\varphi_4 \tag{7.57}$$

$$\psi_E{}' = \varphi_1 - i\,\varphi_2 - \varphi_3 + i\,\varphi_4 \tag{7.58}$$

Normalization of above equation gives:

$$\psi_A = \frac{1}{2}(\varphi_1 + \varphi_2 + \varphi_3 + \varphi_4) \tag{7.59}$$

$$\psi_B = \frac{1}{2}(\varphi_1 - \varphi_2 + \varphi_3 - \varphi_4) \tag{7.60}$$

Equations (7.57) and (7.58), represent two dimension representations. These are combined and then normalized.

$$\begin{aligned}\psi_{E_1} &= \psi_E + \psi_E{}' \\ &= 2\,\phi_1 - 2\,\phi_3 \cong \phi_1 - \phi_3 \\ &= \frac{1}{\sqrt{2}}(\varphi_1 - \varphi_3)\end{aligned} \tag{7.61}$$

Subtracting Eq. (7.58) from Eq. (7.57) and dividing by i, and then on normalization, we get:

$$\begin{aligned}\psi_{E2} &= \frac{\psi_E - \psi_E{}'}{i} \\ &= 2\,\phi_2 - 2\,\phi_4 \cong \phi_2 - \phi_4 \\ &= \frac{1}{\sqrt{2}}(\varphi_2 - \varphi_4)\end{aligned} \tag{7.62}$$

Thus, the four π-MO's wave functions are represented by Eqs. (7.59)–(7.62). The secular determinant for this system is:

$$\begin{bmatrix} H_{11} - ES_{11} & H_{12} - ES_{12} & H_{13} - ES_{13} & H_{14} - ES_{14} \\ H_{21} - ES_{21} & H_{22} - ES_{22} & H_{23} - ES_{23} & H_{24} - ES_{24} \\ H_{31} - ES_{31} & H_{32} - ES_{32} & H_{33} - ES_{33} & H_{34} - ES_{34} \\ H_{41} - ES_{41} & H_{42} - ES_{42} & H_{43} - ES_{45} & H_{44} - ES_{44} \end{bmatrix} = 0$$

Using Huckel approximations

$$H_{11} = \frac{1}{2} \cdot \frac{1}{2} \left[\int (\varphi_1 + \varphi_2 + \varphi_3 + \varphi_4) H (\varphi_1 + \varphi_2 + \varphi_3 + \varphi_4) \right] d\tau$$

$$= \alpha + 2\beta$$

Similarly, $H_{22} = \alpha - 2\beta$

$$H_{23} = H_{44} = \alpha$$

$$H_{34} = H_{43} = 0$$

Step 4: The secular determinant for this system is:

$$\begin{bmatrix} \alpha + 2\beta - E & 0 & 0 & 0 \\ 0 & \alpha - 2\beta + E & 0 & 0 \\ 0 & 0 & \alpha - E & 0 \\ 0 & 0 & 0 & \alpha - E \end{bmatrix} = 0$$

The determinant of 4 × 4 dimension can be reduced into determinant 1 × 1 dimensions by block out method.

$$\alpha + 2\beta - E_1 = 0$$

or $$E_1 = \alpha + 2\beta$$

$$\alpha - 2\beta - E_2 = 0$$

or $$E_2 = \alpha - 2\beta$$

$$\alpha - E_3 = 0$$

or $$E_3 = \alpha = E_4$$

and, therefore, four energy levels are obtained as:

$$E_1 = \alpha + 2\beta$$

$$E_2 = \alpha - 2\beta$$

$$E_3 = E_4 = \alpha$$

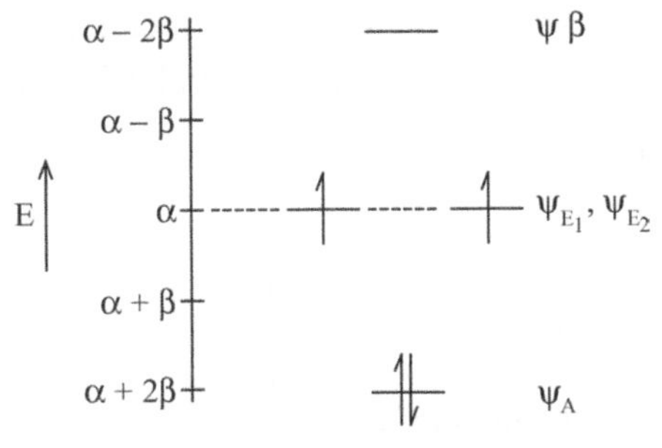

Total energy (E_π) of the system is:

$$E_\pi = 2\,(\alpha + 2\,\beta) + \alpha + \alpha$$
$$= 4\,\alpha + 4\beta$$

Delocalization energy (E_D) is:

$$E_D = E_\pi - 4\,(\alpha + \beta)$$

Here, $4(\alpha + \beta)$ is π-electron energy of two isolated ethylenic linkage.

$$E_D = 4\,\alpha + 4\,\beta - 4\,(\alpha + \beta) = 0$$

The HMO's of C_4H_4 and its nodal characteristics are:

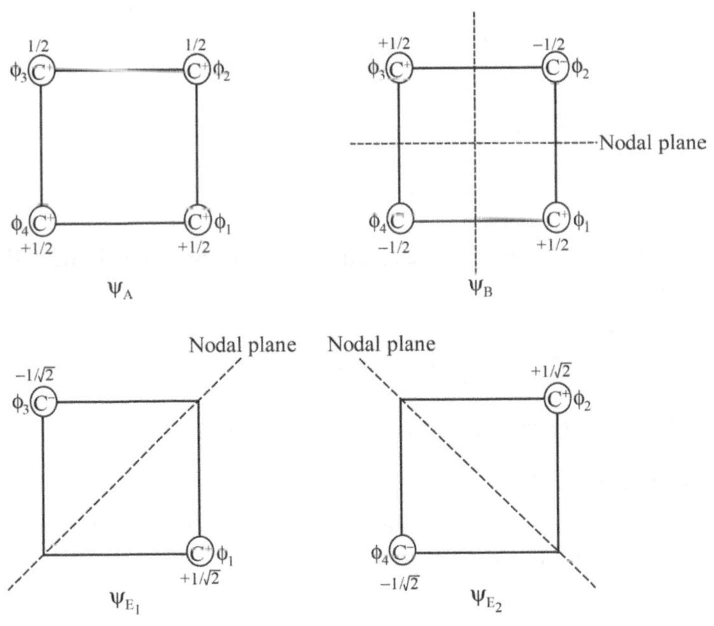

Using coefficient of AO's in wave function, Huckel parameters like electron density, charge density, bond order and free valence can be determined.

(i) Electron density

$$ED_r = \sum_{i}^{Occ.MOs} n_i . c_{ij}^2$$

$$ED_1 \text{ at carbon } (1) = 2 \times \left(\frac{1}{2}\right)^2 + 1 \times \left(\frac{1}{\sqrt{2}}\right)^2 + 0$$

$$= \frac{1}{2} + \frac{1}{2} = 1.00$$

$$ED_2 \text{ at carbon } (2) = 2 \times \left(\frac{1}{2}\right)^2 + 0 + 1 \times \left(\frac{1}{\sqrt{2}}\right)^2$$

$$= 1.00$$

$$ED_3 \text{ at carbon } (3) = 2 \times \left(\frac{1}{2}\right)^2 + 1 \times \left(-\frac{1}{\sqrt{2}}\right)^2 + 0$$

$$= 1.00$$

$$ED_3 \text{ at carbon } (4) = 2 \times \left(\frac{1}{2}\right)^2 + 0 + 1 \times \left(-\frac{1}{\sqrt{2}}\right)^2$$

$$= 1.00$$

(ii) Charge density (q_r)

$$q_r = 1 - E\,D_r$$

For all the four carbon atoms, electron density is zero and hence, the charge density will be

$$q = 1 - 1.00 = 0$$

It means, that each carbon atom have zero charge density.

(iii) Bond order

The bond order lies between 1 and 2, because ψ_A have two π electrons and ψ_{E_1} and ψ_{E_2} have one π-electron in each.

$$P_{12} = 2 \times \left(\frac{1}{2} \times \frac{1}{2}\right) + 1\left(\frac{1}{\sqrt{2}} \times 0\right) + 1\left(0 \times \frac{1}{\sqrt{2}}\right) = \frac{1}{2}$$

$$P_{23} = 2 \times \left(\frac{1}{2} \times \frac{1}{2}\right) + 1\left(0 \times -\frac{1}{\sqrt{2}}\right) + 1\left(\frac{1}{\sqrt{2}} \times 0\right) = \frac{1}{2}$$

$$P_{34} = 2 \times \left(\frac{1}{2} \times \frac{1}{2}\right) + 1\left(-\frac{1}{\sqrt{2}} \times 0\right) + 1\left(0 \times -\frac{1}{\sqrt{2}}\right) = \frac{1}{2}$$

and
$$P_{14} = 2 \times \left(\frac{1}{2} \times \frac{1}{2}\right) + 1\left(\frac{1}{\sqrt{2}} \times 0\right) + 1\left(0 \times -\frac{1}{\sqrt{2}}\right) = \frac{1}{2}$$

It shows that all the 4 $C-C$ bond are equivalent in C_4H_4 molecule, i.e., each carbon atom is bonded to two other carbon by partial double bond.

(iv) Free valence (F_r)

Total π bond order for C_1 is:

$$P_{12} + P_{14} = 1/2 + 1/2 = 1.0$$

Therefore
$$F_r = N_{max} - N_r$$

$$F_1 = 4.732–4.0$$

$$= 0.732$$

In the same manner, π bond order for C_3 is:

$$P_{23} + P_{34} = 1/2 + 1/2 = 1.0$$

$$F_3 = 4.732–4.0$$

$$= 0.732$$

Similarly for C_2 and C_4 carbon atoms, free valence is:

$$F_2 = F_4 = 4.732–4.0 = 0.732$$

Thus, in cyclobutadiene molecule, all the 4 $C-C$ bonds are equivalent and there are two true double bonds. Each carbon atom is equally reactive for attack by any kind of reagent as all the four atoms are having same free valence.

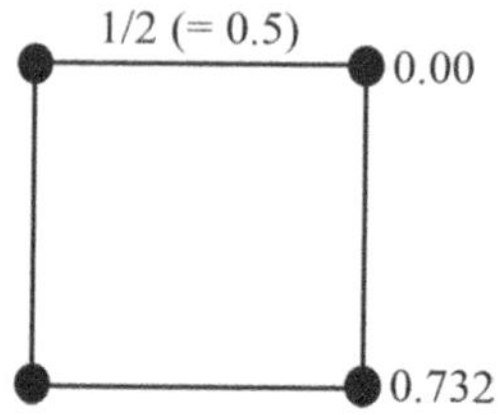

Therefore, change density in C_4H_4 molecule at each carbon atom is 0.0, the π-bond order for each bond is 0.5, and free valence at each C is 0.732.

7.5 BENZENE

Step 1: Benzene belongs to D_{6h} point group. Γ_π can be obtained using a simpler point group C_6.

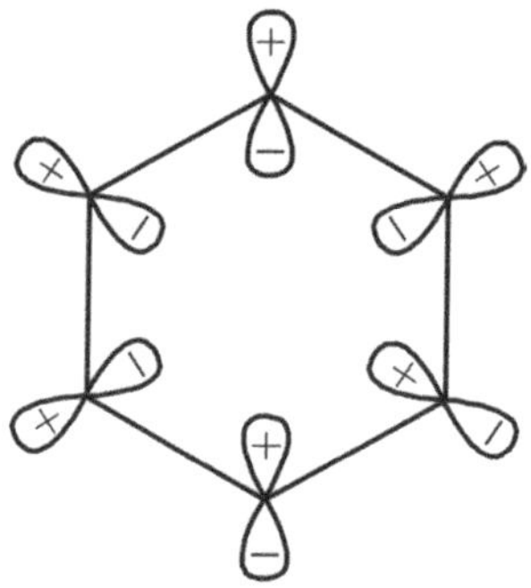

C_6	E	C_6	$C_6^2 = C_3$	$C_6^3 = C_2$	$C_6^4 = C_3$	C_6^5
A	1	1	1	1	1	1
B	1	−1	1	−1	1	−1
E_1	1	ε	$-\varepsilon^*$	−1	$-\varepsilon$	ε^*
	1	ε^*	$-\varepsilon$	−1	$-\varepsilon^*$	ε
E_2	1	$-\varepsilon^*$	$-\varepsilon$	1	$-\varepsilon^*$	$-\varepsilon$
	1	$-\varepsilon$	$-\varepsilon^*$	1	$-\varepsilon$	$-\varepsilon^*$
Γ_π	6	0	0	0	0	0

Step 2: Using reduction formula $a_i = \frac{1}{h}\sum \chi(R)\, n_R\, \chi_i(R)$

We have the contribution of irreducible representation in reducible representation as:

$$a_A = \frac{1}{6}[1 \times 1 \times 6 + 1 \times 1 \times 0 + 1 \times 1 \times 0 + 1 \times 1 \times 0 + 1 \times 1 \times 0 + 1 \times 1 \times 0] = \frac{6}{6} = 1$$

$$a_B = \frac{1}{6}[1 \times 1 \times 6 + 1 \times (-1) \times 0 + 1 \times 1 \times 0 + 1 \times (-1) \times 0 + 1 \times 1 \times 0 + 1 \times (-1) \times 0] = \frac{6}{6} = 1$$

Step 3: Here, number of carbon atoms in benzene are equal to order of C_6 group and hence, wave function is given by:

$$\psi_A = \phi_1 + \phi_2 + \phi_3 + \phi_4 + \phi_5 + \phi_6 \tag{7.63}$$

$$\psi_B = \phi_1 - \phi_2 + \phi_3 - \phi_4 + \phi_5 - \phi_6 \tag{7.64}$$

$$\psi_{E1} = \phi_1 + \varepsilon\phi_2 - \varepsilon^*\phi_3 - \phi_4 - \varepsilon\phi_5 - \varepsilon^*\phi_6 \tag{7.65}$$

$$\psi_{E1}' = \phi_1 + \varepsilon^*\phi_2 - \varepsilon\phi_3 - \phi_4 - \varepsilon^*\phi_5 - \varepsilon\phi_6 \tag{7.66}$$

$$\psi_{E2} = \phi_1 - \varepsilon^*\phi_2 - \varepsilon\phi_3 + \phi_4 - \varepsilon^*\phi_5 - \varepsilon\phi_6 \tag{7.67}$$

$$\psi_{E2}' = \phi_1 \quad \varepsilon\phi_2 - \varepsilon^*\phi_3 + \phi_4 - \varepsilon\phi_5 - \varepsilon^*\phi_6 \tag{7.68}$$

Normalization of Eqs. (7.63) and (7.64) gives:

$$\psi_A = \frac{1}{\sqrt{6}}(\phi_1 + \phi_2 + \phi_3 + \phi_4 + \phi_5 + \phi_6)$$

$$\psi_B = \frac{1}{\sqrt{6}}(\phi_1 - \phi_2 + \phi_3 - \phi_4 + \phi_5 - \phi_6)$$

Equations (7.65) and (7.66) represent two dimension representations. Firstly, these are combined and then normalized.

$$\psi_{E_1} + \psi_{E_1}' = 2\phi_1 + (\varepsilon + \varepsilon^*)\,\phi_2 - (\varepsilon + \varepsilon^*)\,\phi_{3-} 2\phi_4 - (\varepsilon + \varepsilon^*)\,\phi_5 + (\varepsilon + \varepsilon^*)\,\phi_6$$

but $\varepsilon + \varepsilon^* = 1$ and hence,

$$\psi_{E_1} + \psi_{E_1}' = 2\phi_1 + \phi_2 - \phi_3 \_ 2\phi_4 - \phi_5 + \phi_6 \tag{7.69}$$

$$\psi_{E_1} - \psi_{E_1}' = (\varepsilon - \varepsilon^*)\,\phi_2 + (\varepsilon - \varepsilon^*)\,\phi_3 - (\varepsilon - \varepsilon^*)\,\phi_5 + (\varepsilon^* - \varepsilon)\,\phi_6$$

Above equation is divided by i

$$\frac{\psi_{E_1} - \psi_{E_1}'}{i} = \frac{(\varepsilon - \varepsilon^*)}{i}\,\phi_2 + \frac{(\varepsilon - \varepsilon^*)}{i}\,\phi_3 - \frac{(\varepsilon - \varepsilon^*)}{i}\,\phi_5 + \frac{(\varepsilon^* - \varepsilon)}{i}\,\phi_6$$

but $$= \frac{(\varepsilon - \varepsilon^*)}{i} = \sqrt{3}$$

$$\frac{\psi_E - \psi_E'}{i} = \sqrt{3}\,\phi_2 + \sqrt{3}\,\phi_2 - \sqrt{3}\,\phi_5 - \sqrt{3}\,\phi_6 \tag{7.70}$$

Normalization of Eqs. (7.69) and (7.70) gives:

$$\psi_{E1} + \psi_{E1}' = \frac{1}{\sqrt{12}}\,(2\phi_1 + \phi_2 - \phi_3 - 2\phi_4 - \phi_5 + \phi_6)$$

$$\psi_{E1} + \psi_{E1}' = \frac{1}{2}(\phi_2 + \phi_3 - \phi_5 - \phi_6)$$

Similarly, Eqs. (7.67) and (7.68) gives:

$$\begin{aligned}\psi_{E_2} + \psi_{E_2}' &= 2\phi_1 - (\varepsilon + \varepsilon^*)\,\phi_2 - (\varepsilon + \varepsilon^*)\,\phi_3 + 2\phi_4 - (\varepsilon + \varepsilon^*)\,\phi_5 - \\ &\quad (\varepsilon + \varepsilon^*)\,\phi_6 \\ &= 2\phi_1 - \phi_2 - \phi_3 + 2\phi_4 - \phi_5 - \phi_6 \end{aligned} \tag{7.71}$$

$$\begin{aligned}\psi_{E_2} + \psi_{E_2}' &= (\varepsilon - \varepsilon^*)\,\phi_2 + (\varepsilon^* - \varepsilon)\,\phi_3 - (\varepsilon^* - \varepsilon)\,\phi_5 - (\varepsilon - \varepsilon^*)\,\phi_6 \\ &= \phi_2 - \phi_3 + \phi_5 - \phi_6 \end{aligned} \tag{7.72}$$

Normalization of Eqs. (7.71) and (7.72) gives:

$$\psi_{E2} + \psi_{E2}' = \frac{1}{\sqrt{12}}\,(2\phi_1 - \phi_2 - \phi_3 + 2\phi_4 - \phi_5 - \phi_6)$$

$$\psi_{E2} - \psi_{E2}' = \frac{1}{2}\,(\phi_2 - \phi_3 + \phi_5 - \phi_6)$$

The secular determinant obtained using the SALS's is:

$$\begin{bmatrix} H_{11} - ES_{11} & H_{12} - ES_{12} & H_{13} - ES_{13} & H_{14} - ES_{14} & H_{15} - ES_{15} & H_{16} - ES_{16} \\ H_{21} - ES_{21} & H_{22} - ES_{22} & H_{23} - ES_{23} & H_{24} - ES_{24} & H_{25} - ES_{25} & H_{26} - ES_{26} \\ H_{31} - ES_{31} & H_{32} - ES_{32} & H_{33} - ES_{33} & H_{34} - ES_{34} & H_{35} - ES_{35} & H_{36} - ES_{36} \\ H_{41} - ES_{41} & H_{42} - ES_{42} & H_{43} - ES_{43} & H_{44} - ES_{44} & H_{45} - ES_{45} & H_{46} - ES_{46} \\ H_{51} - ES_{51} & H_{52} - ES_{52} & H_{53} - ES_{53} & H_{54} - ES_{54} & H_{55} - ES_{55} & H_{56} - ES_{56} \\ H_{61} - ES_{61} & H_{62} - ES_{62} & H_{63} - ES_{63} & H_{64} - ES_{64} & H_{65} - ES_{65} & H_{66} - ES_{66} \end{bmatrix} = 0$$

Using Huckel Approximation

$$H_{11} = \frac{1}{\sqrt{6}} \cdot \frac{1}{\sqrt{6}} \int (\phi_1 + \phi_2 + \phi_3 + \phi_4 + \phi_5 + \phi_6)\, H\, (\phi_1 + \phi_2 + \phi_3 + \phi_4 + \phi_5 + \phi_6)\, d\tau$$

$$H_{11} = \frac{1}{6} [H_{11'} + H_{12'} + H_{13'} + H_{14'} + H_{15'} + H_{16'} + H_{21'} + H_{22'} + H_{23'} + H_{24'} + H_{25'} + H_{26'} + H_{31'} + H_{32'} + H_{33'} + H_{34'} + H_{35'} + H_{36'} + H_{31'} + H_{32'} + H_{33'} + H_{34'} + H_{35'} + H_{36'} + H_{31'} + H_{32'} + H_{33'} + H_{34'} + H_{35'} + H_{36'} + H_{31'} + H_{32'} + H_{33'} + H_{34'} + H_{35'} + H_{36'} + H_{41'} + H_{42'} + H_{43'} + H_{44'} + H_{45'} + H_{46'} + H_{51'} + H_{52'} + H_{53'} + H_{54'} + H_{55'} + H_{56'} + H_{61'} + H_{62'} + H_{63'} + H_{64'} + H_{65'} + H_{66'}]$$

$$H_{11} = \frac{1}{6} [\alpha + \beta + 0 + 0 + 0 + \beta + \beta + \alpha + \beta + 0 + 0 + 0 + 0 + \beta + \alpha + \beta + 0 + 0 + 0 + 0 + \beta + \alpha + \beta + 0 + 0 + 0 + 0 + \beta + \alpha + \beta + \beta + 0 + 0 + 0 + \beta + \alpha]$$

$$= \frac{1}{6} [6\alpha + 12\beta] = \alpha + 2\beta$$

Similarly,

$$H_{22} = \frac{1}{\sqrt{6}} \cdot \frac{1}{\sqrt{6}} \int (\phi_1 - \phi_2 + \phi_3 - \phi_4 + \phi_5 - \phi_6)\, H\, (\phi_1 - \phi_2 + \phi_3 - \phi_4 + \phi_5 - \phi_6)\, d\tau$$

$$= \alpha - 2\beta$$

$$H_{33} = \alpha + \beta$$

$H_{44} = \alpha + \beta$ and

$$H_{55} = \alpha - \beta$$

Similarly, $H_{12} = H_{13} = H_{14} = H_{15} = H_{16} = 0$

Secular determinant is:

$$\begin{bmatrix} \alpha+2\beta-E & 0 & 0 & 0 & 0 & 0 \\ 0 & \alpha-2\beta-E & 0 & 0 & 0 & 0 \\ 0 & 0 & \alpha+\beta-E & 0 & 0 & 0 \\ 0 & 0 & 0 & \alpha+\beta-E & 0 & 0 \\ 0 & 0 & 0 & 0 & \alpha-\beta-E & 0 \\ 0 & 0 & 0 & 0 & 0 & \alpha-\beta-E \end{bmatrix} = 0$$

The determinant of 6 × 6 dimension can be reduced into determinants of 1 × 1 dimensions by block out method.

$$\alpha + 2\,\beta - E = 0 \tag{7.73}$$

$$\alpha - 2\,\beta - E = 0 \tag{7.74}$$

$$\alpha - \beta - E = 0 \tag{7.75}$$

$$\alpha + \beta - E = 0 \tag{7.76}$$

and therefore six energy levels are obtained as

$E_A = E_1 = \alpha + 2\,\beta$

$E_E = E_2 = \alpha + \beta$ (Doubly degenerate)

$E_E = E_3 = \alpha - \beta$ (Doubly degenerate)

$E_B = E_4 = \alpha - 2\,\beta$

These are diagrammatically represented as:

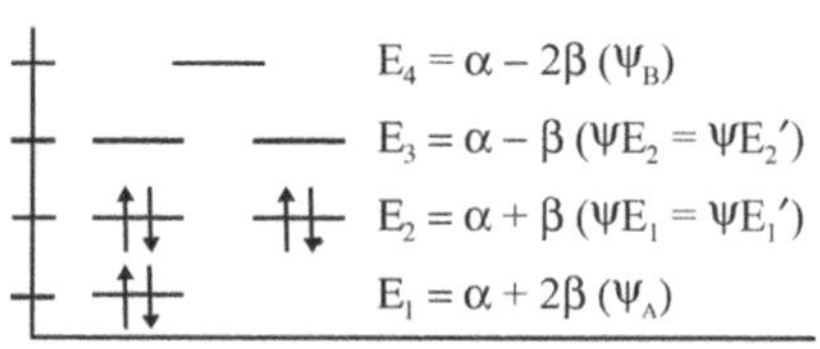

Energy level diagram

As the six electrons are filled in three levels, E_1 and E_2 (doubly degenerate), total energy E of the system will be

$$E = 2\,(\alpha + 2\,\beta) + 4\,(\alpha + \beta)$$
$$= 6\,\alpha + 8\,\beta$$

Benzene is made up of three ethylene units and if there is no resonance, it energy will be 3 times of ethylene, i.e., $3\,(2\,\alpha + 2\,\beta) = 6\,\alpha + 6\,\beta$. But the energy of benzene is $6\,\alpha + 8\,\beta$.

So, resonance Energy = Actual energy – Energy of three ethylene
$$= (6\,\alpha + 8\,\beta) - 3\,(2\,\alpha + 2\,\beta)$$

Resonance Energy of benzene $= 2\,\beta$

The SALC's themselves are the HMO'S. The electron configuration of benzene is ground state (G.S.) may be thus denoted as $A^2\,E_1{}^4$ (when full symmetry of benzene is considered, it is $A_{1u}{}^2E_{1g}{}^4$).

The HMO's of benzene and their nodal characteristics are represented.

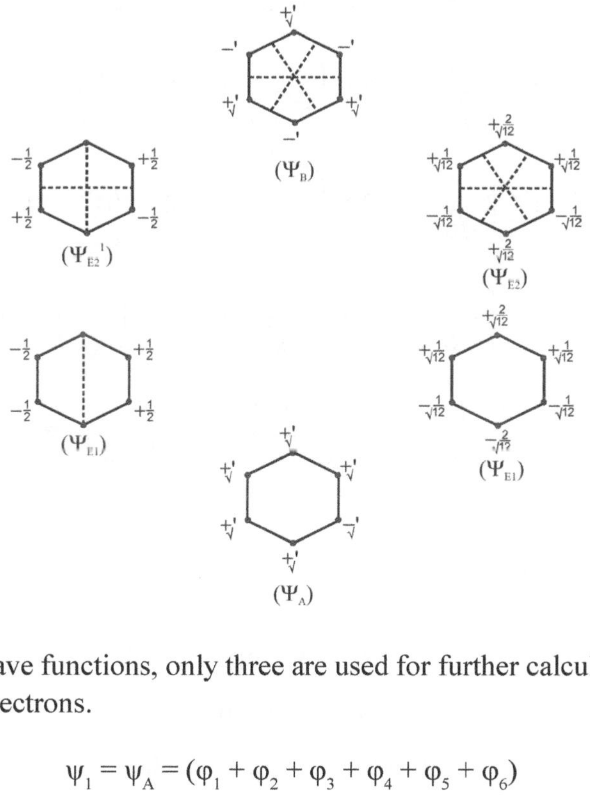

Out of six wave functions, only three are used for further calculation, which are having electrons.

$$\psi_1 = \psi_A = (\varphi_1 + \varphi_2 + \varphi_3 + \varphi_4 + \varphi_5 + \varphi_6)$$

$$\psi_2 = \psi_{E1} = (2\varphi_1 + \varphi_2 - \varphi_{3-}2\varphi_4 - \varphi_5 + \varphi_6)$$

$$\psi_3 = \psi_{E1} = (\varphi_2 + \varphi_3 - \varphi_5 - \varphi_6)$$

(i) Electron density

$$E\,D_i = \sum_{j}^{\text{occ MOs}} n_j\, C_{ij}^2$$

n_j = Number of electrons in j[th] energy level;
C_{ij} = Coefficient of ith atom in j[th] energy level.

$$E\,D_1 = 2 \times \left(\frac{1}{\sqrt{6}}\right)^2 + 2 \times \left(\frac{2}{\sqrt{12}}\right)^2 + 2 \times (0)^2 \quad = 1$$

$$E\,D_2 = 2 \times \left(\frac{1}{\sqrt{6}}\right)^2 + 2 \times \left(\frac{2}{\sqrt{12}}\right)^2 + 2 \times \left(\frac{1}{2}\right)^2 = 1$$

$$E\,D_3 = 2 \times \left(\frac{1}{\sqrt{6}}\right)^2 + 2 \times \left(-\frac{1}{\sqrt{12}}\right)^2 + 2 \times \left(\frac{1}{2}\right)^2 = 1$$

$$E\,D_4 = 2 \times \left(\frac{1}{\sqrt{6}}\right)^2 + 2 \times \left(-\frac{2}{\sqrt{12}}\right)^2 + 2 \times (0)^2 \quad = 1$$

$$E\,D_5 = 2 \times \left(\frac{1}{\sqrt{6}}\right)^2 + 2 \times \left(-\frac{1}{\sqrt{12}}\right)^2 + 2 \times \left(-\frac{1}{2}\right)^2 = 1$$

$$E\,D_6 = 2 \times \left(\frac{1}{\sqrt{6}}\right)^2 + 2 \times \left(\frac{1}{\sqrt{12}}\right)^2 + 2 \times \left(-\frac{1}{2}\right)^2 = 1$$

(ii) Charge density

$$q_i = 1 - ED_i$$

$$q_1 = 1 - 1 = 0$$

Similarly, $q_2 = q_3 = q_4 = q_5 = q_6 = 0$

(iii) Bond order

$$p^{J}_{kl} = \sum_{J}^{\text{occ MOs}} n_j\, C_{kj}\, C_{ej}$$

n_j = Number of electrons in jth energy level;
C_{kj} = Coefficient of kth atom in jth energy level.

$$p_{12} = 2 \times \frac{1}{\sqrt{6}} \times \frac{1}{\sqrt{6}} + 2 \times \frac{2}{\sqrt{12}} \times \frac{1}{\sqrt{12}} + 0 = \frac{2}{3} = 0.66$$

$$p_{23} = 2 \times \frac{1}{\sqrt{6}} \times \frac{1}{\sqrt{6}} + 2 \times \frac{1}{\sqrt{12}} \times -\frac{1}{\sqrt{12}} + 2 \times \frac{1}{\sqrt{2}} \times \frac{1}{\sqrt{2}} = \frac{2}{3} = 0.66$$

$$p_{34} = 2 \times \frac{1}{\sqrt{6}} \times \frac{1}{\sqrt{6}} + 2 \times -\frac{1}{\sqrt{12}} \times -\frac{2}{\sqrt{12}} + 2 \times \frac{1}{2} \times 0 = \frac{2}{3} = 0.66$$

$$p_{45} = 2 \times \frac{1}{\sqrt{6}} \times \frac{1}{\sqrt{6}} + 2 \times -\frac{2}{\sqrt{12}} \times -\frac{1}{\sqrt{12}} + 2 \times 0 \times -\frac{1}{2} = \frac{2}{3} = 0.66$$

$$p_{56} = 2 \times \frac{1}{\sqrt{6}} \times \frac{1}{\sqrt{6}} + 2 \times -\frac{1}{\sqrt{12}} \times -\frac{1}{\sqrt{12}} + 2 \times -\frac{1}{2} \times -\frac{1}{2} = \frac{2}{3} = 0.66$$

$$p_{61} = 2 \times \frac{1}{\sqrt{6}} \times \frac{1}{\sqrt{6}} + 2 \times \frac{2}{\sqrt{12}} \times \frac{1}{\sqrt{12}} + 2 \times 0 \times -\frac{1}{2} = \frac{2}{3} = 0.66$$

This show all the six C – C bonds in benzene are equivalent and the bonds are not true double bonds.

(iv) Free valence

In benzene, all the six position are equivalent and each carbon atom is joined to two others by partial double bond. e.g., For carbon atom 1, the total π-bond order will be

$$p_{12} + p_{16} = \frac{2}{3} + \frac{2}{3} = \frac{4}{3}$$

Free valence at position 1 is:

$$F_1 = 4.73 – 4.33$$
$$= 0.40$$

Similarly, all the six carbon atoms have the same free valence:

$$F_2 = F_3 = F_4 = F_5 = F_6 = 0.40$$

Therefore, all the six positions in benzene are equally reactive.

Thus, in benzene molecule, charge density at each carbon is 0.0, the π-bond order for each bond is 0.66 and the free valence at each C is 0.40.

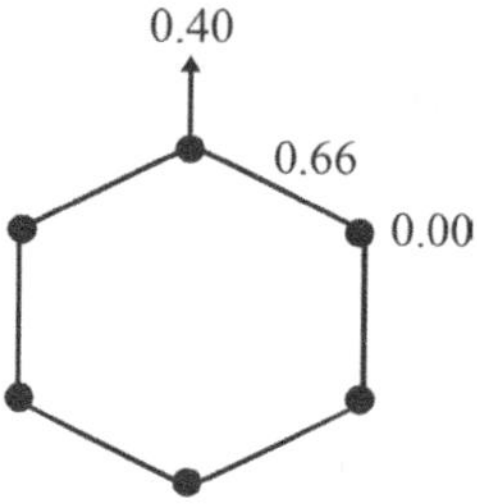

7.6 CYCLOPROPENYL GROUP (C_3H_3)

This is the simplest carbocycle with a delocalized π-system. The carbocyclic system is characterired by a general formula $(CH)_n$. It is assumed that in the carbocyclic system, the carbon atom uses sp^2 hybrids to form σ bonds. If the molecular plane is xy, it contains p orbitals specifically p_x and p_y. There remains one p_z-orbital on each carbon atom, which is perpendicular to the molecular plane. These p_z-orbitals may combine to form π-molecular orbitals.

Step 1: This molecule belongs to point group D_{3h}.

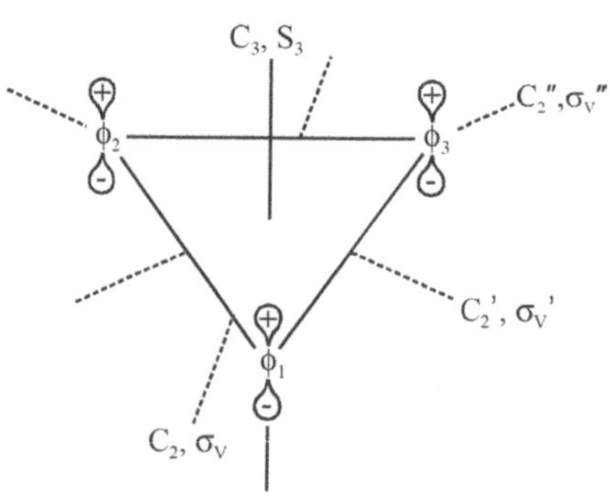

Γ_π can be obtained by C_3 group, (a simpler group).

C_3	E	C_3	C_3^2
A	1	1	1
E	1	ε	ε^*
	1	ε^*	ε
Γ_π	3	0	0

Step 2: Using reduction formula $a_i = \frac{1}{h}\sum \chi(R)\, n_R\, \varsigma_i(R)$ we have:

$$\Gamma_\pi = A + E \tag{7.77}$$

Step 3: Here, number of carbon atoms in cyclopropenyl system equal to order C_3 group, hence, wave function is given by:

$$\psi_A = \phi_1 + \phi_2 + \phi_3 \tag{7.78}$$

$$\psi_E = \phi_1 + \varepsilon\,\phi_2 + \varepsilon^*\,\phi_3 \tag{7.79}$$

$$\psi_E' = \phi_1 + \varepsilon^*\,\phi_2 + \phi_3 \tag{7.80}$$

Normalization of Eq. (7.78) gives:

$$\psi_1 = \frac{1}{\sqrt{3}}(\phi_1 + \phi_2 + \phi_3) \tag{7.81}$$

Equations (7.79) and (7.80) represent two dimension representation. They are first combined and then normalized.

$$\psi_E - \psi_E' = 2\,\phi_1 + (\varepsilon - \varepsilon^*)\,\phi_2 + (\varepsilon^* - \varepsilon)\,\phi_3$$

But $\quad \varepsilon - \varepsilon^* = \cos\frac{2\pi}{3} + \sin\frac{2\pi}{3} - \sin\frac{2\pi}{3}$

$$= 2\cos\frac{2\pi}{3} = 2\cos 120^\circ = 2\left(-\frac{1}{2}\right) = -1$$

$$\psi_E - \psi_E{}' = 2\,\phi_1 - \phi_2 - \phi_3 \tag{7.82}$$

Then it is normalized to give:

$$\psi_{Ea} = \frac{1}{\sqrt{6}}(2\,\phi_1 - \phi_2 - \phi_3) \tag{7.83}$$

Then Eqs. (7.79) and (7.80) are subtracted and normalized.

$$\frac{\psi_E - \psi_E{}'}{-i} = \frac{(\varepsilon - \varepsilon^*)}{-i}\,\phi_2 + \frac{(\varepsilon^* - \varepsilon)}{-i}\,\phi_3$$

$$\frac{(\varepsilon - \varepsilon^*)}{-i} = \frac{1}{-i}\,[\cos\frac{2\pi}{3} - i\sin\frac{2\pi}{3} - \cos\frac{2\pi}{3} + i\sin\frac{2\pi}{3}] = \sqrt{3}$$

$$\frac{\psi_E - \psi_E{}'}{-i} = \sqrt{3}\,(\phi_2 - \phi_3)$$

$$\psi_{Eb} = (\phi_2 - \phi_3) \tag{7.84}$$

After normalization it gives:

$$\psi_{Eb} = \frac{1}{\sqrt{2}}\phi_2 - \frac{1}{\sqrt{2}}\phi_3 \tag{7.85}$$

Secular determinant obtained using the SALS's takes this form:

$$\begin{bmatrix} H_{11} - ES_{11} & 0 & 0 \\ 0 & H_{22} - ES_{22} & H_{23} - ES_{23} \\ 0 & H_{32} - ES_{32} & H_{33} - Es_{33} \end{bmatrix} = 0$$

$$\begin{bmatrix} \alpha + 2\beta - E & 0 & 0 \\ 0 & \alpha - \beta - E & 0 \\ 0 & 0 & \alpha - \beta - E \end{bmatrix} = 0$$

Using Huckel approximation

$$H_{11} = \frac{1}{\sqrt{3}} \cdot \frac{1}{\sqrt{3}}\,[\int(\phi_1 + \phi_2 + \phi_3)\,H\,(\phi_1 + \phi_2 + \phi_3)]\,d\tau$$

$$= \alpha + \beta$$

$$H_{22} = \frac{1}{\sqrt{6}} \cdot \frac{1}{\sqrt{6}} [\int (\phi_1 - \phi_2 - \phi_3) H (2\phi_1 - \phi_2 - \phi_3)] d\tau$$
$$= \alpha + \beta$$

$$[H_{33} = \frac{1}{\sqrt{2}} \cdot \frac{1}{\sqrt{2}} [\int (\phi_2 - \phi_3) H (\phi_2 - \phi_3)] d\tau$$
$$= \alpha]$$

Step 4: The secular determinant for this system is:

$$\begin{bmatrix} H_{11} - ES_{11} & H_{12} - ES_{12} & H_{13} - ES_{13} \\ H_{21} - ES_{21} & H_{22} - ES_{22} & H_{23} - ES_{23} \\ H_{31} - ES_{31} & H_{32} - Es_{32} & H_{33} - Es_{33} \end{bmatrix} = 0$$

Now $$H_{11} = H_{22} = H_{33} = \alpha$$

$$S_{11} = S_{22} = S_{33} = 1$$

$$S_{12} = S_{21} = S_{13} = S_{31} = S_{23} = S_{32} = 0$$

and $$H_{12} = H_{21} = H_{13} = H_{31} = H_{23} = H_{32} = \beta$$

(because here C_1 and C_3 are also neighbors)

$$\begin{vmatrix} \alpha - E & \beta & \beta \\ \beta & \alpha - E & \beta \\ \beta & \beta & \alpha - E \end{vmatrix} = 0$$

Dividing all the elements by β and putting $\frac{\alpha - E}{\beta} = x$, we get

$$\begin{vmatrix} x & 1 & 1 \\ 1 & x & 1 \\ 1 & 1 & x \end{vmatrix} = 0$$

$$x\begin{vmatrix} x & 1 \\ 1 & x \end{vmatrix} - 1\begin{vmatrix} 1 & 1 \\ 1 & x \end{vmatrix} + 1\begin{vmatrix} 1 & x \\ 1 & 1 \end{vmatrix} = 0$$

or $$x\,(x^2 - 1) - 1\,(x - 1) + 1\,(1 - x) = 0$$

or $$x^3 - x - x + 1 + 1 - x = 0$$

or $$x^3 - 3\,x + 2 = 0$$

It can be rewritten as

or $$x^3 - 2\,x^2 + x + 2 + 2\,x^2 - 4\,x = 0$$

or $$x^2\,(x + 2) - 2\,x\,(x + 2) + 1\,(x + 2) = 0$$

or $$(x + 2)\,(x^2 - 2\,x + 1) = 0$$

or $$(x + 2)\,(x - 1)^2 = 0$$

Therefore, the roots of equation are $x_1 = -2$ and $x_2 = x_3 = 1$

The corresponding energy levels are

$$\frac{(\alpha - E)}{\beta} = -2$$

or $$E_1 = \alpha + 2\beta$$

$$\frac{(\alpha - E)}{\beta} = 1$$

or $$E_2 = E_3 = \alpha - \beta$$

Thus, two levels (E_2 and E_3) are degenerate.

The ground state π-electron distribution in the three HMO'S of the cyclopropenyl carbonuim ion, radical and the carbonuim is shown diagrammatically as:

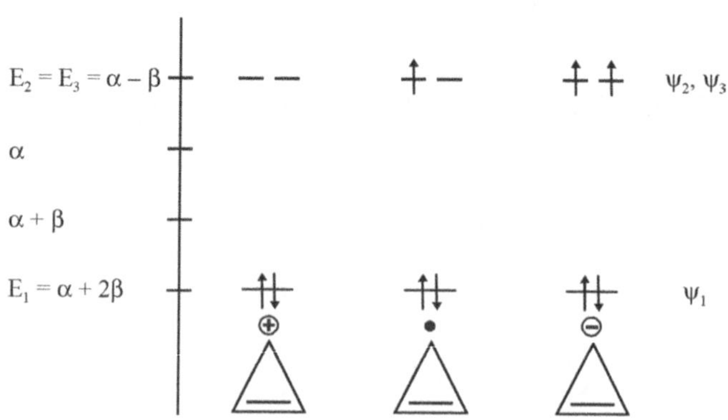

The total energy E_π and the delocalization energy for these three systems is:

Cation $E_\pi = 2(\alpha + 2\beta) = 2\alpha + 4\beta$

$$\text{D.E.} = 2(\alpha + 2\beta) - 2(\alpha + \beta) = 2\beta$$

Radical $E_\pi = 2(\alpha + 2\beta) + (\alpha - \beta) = 3\alpha + 3\beta$

$$\text{D.E.} = (3\alpha + 3\beta) - 2(\alpha + \beta) - \alpha = \beta$$

Anion $E_\pi = 2(\alpha + 2\beta) + 2(\alpha - \beta) = 4\alpha + 2\beta$

$$\text{D.E.} = (4\alpha + 2\beta) - 2(\alpha + \beta) - \alpha - \alpha = 0$$

Out of three wave functions, two are used for anion and radical cyclopropenyl and one for cationic system.

$$\psi_1 = \frac{1}{\sqrt{3}}(\phi_1 + \phi_2 + \phi_3)$$

$$\psi_3 = \psi_{EA} = \frac{1}{\sqrt{6}}(2\phi_1 - \phi_2 - \phi_3)$$

$$\psi_2 = \psi_{EB} = \frac{1}{\sqrt{2}}(\phi_2 - \phi_3)$$

(i) Electron density

$$ED_i = \sum_{J}^{\text{occ MOs}} n_j C_{ij}^2$$

Cation

It has 2 e^- in ψ_1 and no electron in ψ_2 and ψ_3.

$$ED_1 = 2 \times \left(\frac{1}{\sqrt{3}}\right)^2 = \frac{2}{3}$$

$$ED_2 = 2 \times \left(\frac{1}{\sqrt{3}}\right)^2 = \frac{2}{3}$$

$$ED_3 = 2 \times \left(\frac{1}{\sqrt{3}}\right)^2 = \frac{2}{3}$$

Radical

In cyclopropenyl radical, two-electrons are in ψ_1 and third electron may be placed in either ψ_2 or ψ_3 (degenerate). In such a case, the electron density is calculated by assuming that half of the available electron is in each of the degenerate MO's ψ_2 and ψ_3.

$$ED_1 = 2 \times \left(\frac{1}{\sqrt{3}}\right)^2 + \frac{1}{2}(0)^2 + \frac{1}{2} \times \left(\frac{2}{\sqrt{6}}\right)^2 = 1$$

$$ED_2 = 2 \times \left(\frac{1}{\sqrt{3}}\right)^2 + \frac{1}{2}\left(\frac{1}{\sqrt{2}}\right)^2 + \frac{1}{2} \times \left(-\frac{1}{\sqrt{6}}\right)^2 = 1$$

$$ED_3 = 2 \times \left(\frac{1}{\sqrt{3}}\right)^2 + \frac{1}{2}\left(-\frac{1}{\sqrt{2}}\right)^2 + \frac{1}{2}\left(-\frac{1}{\sqrt{6}}\right)^2 = 1$$

Anion

Like radical, considering ψ_2 or ψ_3 (degenerate).

$$ED_1 = 2 \times \left(\frac{1}{\sqrt{3}}\right)^2 + 1 \times (0)^2 + 1 \times \left(\frac{2}{\sqrt{6}}\right)^2$$

$$= \frac{2}{3} + 0 + \frac{4}{6} = \frac{4}{3}$$

$$ED_2 = 2 \times \left(\frac{1}{\sqrt{3}}\right)^2 + 1 \times \left(\frac{1}{\sqrt{2}}\right)^2 + 1 \times \left(-\frac{1}{\sqrt{6}}\right)^2$$

$$= \frac{2}{3} + \frac{1}{2} + \frac{1}{6} = \frac{4}{3}$$

$$ED_3 = 2 \times \left(\frac{1}{\sqrt{3}}\right)^2 + 1 \times \left(-\frac{1}{\sqrt{2}}\right)^2 1 \times \left(-\frac{1}{\sqrt{6}}\right)^2$$

$$= \frac{2}{3} + \frac{1}{2} + \frac{1}{6} = \frac{4}{3}$$

(ii) Charge density

$$q_i = 1 - ED_i$$

For cyclopropenyl cation

$$q_1 = q_2 = q_3 = 1 - \frac{2}{3} = \frac{1}{3}$$

For cyclopropenyl radical

$$q_1 = q_2 = q_3 = 1 - 1 = 0$$

For cyclopropenyl anion

$$q_1 = q_2 = q_3 = 1 - \frac{4}{3} = -\frac{1}{3}$$

(iii) Bond order

$$p_{kl}^{J} = \sum_{J}^{\text{occ MOs}} n_j \, C_{kj} \, C_{ej}$$

where n_j = number of electron in j[th] energy level; C_{kj} = coefficient of k atom in j[th] energy level.

For cyclopropenyl cation

$$p_{12} = 2 \times \frac{1}{\sqrt{3}} \times \frac{1}{\sqrt{3}} = \frac{2}{3} = 0.666$$

Similarly,

$$p_{23} = p_{31} = \frac{2}{3} = 0.666$$

For cyclopropenyl radical

$$p_{12} = 2 \times \frac{1}{\sqrt{3}} \times \frac{1}{\sqrt{3}} + \frac{1}{2} \times 0 \times \frac{1}{\sqrt{2}} + \frac{1}{2} \times \frac{2}{\sqrt{6}} \times \left(-\frac{1}{\sqrt{6}}\right)$$

$$= \frac{2}{3} + 0 - \frac{1}{6} = \frac{3}{6} = \frac{1}{2} = 0.50$$

$$P_{23} = 2 \times \frac{1}{\sqrt{3}} \times \frac{1}{\sqrt{3}} + \frac{1}{2} \times \frac{1}{\sqrt{2}} \times \left(-\frac{1}{\sqrt{2}}\right) + \frac{1}{2} \times \left(-\frac{1}{\sqrt{6}}\right) \times \left(-\frac{1}{\sqrt{6}}\right)$$

$$= \frac{2}{3} - \frac{1}{4} + \frac{1}{12}$$

$$= \frac{8-3+1}{12} = \frac{6}{12} = \frac{1}{2} = 0.50$$

$$P_{31} = 2 \times \frac{1}{\sqrt{3}} \times \frac{1}{\sqrt{3}} + \frac{1}{2} \times \left(-\frac{1}{\sqrt{2}}\right) \times (0) + \frac{1}{2} \times -\frac{1}{\sqrt{6}} \times \frac{2}{\sqrt{6}}$$

$$= \frac{2}{3} + 0 - \frac{1}{6} = 0.50$$

$$= \frac{4-1}{6} = \frac{3}{6} = \frac{1}{2}$$

For cyclopropenyl anion

$$p_{12} = 2 \times \frac{1}{\sqrt{3}} \times \frac{1}{\sqrt{3}} + 1 \times \frac{2}{\sqrt{6}} \times \frac{1}{\sqrt{6}} + 1 \times (0) \times \frac{1}{\sqrt{2}}$$

$$= \frac{2}{3} - \frac{2}{6} = \frac{1}{3} = 0.33$$

$$p_{23} = 2 \times \frac{1}{\sqrt{3}} \times \frac{1}{\sqrt{3}} + 1 \times \frac{1}{\sqrt{2}} \times -\frac{1}{\sqrt{2}} + 1 \times -\frac{1}{\sqrt{6}} \times -\frac{1}{\sqrt{6}}$$

$$= \frac{2}{3} - \frac{1}{2} = \frac{1}{6}$$

$$= \frac{8-3+1}{12} = \frac{2}{6} = \frac{1}{3} = 0.333$$

$$p_{31} = 2 \times \frac{1}{\sqrt{3}} \times \frac{1}{\sqrt{3}} + 1 \times -\frac{1}{\sqrt{2}} \times (0) + 1 \times -\frac{1}{\sqrt{6}} \times \frac{2}{\sqrt{6}}$$

$$= \frac{2}{3} + 0 - \frac{2}{6} = \frac{2}{6} = \frac{1}{3} = 0.333$$

(iv) Free valence

In $C_3H_3^+$ ion, all three positions are equivalent and each carbon join another carbon atom by a partial double bond.

For cyclopropenyl cation

Thus, the total π-bond order for carbon will be:

$$= P_{12} + P_{31}$$

$$= 0.666 + 0.666 = 1.332$$

So, Fr $= 4.732 - 4.332$

$$= 0.40$$

Similarly,

For cyclopropenyl radical

Total bond order = 0.50 + 0.50 = 1.0

So free valence = 4.732 – 4.0 = 0.732

For cyclopropenyl anion

Total bond order = 0.333 + 0.333 = 0.666

Hence, free valence = 4.732 – 3.666 = 1.066

It may thus be concluded that reactivity of these species will be in the order:

Cyclopropenyl anion > Cyclopropenyl radical > Cyclopropenyl cation

Reverse will be the order of their stability-

Cation > Radical > Anion

Cyclopropenyl cation is stabilized because of resonance.

7.7 CYCLOPENTADIENYL GROUP

Such molecules have planar geometry, where each carbon atom is sp^2 hybridized. All carbon atoms have p_z orbital perpendicular to the plane of the molecule. These π-orbitals combine to form 5 π MO's, in which one is non-degenerate and other two sets are doubly degenerate MO's.

Step 1: The C_5H_5 belongs to D_{5h} point group.

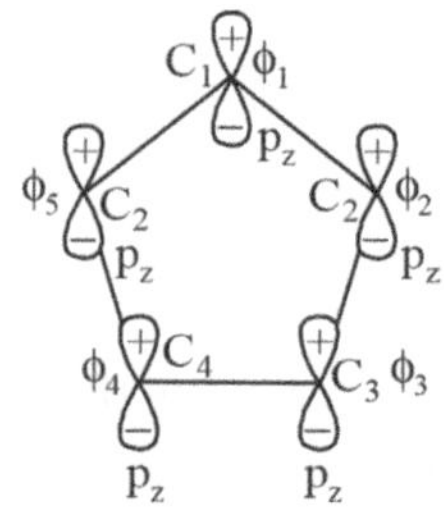

Γ_π can be obtained by using C_5 group

C_5	E	C_5	$C_5^{\,2}$	$C_5^{\,3}$	$C_5^{\,4}$
A	1	1	1	1	1
E_1	$\begin{Bmatrix}1 \\ 1\end{Bmatrix}$	ε ε^*	ε^2 ε^{2*}	ε^{2*} ε^2	ε^* ε
E_2	$\begin{Bmatrix}1 \\ 1\end{Bmatrix}$	ε^2 ε^{2*}	ε^* ε	ε ε^*	ε^{2*} ε^2
$\Gamma_\pi(R)$	5	0	0	0	0

Step 2: Using reduction formula,

$$\Gamma_\pi(R) = A + E_1 + E_2$$

Step 3: Because, number of carbon atoms are equal to order of C_5 group. Therefore, wave function are:

$$\psi_A = \varphi_1 + \varphi_2 + \varphi_3 + \varphi_4 + \varphi_5 \tag{7.86}$$

$$\psi_{E_1}\begin{cases}\psi_{E_1} = \varphi_1 + \mu\varphi_2 + \mu^2\varphi_3 + \mu^{2*}\varphi_4 + \mu^*\varphi_5 \\ \psi_{E_1}{}' = \varphi_1 + \mu^*\varphi_2 + \mu^*\varphi_2 + \mu^{2*}\varphi_3 + \mu^2\varphi_4 + \mu\varphi_5\end{cases} \tag{7.87}$$

$$\psi_{E_2}\begin{cases}\psi_{E_2} = \varphi_1 + \mu^2\varphi_2 + \mu^*\varphi_3 + \mu\varphi_4 + \mu^{2*}\varphi_5 \\ \psi_{E_2}{}' = \varphi_1 + \mu^{2*}\varphi_2 + \mu\varphi_3 + \mu^*\varphi_4 + \mu^2\varphi_5\end{cases} \tag{7.89}$$

On normalization of Eq. (7.86), we get:

$$\psi_A = \frac{1}{\sqrt{5}}(\varphi_1 + \varphi_2 + \varphi_3 + \varphi_4 + \varphi_5) \tag{7.91}$$

On adding Eqs. (7.87) and (7.88) and normalization, it gives:

$$\psi_{E_1} = \psi_{E_1} + \psi_{E_1'}$$

$$= 2\,\phi_1 + (\varepsilon + \varepsilon^*)\,\phi_2 + (\varepsilon^2 + \varepsilon^{2*})\,\phi_3 + (\varepsilon^{2*} + \varepsilon^2)\,\phi_4 + (\varepsilon^* + \varepsilon)\,\phi_5$$

As $$\varepsilon + \varepsilon^* = \left(\cos\frac{2\pi}{5} + i\sin\frac{2\pi}{5}\right) + \left(\cos\frac{2\pi}{5} - i\sin\frac{2\pi}{5}\right)$$

$$= 2\cos\frac{2\pi}{5}$$

and $$\frac{\varepsilon - \varepsilon^*}{i} = 2\sin\frac{2\pi}{5}$$

Also, $$\varepsilon^2 + \varepsilon^{2*} = 2\cos\frac{4\pi}{5}$$

and $$\frac{\varepsilon^2 - \varepsilon^{2*}}{i} = 2\sin\frac{4\pi}{5}$$

Note $$\frac{2\pi}{5} = \frac{2 \times 180^\circ}{5} = 72^\circ$$

$\because$ $$\frac{4\pi}{5} = 2 \times 72 = 144^\circ$$

Therefore,

$$\psi_{E_1} + \psi_{E_1}{}' = 2\,\varphi_1 + 2\,\varphi_2\cos\frac{2\pi}{5} + 2\,\varphi_3\cos\frac{4\pi}{5} + 2\,\varphi_3\cos\frac{4\pi}{5} + 2\,\varphi_5\cos\frac{2\pi}{5}$$

$$= \sqrt{\frac{2}{5}}\left(\varphi_1 + \varphi_2\cos 72^\circ + \varphi_3\cos 144^\circ + \varphi_4\cos 144^\circ + \varphi_5\cos 72^\circ\right) \quad (7.92)$$

and on subtracting, Eqs. (7.87) and (7.88) and dividing by i, then normalization give

$$\frac{\psi_{E_1 - E_1'}}{i} = \sqrt{\frac{2}{5}}\left(\varphi_1\sin 72^\circ + \varphi_3\sin 144^\circ - \varphi_4\sin 144^\circ - \varphi_5\sin 72^\circ\right) \quad (7.93)$$

on ψ_{E_1} and ψ_{E_1}', ψ_{E_2} and ψ_{E_2}' are also operated, which results into:

$$\psi_{E_2} + \psi_{E_2'} = \sqrt{\frac{2}{5}}\left(\varphi_1 + \varphi_2 \cos\ 144^{\circ} + \varphi_3 \cos\ 72^{\circ} + \varphi_4 \cos\ 72^{\circ} + \varphi_5 \cos\ 144^{\circ}\right)$$

(7.94)

$$\frac{\psi_{E_2} - \psi_{E_2'}}{i} = \sqrt{\frac{2}{5}}\begin{pmatrix} \varphi_2 \sin 144^{\circ} - \varphi_3 \sin 72^{\circ} + \varphi_4 \sin 72^{\circ} + \\ \varphi_4 \sin 72^{\circ} - \varphi_5 \sin 144^{\circ} \end{pmatrix} \qquad (7.95)$$

The five π-MO's wave functions are expressed by Eqs. (7.91)–(7.95).

The secular determinant for the system is:

$$\begin{bmatrix} H_{11}-ES_{11} & H_{12}-ES_{12} & H_{13}-ES_{13} & H_{14}-ES_{14} & H_{15}-ES_{15} \\ H_{21}-ES_{21} & H_{22}-ES_{22} & H_{23}-ES_{23} & H_{24}-ES_{24} & H_{25}-ES_{25} \\ H_{31}-ES_{31} & H_{32}-ES_{32} & H_{33}-ES_{33} & H_{34}-ES_{34} & H_{35}-ES_{35} \\ H_{41}-ES_{41} & H_{42}-ES_{42} & H_{43}-ES_{43} & H_{44}-ES_{44} & H_{45}-ES_{45} \\ H_{51}-ES_{51} & H_{52}-ES_{52} & H_{53}-ES_{53} & H_{54}-ES_{54} & H_{55}-ES_{55} \end{bmatrix} = 0$$

Using Huckel approximation, all the off-diagonal elements in the secular determinant will vanish. Integral in which the π-MO's wave functions belonging to same symmetry but orthogonal to each other will also vanish. Thus we get,

$$H_{11} = \alpha + 2\,\beta$$
$$H_{22} = \alpha + (2 \cos 72°).\ \beta$$
$$H_{33} = \alpha + (2 \cos 72°).\ \beta$$
$$H_{44} = H_{55} = \alpha + 2 \cos (144°).\ \beta$$

Step 4: The secular determinant for this system will be:

$$\begin{bmatrix} \alpha+2\beta-E & 0 & 0 & 0 & 0 \\ 0 & \alpha+(2\cos 72^{\circ})\beta-E & 0 & 0 & 0 \\ 0 & 0 & \alpha+(2\cos 72^{\circ})\beta-E & 0 & 0 \\ 0 & 0 & 0 & \alpha+(2\cos 144^{\circ})\beta-E & 0 \\ 0 & 0 & 0 & 0 & \alpha+(2\cos 144^{\circ})\beta-E \end{bmatrix} = 0$$

The determinant of 5 × 5 dimension can be reduced into determinant of 1 × 1 dimensions by block out method.

$$E_1 = \alpha + 2\,\beta - E = 0$$
$$E_2 = E_3 = \alpha + (2\cos 72°)\,\beta - E = 0$$
$$E_4 = E_5 = \alpha + (2\cos 144°)\,\beta - E = 0$$

and therefore, five energy levels are:

$$E_1 = \alpha + 2\,\beta$$
$$E_2 = E_3 = \alpha + (2\cos 72°)\,\beta$$
$$E_4 = E_5 = \alpha + (2\cos 144°)\,\beta$$

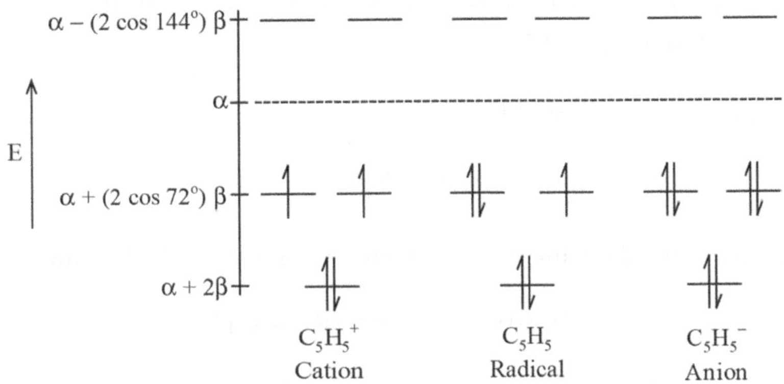

$C_5H_5^+$ cation

The total π-electron energy of $C_5H_5^+$ is:

$$E_\pi = 2\,(\alpha + 2\,\beta) + 1\,(\alpha + 2\,\beta\cos 72°) + 1\,(\alpha + 2\,\beta\cos 72°)$$
$$= 4\,\alpha + 5.236\,\beta$$

Delocalization energy is:

$$E_\pi(0) = E_\pi - 4\,(\alpha + \beta)$$
$$= 4\,\alpha - 5.236\,\beta - 4\,\alpha - 4\,\beta$$
$$= 1.236\,\beta \cong (4\cos w)\,\beta$$

C_5H_5 radical

The total energy (E_π) of the radical C_5H_5 system is:

$$E_\pi = 2\,(\alpha + 2\,\beta) + 2\,(\alpha + 2\,\beta\cos 72°) + 1\,(\alpha + 2\,\beta\cos 75°)$$
$$= 2\,\alpha + 4\,\beta + 3\alpha + 6\,\beta\cos 72\ (\cos 72 = 0.3090)$$
$$= 5\,\alpha + 5.854$$

Delocalization energy (E_D) for C_5H_5 radical is:

$$E_\pi(0) = E_A - [4(\alpha + \beta) + \alpha] = 5\alpha + 5.854\beta - 4\alpha - 4\beta - \alpha$$
$$= 1.854\beta \cong (6\cos w)\beta$$

where w = 72°.

$C_5H_5^-$ anion
E_π for $C_5H_5^-$ anion:

$$E_\pi = 2(\alpha + 2\beta) + 2(\alpha + 2\beta\cos 72°) + 2(\alpha + 2\beta\cos 72°)$$
$$= 6\alpha + 6.472\beta$$

Delocalization energy of $C_5H_5^-$ is:

$$E_\pi(0) = 6\alpha + 6.472\beta - 4(\alpha + \beta) - 2\alpha$$

Here, $4(\alpha + \beta) - 2\alpha$ is the π-electron energy of localized ethylenic linkage.

$$E_\pi(0) = 2.472\beta \cong (8\cos w)\beta$$

The HMO's of C_5H_5 and its nodel characteristics are:

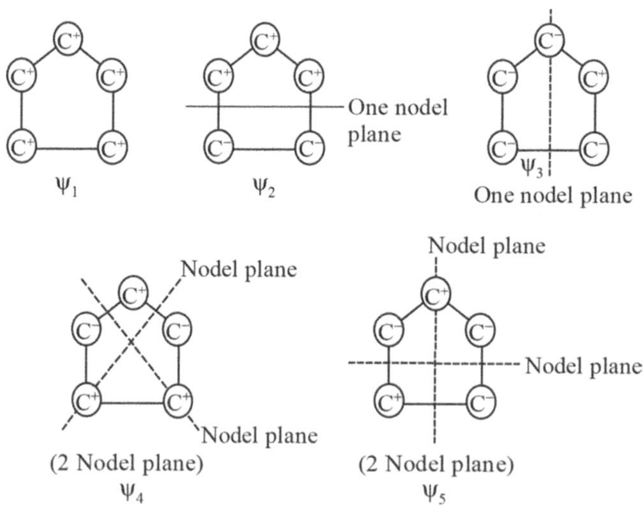

Huckel parameters can be determined by using coefficient of AO's in wave function.

(i) Electron density

$$ED_r = \sum_{j}^{Occ.MOs} n_j.c_{ij}^2$$

$$ED_1 \text{ at carbon (1)} = 2 \times \left(\frac{1}{\sqrt{5}}\right)^2 + 1 \times \left(\sqrt{\frac{1}{5}}\right)^2 + 2(0)$$

$$= \frac{2}{5} + \frac{4}{5} = 1.2$$

$$ED_2 \text{ at carbon (2)} = 2 \times \left(\frac{1}{\sqrt{5}}\right)^2 + 2\left(\sqrt{\frac{2}{5}}\cos 72^\circ\right)^2 + 2\left(\sqrt{\frac{2}{5}}\sin 72^\circ\right)^2$$

$$= \frac{2}{5} + \frac{4}{5} = 1.2$$

$$ED_3 \text{ at carbon (3)} = 2 \times \left(\frac{1}{\sqrt{5}}\right)^2 + 2\left(\sqrt{\frac{2}{5}}\cos 144^\circ\right)^2 + 2\left(\sqrt{\frac{2}{5}}\sin 144^\circ\right)^2$$

$$+2\left(\sqrt{\frac{2}{5}}\sin 144^\circ\right)^2 = 1.2$$

$$ED_4 \text{ at carbon (4)} = 2 \times \left(\frac{1}{\sqrt{5}}\right)^2 + 2\left(\sqrt{\frac{2}{5}}\cos 144^\circ\right)^2 + 2\left(-\sqrt{\frac{2}{5}}\sin 144^\circ\right)^2 = 1.2$$

$$ED_d \text{ at carbon (5)} = 2 \times \left(\frac{1}{\sqrt{5}}\right)^2 + 2\left(\sqrt{\frac{2}{5}}\cos 72^\circ\right)^2 + 2\left(-\sqrt{\frac{2}{5}}\sin 72^\circ\right)^2 = 1.2$$

It shows that total electron density of $C_5H_5^-$ anion is:

$$1.2 + 1.2 + 1.2 + 1.2 + 1.2 = 6.0$$

(ii) Charge density (q_r)

In $C_5H_5^-$ ion

$$q_r = 1 - E\,D_r$$

As the electron density of all the 5 carbon atoms are equal

$$q_1 = q_2 = q_3 = q_4 = q_5 = 1.2$$

Charge density of each carbon atom will be:

$$= 1 - 1.2$$

$$= -0.2 = -1/5$$

(iii) Bond order

Bond order for $C_5H_5^-$ will be calculated from π-bond order between any two adjacent carbon atoms:

$$P_{12} = 2 \times \left(\frac{1}{\sqrt{5}} \times \frac{1}{\sqrt{5}} \right) + 2 \left(\sqrt{\frac{2}{5}} \times \sqrt{\frac{2}{5}} \text{ con } 72^{\circ} \right) + 2 \left(0 \times \sqrt{\frac{2}{5}} \sin 72^{\circ} \right)$$

$$= \frac{2}{5} + \frac{2 \times 2}{5} \times 0.309$$

$$= 0.647$$

$$P_{23} = 2 \times \left(\frac{1}{\sqrt{5}} \times \frac{1}{\sqrt{5}} \right) + 2 \left(\sqrt{\frac{2}{5}} \cos 72^{o} \times \sqrt{\frac{2}{5}} \cos 144^{\circ} \right) + 2 \left(\sqrt{\frac{2}{5}} \sin 72^{\circ} \times \sqrt{\frac{2}{5}} \cos 144^{\circ} \right)$$

$$= \frac{2}{5} + \frac{2 \times 2}{5} \times 0.309$$

$$= 0.647$$

In same way, $P_{34} = P_{45} = P_{15}$ or $P_{51} = 0.647$

It shows that all five $C - C$ bonds are equivalent, but these are not true double bonds.

(iv) Free valence

In $C_5H_5^-$ ion, all five positions are equivalent and each carbon join another carbon atom by a partial double bond.

Thus, the total π-bond order for carbon will be:

$$= P_{12} + P_{51}$$

$$= 0.647 + 0.647 = 1.294$$

So, $$F_1 = 4.732 - 4.294$$

$$= 0.438$$

Similarly, the free valence at all the carbon atoms in $C_5H_5^-$ anion is 0.438. It means all five carbons will be equally reactive.

KEYWORDS

- **Bond order**
- **Charge density**
- **Delocalization energy**
- **Electron density**
- **Face valence**
- **Molecular orbital theory**
- **Secular determinan**

CHAPTER 8

MOLECULAR VIBRATIONS

CONTENTS

8.1 NORMAL MODES OF VIBRATION

The complex vibrations of a molecule are the superposition of relatively simple vibrations, which are called the normal mode of vibration. Suppose a molecule has N number of atoms, then total degrees of freedom for that molecule is 3 N. The normal mode corresponds to coordinates at X-, Y- and Z-axes.

Each normal mode of vibration has a fixed frequency. Molecules have translational, rotational and vibrational motions, and therefore, total degrees of freedom (3 N) can be determined for each motion depending on type of motion.

8.1.1 TRANSLATIONAL MOTION

In this motion, molecule moves from one place to other but without changing its shape. It means that molecule moves as a whole unit. Translational motion uses all the three coordinates. Therefore, number of translational degrees of freedom is 3.

8.1.2 ROTATIONAL MOTION

Rotational degrees of freedom depend on shape of molecule. In linear molecules, rotation occurs about X- and Y-axes, whereas in non-linear molecules, rotation occur along all the three axes X-, Y- and Z-. Therefore, linear molecules have two rotational degrees of freedom while non-linear molecules have three rotational degrees of freedom.

8.1.3 VIBRATIONAL MOTION

It is determined by the difference between total degrees of freedom and the sum of translational and rotational degrees of freedom, i.e., [3 N – (Trans + Rot.)]. Thus,

$$\text{Normal mode for a linear molecule} = 3\text{ N} - (3 + 2) = 3\text{ N} - 5$$
$$\text{Normal mode for a non-linear molecule} = 3\text{ N} - (3 + 3) = 3\text{ N} - 6$$

At room temperature, all the molecules are in their lowest vibrational energy level with quantum number equal to zero for each mode. The most probable vibrational transition is form $v = 0$ to $v = 1$. This transition, i.e., $v = 0 \rightarrow 1$ gives strong IR and Raman bands and is called a fundamental normal mode.

8.2 MOLECULAR VIBRATIONS

The vibrations in molecule can be classified into two types:

- Bond stretching vibrations and
- Bending or deformation vibrations.

8.2.1 BOND STRETCHING VIBRATIONS

Stretching vibrations occur due to displacement of atoms along the bond and it leads to change in bond length. This mode is represented by a change in bond length while keeping the bond angle fixed showing the direction of the movement of atoms along the bond. The change in bond length is given by the symbol ν or (strictly speaking Δr) and represents internal coordinate of bond vector. The stretching of the bond has been conventionally designated as ν. There are two types of stretching modes/coordinates. These are:

8.2.1.1 Symmetric Stretching Vibration (ν_s)

It involves stretching or compressing of bond from both the sides simultaneously, i.e., together movement of atoms in the same direction along the bond. For example in xy_2 molecule:

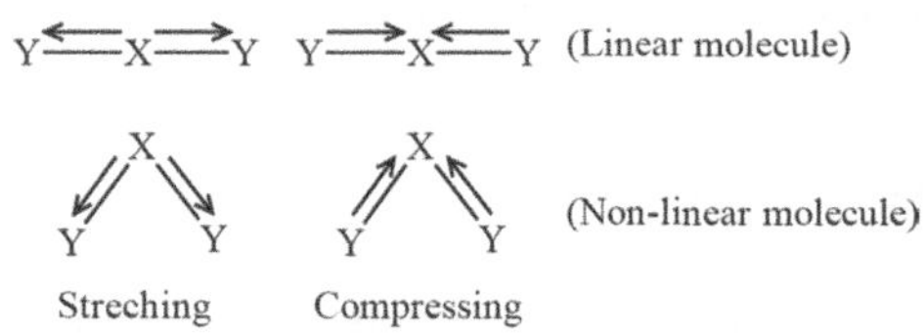

Symmetric streching

8.2.1.2 Asymmetric Stretching Vibration (ν_{as})

It involves simultaneous movement of atoms in the different directions along the bond. It means that when one bond is being stretched, then the other bond is compressed.

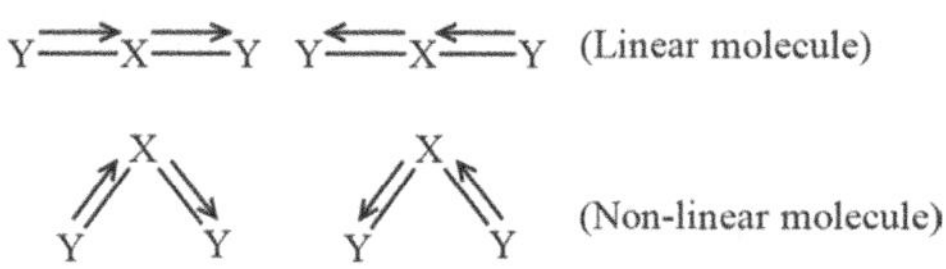

Asymmetric stretching

Asymmetric stretching vibrations have greater energy than energy of symmetric stretching vibrations.

8.2.2 BENDING VIBRATIONS

This mode represents a change in the angle between two bonds while keeping the bond length constant (unaltered). Bond angle changes because of the movement of atoms in-plane or out-of-plane of the molecule. The individual angle bending coordinates are represented by α, β, γ, etc. For the angle bending, the atoms connected by the bonds move in such a way that the direction of displacement of these atoms is perpendicular to the bonds. Such an angle bending has been represented by δ and it is represented as:

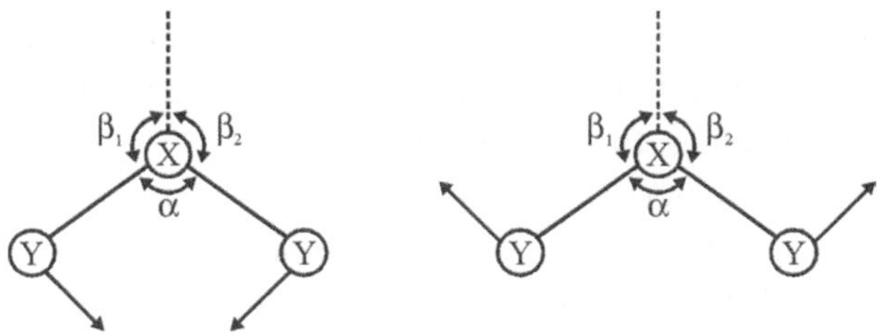

Two types of angle bending deformations are therefore called as in-plane mode and out-of-plane mode.

8.2.2.1 In-Plane Mode

All the atoms lie in the same plane during in-plane bending vibrations (plane mode). Plane mode can be of two types:

(i) Scissoring mode (Symmetric) – It involves change in both; the internal coordinates separated by α and β. It is designated as δ_s.
(ii) Rocking mode (Asymmetric) – It involves changes only in β type coordinates while α remains constant. It is designated as δ_r.

The scissoring and rocking vibrations are represented as:

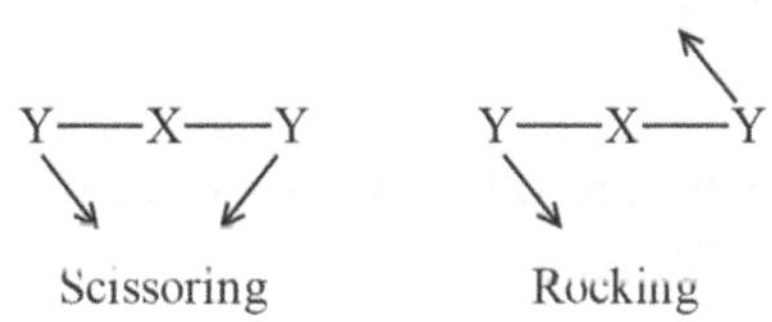

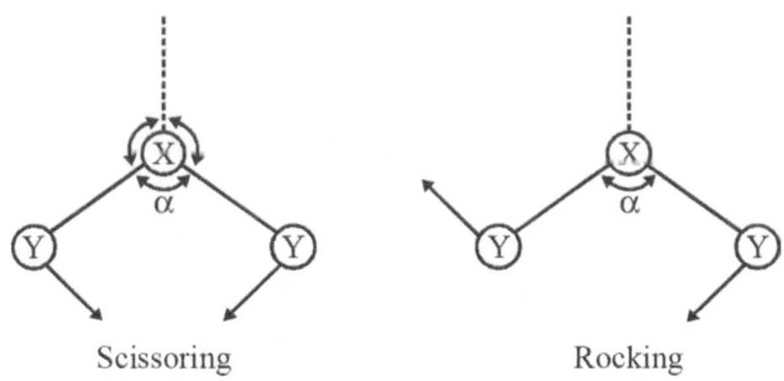

8.2.2.2 Out-of-Plane Mode

In out-of-plane bending vibration, atoms do not remain in the plane but they move out-of-plane.

(i) Wagging mode (Symmetric) – This type of vibrational motion results, where both the X-Y bonds go out-of-plane and come back also simultaneously and it is therefore, designated as δ_W.
Out-of-plane movement can be represented by ⊕ and ⊖ sign. ⊕ Sign shows atom is moving above-the-plane of the molecule. On the other hand, ⊖ sign indicates that the atoms are moving below-the-plane of the molecule.

(ii) Twisting mode (Asymmetric) – In this case, one X-Y bond moves above the plane (⊕) and the other bond move below the plane (⊖). This coordinate results in the asymmetric displacement of XY_2 unit and pushes both the X-Y bonds out of the main frame of molecular plane.
It is represented by δ_t. The two kinds of out-of-plane bending vibrations are:

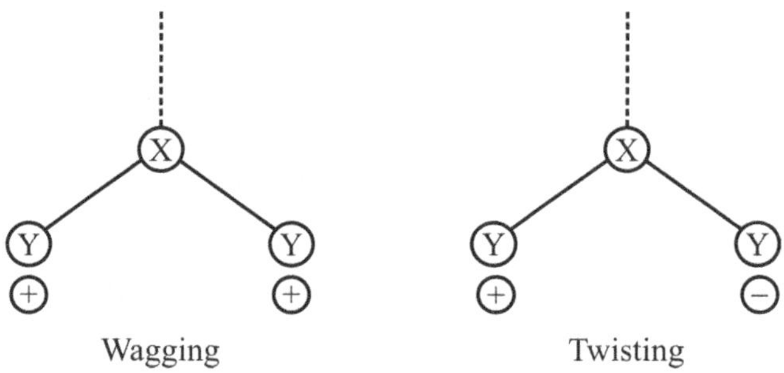

In addition to these, there are other types of out-of-plane modes in planar molecules (tetratomic XY_{4-}) as:

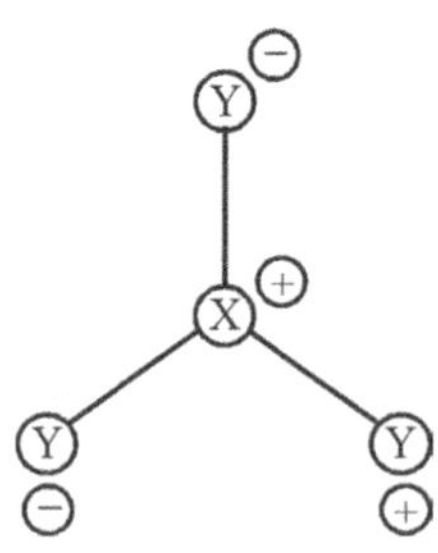

Here, central atom (X) is pulled out of the plane and the final configuration of the molecule resembles a pyramid. Thus, this is a type of 'breathing mode,' labeled as π.

Thus, it can be concluded that out of six vibrational modes in a linear molecule, rocking and twisting are rotating around *Y*- and *Z*-axes and thus, they do not absorb in IR region. In contrary to non-linear molecule (bent molecule), wagging is partly rotating around *X*-axis in linear molecule. It is so, as in rotational mode in a linear molecule, rotation around the molecular axis is not considered. Hence, the total number of vibration in linear tri-atomic molecule is 3 N – 5 = (3 x 3) – 5 = 4.

In non-linear (bent) molecules, rocking, wagging and twisting vibrations are part of rotation around *X*-, *Y*- and *Z*-axes, respectively, This means these three vibrations are inactive in IR region and do not absorb in the IR region. Therefore, symmetric, asymmetric stretching and scissoring bending are possible mode of vibrations.

Hence, total number of fundamental vibrations for a triatomic (XY_2) non-linear molecule is 3 N – 6 = 3 x 3) – 6 = 3, while it was 4 for linear molecule.

The bending vibration has lower energy than a stretching vibration. The order of energies for the vibrations are:

$$\nu_{as} > \nu_s > \delta$$

8.3 SELECTION RULES FOR IR AND RAMAN SPECTRA

The selection rule/transition rule constrains the possible transition from one quantum state to another. According to quantum mechanics and as per selection rule, vibrational transitions are allowed when $\Delta v = +1$ and $\Delta J = +1$. Vibrational spectroscopy (IR and Raman) depends on two concepts. Firstly, change in dipole moment (for IR) and second, change in polarizability (for Raman) of molecule.

It means, when dipole moment changes during the vibration, the vibrational transition will be IR active, but when polarizability of molecule (α) changes, then vibrational transition will be Raman active. It can be expressed as:

$$\frac{\partial \mu_i}{\partial \varphi} \neq 0 \qquad \text{(IR active)}$$

and

$$\frac{\partial \alpha_{ij}}{\partial \varphi} \neq 0 \qquad \text{(Raman active)}$$

Here, μ_i is the one of the component of dipole moment, ϕ is normal coordinate with internuclear distance (r) and α_{ij} is one of the component of the polarizability tensor.

Dipole moment has three components μ_x, μ_y and μ_z along *X*-, *Y*- and *Z*-axes and polarizability has six components, i.e., α_{xx}, α_{yy}, α_{zz}, α_{xy}, α_{yz}, and α_{xz}. Majority of the molecules with $\mu = 0$ are IR inactive but it is not always true. In case of homonuclear diatomic molecule (such as N_2, O_2, H_2, Cl_2), there will be no permanent dipole. So, they will not give any IR spectrum but hetronuclear diatomic molecule (such as CO, NO) will be IR active. When molecules do not have permanent dipole moment, even then they may produce change in dipole moment during vibration because they contain bonds, which have dipole moments. For example CO_2.

In equilibrium position, symmetric stretching of CO_2 has $\mu = 0$ but during asymmetric stretching and bending, $\mu \neq 0$ and therefore, in these two cases, CO_2 will be IR active.

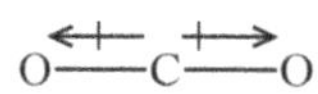

Equilibrium position

$\mu = 0$

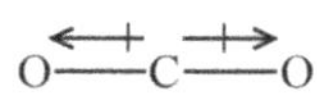

Symmetric stretching (ν_s)

$\mu = 0$ (IR inactive)

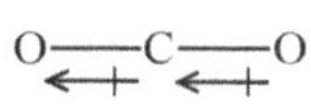

Asymmetric stretching

$\mu \neq 0$

IR active

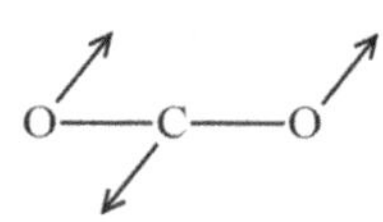

Bending vibration

$\mu \neq 0$

IR active

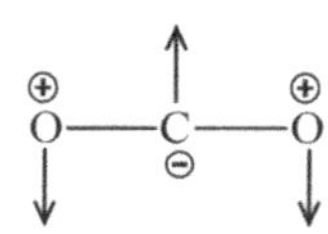

Bending vibration

$\mu \neq 0$

IR active

The allowed and forbidden transitions in term of the symmetry of ground and excited wave functions and the symmetry of the operation can be expressed as:

$$P = \int \psi_{ground} \text{ (Operator) } \psi_{excited} \, d\tau$$

Here, P = Transition moment integral.

P —
- Zero (Forbidden transition)
- Non-zero (Allowed transition)

A molecule need not possess a permanent dipole moment for absorption of infrared radiation. Only a change in dipole moment is necessary during a vibration. Using the dipole moment operator μ for IR and polarizability tensor α for Raman, the transition moment integral (P_{gi}) for absorption in IR can be written as:

$$P_{gi} = \int \varphi_v^g \mu \varphi_v^i \, d\tau \tag{8.1}$$

where, φ_v^g refers to the vibrational wave function in the ground state and φ_v^i refers to the vibrational wave function in the ith excited state. μ can be written as $\mu = \mu_x + \mu_y + \mu_z$, where these refer to the components of the dipole moment operator along the three axes X-, Y- and Z-, respectively. They can be written as:

$$\mu_x = e.x; \ \mu_y = e.y; \text{ and } \mu_z = e.z$$

where e refers to the electronic charge and *x*, *y*, and *z* refer to the Cartesian coordinates. Thus, if any one of these components changes during the vibration, then transition moment integrals can be written as:

$$P_{gi}(x) = e \int \varphi_v^g \mu_x \varphi_v^i \, d\tau \tag{8.2}$$

$$P_{gi}(y) = e \int \varphi_v^g \mu_y \varphi_v^i \, d\tau \tag{8.3}$$

$$P_{gi}(z) = e \int \varphi_v^g \mu_z \varphi_v^i \, d\tau \tag{8.4}$$

The integral $e\int_{-\infty}^{\infty} \varphi_v^g \, x \, \varphi_v^i \, d\tau$ is non-zero, if the direct product representation is totally symmetric. The same procedure can be used to find out other integrals (i.e., these may have non-zero value or not). For a fundamental transition ($v = 0 \rightarrow 1$) to occur by absorption of infrared radiation, it is necessary that one of the integrals be non-zero. Let us consider the integral:

$$c\int_{-\infty}^{\infty} \varphi_v^g \, x \, \varphi_v^i \, d\tau \tag{8.5}$$

The direct product of the symmetry species of the integrand is totally symmetric only, if the integral $\int e\ \psi_0^g\ \hat{x}\ \psi_1^i\ d\tau$ is not equal to zero (non-zero). This condition occurs only, when $\psi_1{}^i$ function has the same symmetry as x.

In this integral, $\psi_0{}^g$ is ground state vibration wave function, which is totally symmetric for all molecules except free radicals and $\psi_1{}^i$ is excited wave function, which has all symmetry of the normal mode.

Here $\varphi_v^g \sin v = 0$ and

$$\varphi_i^g \sin v = 1$$

Thus, a fundamental will be IR active, if the excited normal model modes have the same symmetry as one of the Cartesian coordinates.

A vibrational mode in a molecule will be Raman active, if the polarizability of the molecule changes during the vibration. Using the polarizability operator α, the transition moment integral can be written as:

$$P_{gi} = \int_{-\infty}^{\infty} \varphi_v^g\ \alpha\ \varphi_v^i\ d\tau \tag{8.6}$$

The integral $\int_{-\infty}^{+\infty} e\ \psi_v^o\ \hat{\alpha}\ \psi_v^i\ d\tau$ must be non-zero for a vibrational mode to be Raman active and it is possible only, when direct product representation of the integrand leads to totally symmetric representation or the direct product representation containing a totally symmetric irreducible representation. The polarizability operation has axis component and the symmetric species of the component is same as the binary product of Cartesian coordinates, i.e., x^2, y^2, z^2 xy, yz and xz. Other combinations like $x^2 - y^2$, $y^2 + z^2$, etc., are presented in the character table of a point group, to which the molecule belongs. Therefore, by using character table, the symmetry species of polarizability operator can be identified.

Thus, a fundamental transition will be Raman active, if the normal mode involved belongs to same representation as one or more components of the polarizability tensor of the molecule. Here α's are components of the polarizability tensor or we can say fundamental transition is Raman active, if the vibration has the same irreducible representation as one of the quadratic or binary Cartesian coordinates. If the electrical vector of the electromagnetic radiation can interact with the oscillating dipole moment of the molecule resulting in resonance, then the molecule gives rise to the spectrum during that vibration.

When a molecule contains two equal but opposite charges (± q), which are separated by a finite distance r, then the electric dipole moment may be given by the relation $\mu = qr$, (μ is a vector quantity and its unit is Debye D ($1D = 10^{-18}$ e.s.u.). It has three components μ_x, μ_y and μ_z.

$$\mu_t = \mu_p + \mu_i \tag{8.7}$$

The condition for infrared activity is that at least one of the dipole moment component derivatives ($\mu_1 = \mu_x$, μ_y or μ_z) with respect to the normal coordinate ϕ, measured at equilibrium position, should be non-zero.

$$\left(\frac{\partial \mu_i}{\partial \varphi}\right)_o \neq 0 \tag{8.8}$$

Atoms and molecules consist of collections of oppositely charged particles, whose relative positions can be altered by the application of an external electric field. This alteration leads to an electric dipole moment (μ_{ind}) being induced in the system. The magnitude of this induced moment will be proportional to the applied electric field (E), i.e., $\mu_{ind} = \alpha E$. The proportionality constant α is called the polarizability of the molecule. As molecules vibrate, the polarizability (α) of the molecule changes. The normal mode is Raman active only, when the polarizability change with the normal coordinate at the equilibrium configuration is non-zero, i.e.,

$$\frac{\partial \alpha}{\partial \varphi} \neq 0$$

For atoms, where the symmetry is spherical, the polarizability will be same in all directions (isotropic). This can be expressed by scalar quantity whereas for molecules with less than spherical symmetry, the polarizability in all directions is not same (anisotropic) and it is described by a tensor (a square matrix). thus, one can write

$$\begin{bmatrix} \mu_{ind}(x) \\ \mu_{ind}(y) \\ \mu_{ind}(z) \end{bmatrix} = \begin{bmatrix} \alpha_{xx} & \alpha_{xy} & \alpha_{xz} \\ \alpha_{yx} & \alpha_{yy} & \alpha_{yx} \\ \alpha_{zx} & \alpha_{zy} & \alpha_{zz} \end{bmatrix} \begin{bmatrix} E_x \\ E_y \\ E_z \end{bmatrix}$$

where
$$\alpha_{xx} = \left(\frac{\partial \mu_x}{\partial E_x}\right)$$

$$\alpha_{xy} = \left(\frac{\partial \mu_x}{\partial E_y}\right)$$

$$\alpha_{xz} = \left(\frac{\partial \mu_x}{\partial E_z}\right)$$

Since polarizability tensor is symmetric, i.e., $\alpha_{ij} = \alpha_{ji}$ (such as $\alpha_{xy} = \alpha_{yx}$, $\alpha_{yz} = \alpha_{zy}$, etc.), only six of the nine components are distinct, i.e., α_{xx}, α_{yy}, α_{zz}, α_{xy}, α_{yz} and α_{zx}. In order for the vibration to be Raman active, the change is polarizability of the molecule with respect to vibrational motion must not be zero at equilibrium position of the vibration.

8.4 THE MUTUAL EXCLUSION RULE

Consider a molecule, which has a center of symmetry (i). Point group of molecules with this element of symmetry has two sets of irreducible representations. The representations, which are symmetric with respect to inversion are called g representation. The representations, which are antisymmetric to inversion are called u representations. Let us consider the inversion of a Cartesian coordinate *x* through the center of inversion. The coordinate *x* becomes $-x$. Therefore, all representations generated by *x*, *y* or *z* must belong to a u representation. On the other hand, the product of two coordinates, (i.e., xy, yz, zx) does not change sign on applying inversion operation. Therefore, it follows that all such binary products, which represent the components of the polarizability tensor, belong to g representations.

From these rules, we can conclude that in centrosymmetric molecules, only fundamental modes belonging to g representations can be Raman active and only fundamental modes belonging to u representations can be infra-red active. This rule is called mutual exclusion rule. It is also obvious that the same must be true for other transitions besides fundamentals, since the reasoning is completely general. Another way of explaining this rule is as follows:

If a molecule has a center of symmetry, then any vibration that is active in the IR, will be inactive in the Raman and vice versa. Therefore, one can infer that a molecule has no center of symmetry, if the same vibration appears in both; IR and Raman spectra.

This rule is very important to get structural details. If center of symmetry is present, then there will be no common band in IR and Raman spectra. This is because Raman active vibration may be very weak to be noticed. But if in a case, center of symmetry is absent in the molecule, then certainly there will be some common vibration (band) in IR and Raman spectra.

Let us take examples of CO_2 and N_2O. Symmetric stretching vibration is IR inactive, but Raman active in CO_2, which clearly indicates the presence of center of symmetry in the molecule. In N_2O, there is a uncertainty to arrive at any conclusion related to center of symmetry, because some vibrations are present in both; Raman and IR spectra. It means, N_2O molecule is non-linear.

Molecule	**Point group**	**Symmetry species**	**IR Active**	**Raman active**
CO_2	$D_{\infty h}$	A_{1g}, A_{1u}, E_{1u}	A_{1u}, E_{1u}	A_{1g}
C_2H_2	$D_{\infty h}$	$2A_{1g}$, E_{1g}, E_{1u}, A_{1u}	E_{1u}, A_{1u}	A_{1g}, E_{1g}
N_2F_2	C_{2h}	$3A_g$, A_u, $2B_u$	A_u, B_u	Ag

Irreducible representation

Subscript u → IR Active

Subscript g → Raman active

According to selection rule

$$\frac{\partial \mu_i}{\partial \varphi} \neq 0 \qquad \frac{\partial \alpha_{ij}}{\partial \varphi} \neq 0$$

$$(\Delta v = \pm 1) \qquad \Delta v = \pm 1$$

(For IR spectra) (For Raman spectra)

Here ϕ represents displacement coordinate. ϕ is a measure of change in bond angle from equilibrium position for bending vibration and extension or compression of bond during stretching vibration.

8.5 NORMAL MODE ANALYSIS

A molecule with N atoms has a total of 3 N degrees of freedom, out of which 3 degrees of freedom correspond to translational motion and 3 more degrees of freedom for non-linear and 2 for linear molecule correspond to rotational motion. The remaining, (3 N – 6) for non-linear and (3 N – 5) for linear molecules, degrees of freedom is due to vibrational motion, and these are called normal modes. These are further divided into stretching and bending modes. In order to determine the symmetry of these normal modes of vibration, a kind of representation is considered for the molecule by choosing a variety of basis systems. There are two group theoretical methods

- Cartesian coordinate method; and
- Internal coordinate method.

In Cartesian coordinate method, the complete set of 3 N vectors are considered for representation in the molecular point group, in which a set of three will be located along *X*-, *Y*- and *Z*-axes on each atom of the molecule.

Firstly, a total reducible representation for 3 N degrees of freedom is determined (Γ_{3N}). It is obtained by performing symmetry operation on this vector; which forms representation basis and represent vibrational mode. The character of particular symmetry operation in Γ_{3N} representation is determined by product of number of unshifted atom (NUA) and contribution per unshifted atom, i.e., only vector on unshifted atom is considered.

X (R) = NUA × Contribution per unshifted atom (equivalent to character of full matrices for the operation)

Now after getting Γ_{3N}, i.e., total reducible representation, reduction formula is used to obtain irreducible representation corresponding to the symmetries of the 3 N degrees of freedom. Translational and rotational motion can be obtained, using character table of point group to which molecule belongs. Then irreducible representation for 3 N – 6 for non-linear molecule and 3 N – 5 (for linear molecule), normal modes are determined by subtracting transitional and rotational representation from reducible representation (Γ_{3N}).

$$\Gamma_{vib.} = \Gamma_{3N} - \Gamma_{Trans.} - \Gamma_{Rot.} \tag{8.9}$$

The normal mode analysis involves determination of symmetry by taking each normal mode of vibration as basis for irreducible representation of the point group of the molecule.

8.5.1 GENERAL SEQUENCE OF STEPS FOR NORMAL MODE ANALYSIS

Group theoretical analysis of vibrations follows a particular sequence of steps and these are:

(i) The geometry of the given molecule has to be assessed correctly and its point group is determined. Its order and number of classes is also found.

(ii) The total number and the symmetry of the Cartesian coordinates is determined, which should be equivalent to the total degrees of freedom, i.e., 3 N.

(iii) Translational and rotational representations are subtracted from the total representation to obtain vibrational modes.

$$\Gamma_{Vib.} = \Gamma_{3N} - \Gamma_{Trans.} - \Gamma_{Tot.}$$

(iv) The internal coordinates are classified and defined into some sets and their individual symmetry is determined.

(v) The spurious modes are recognized and eliminated, which generally occur in internal coordinates and their place in new coordinates are defined.

(vi) I.R. and Raman spectral activity of all the vibrational modes so obtained is determined.

(vii) The normal modes of vibration based on the skeletal framework of the molecule are written and represented as per their symmetry.

8.6 AB_2 MOLECULES (C_{2v} POINT GROUP)

Examples of this group are H_2O or SO_2.

8.6.1 CARTESIAN COORDINATE METHOD

Water molecule belongs to C_{2v} point group. Each atom of molecule has Cartesian coordinates (x, y and z). Z-axis is considered as the principal axis and yz plane is the molecular plane. Since H_2O has three atoms and each atom has 3 displacement vectors. So, H_2O molecule will have in total 9 vectors.

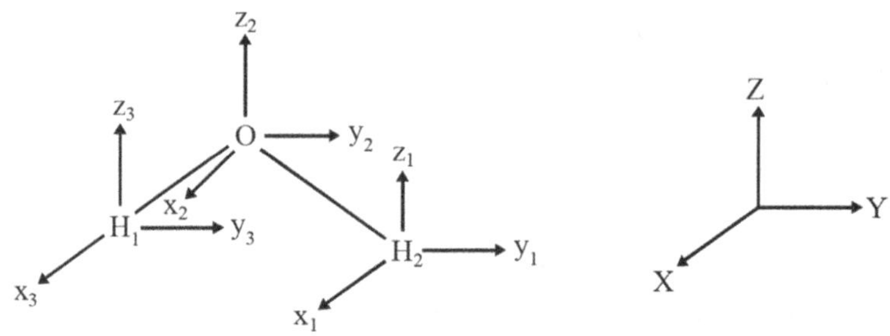

Operation E

The identity operation simply means doing nothing. Therefore, all the coordinates will remain same after operation E also, i.e., x_1, y_1, z_1 x_2, y_2, ..., etc. will be x_1, y_1, z_1, x_2, y_2, ..., etc.

$$
\begin{array}{lll}
x_1' \longrightarrow x_1 & y_1' \longrightarrow y_1 & z_1' \longrightarrow z_1 \\
x_2' \longrightarrow x_2 & y_2' \longrightarrow y_2 & z_2' \longrightarrow z_2 \\
x_3' \longrightarrow x_3 & y_3' \longrightarrow y_3 & z_3' \longrightarrow z_3
\end{array}
$$

Thus, this transformation can be represented as:

$$
E.\begin{bmatrix} x_1' \\ y_1' \\ z_1' \\ x_2' \\ y_2' \\ z_2' \\ x_3' \\ y_3' \\ z_3' \end{bmatrix} = \begin{bmatrix} 1 & 0 & 0 & 0 & 0 & 0 & 0 & 0 & 0 \\ 0 & 1 & 0 & 0 & 0 & 0 & 0 & 0 & 0 \\ 0 & 0 & 1 & 0 & 0 & 0 & 0 & 0 & 0 \\ 0 & 0 & 0 & 1 & 0 & 0 & 0 & 0 & 0 \\ 0 & 0 & 0 & 0 & 1 & 0 & 0 & 0 & 0 \\ 0 & 0 & 0 & 0 & 0 & 1 & 0 & 0 & 0 \\ 0 & 0 & 0 & 0 & 0 & 0 & 1 & 0 & 0 \\ 0 & 0 & 0 & 0 & 0 & 0 & 0 & 1 & 0 \\ 0 & 0 & 0 & 0 & 0 & 0 & 0 & 0 & 1 \end{bmatrix} \begin{bmatrix} x_1 \\ y_1 \\ z_1 \\ x_2 \\ y_2 \\ z_2 \\ x_3 \\ y_3 \\ z_3 \end{bmatrix}
$$

$$\chi\,[E] = 1 + 1 + 1 + 1 + 1 + 1 + 1 + 1 + 1 = 9$$

The character of E is equal to the value of 3 N degrees of freedom

$$\chi\,(E) = 3\text{ N} \quad (\text{N} = \text{Number of atoms})$$

$$9 = 9$$

Operation C_2

On two-fold rotation of molecule along Z-axis, two H atoms change their positions. (They exchange their positions, i.e., H_1 becomes H_2 and vice-versa), but oxygen atom remains unshifted. It leads to following transformation.

$$\begin{array}{lll} x_1' \longrightarrow -x_3 & y_1' \longrightarrow -y_3 & z_1' \longrightarrow z_3 \\ x_2' \longrightarrow -x_2 & y_2' \longrightarrow -y_2 & z_2' \longrightarrow z_2 \\ x_3' \longrightarrow -x_1 & y_3' \longrightarrow -y_1 & z_3' \longrightarrow z_1 \end{array}$$

Thus, the matrix can be shown as:

$$C_2 \cdot \begin{bmatrix} x_1' \\ y_1' \\ z_1' \\ x_2' \\ y_2' \\ z_2' \\ x_3' \\ y_3' \\ z_3' \end{bmatrix} = \begin{bmatrix} 0 & 0 & 0 & 0 & 0 & 0 & -1 & 0 & 0 \\ 0 & 0 & 0 & 0 & 0 & 0 & 0 & -1 & 0 \\ 0 & 0 & 0 & 0 & 0 & 0 & 0 & 0 & 1 \\ 0 & 0 & 0 & -1 & 0 & 0 & 0 & 0 & 0 \\ 0 & 0 & 0 & 0 & -1 & 0 & 0 & 0 & 0 \\ 0 & 0 & 0 & 0 & 0 & 1 & 0 & 0 & 0 \\ -1 & 0 & 0 & 0 & 0 & 0 & 0 & 0 & 0 \\ 0 & -1 & 0 & 0 & 0 & 0 & 0 & 0 & 0 \\ 0 & 0 & 1 & 0 & 0 & 0 & 0 & 0 & 0 \end{bmatrix} \begin{bmatrix} x_1 \\ y_1 \\ z_1 \\ x_2 \\ y_2 \\ z_2 \\ x_3 \\ y_3 \\ z_3 \end{bmatrix}$$

$$\chi(C_2) = (-1) + (-1) + 1 = -1$$

σ_v (xz) Operaton

On σ_v (xz) operation, the two H atoms will exchange their positions, but oxygen atom will remain unshifted. It reverses all the 4 vectors.

Thus, the matrix can be shown as:

$$\sigma_v(xz) \cdot \begin{bmatrix} x_1' \\ y_1' \\ z_1' \\ x_2' \\ y_2' \\ z_2' \\ x_3' \\ y_3' \\ z_3' \end{bmatrix} = \begin{bmatrix} 0 & 0 & 0 & 0 & 0 & 0 & 1 & 0 & 0 \\ 0 & 0 & 0 & 0 & 0 & 0 & 0 & -1 & 0 \\ 0 & 0 & 0 & 0 & 0 & 0 & 0 & 0 & 1 \\ 0 & 0 & 0 & 1 & 0 & 0 & 0 & 0 & 0 \\ 0 & 0 & 0 & 0 & -1 & 0 & 0 & 0 & 0 \\ 0 & 0 & 0 & 0 & 0 & 1 & 0 & 0 & 0 \\ 1 & 0 & 0 & 0 & 0 & 0 & 0 & 0 & 0 \\ 0 & -1 & 0 & 0 & 0 & 0 & 0 & 0 & 0 \\ 0 & 0 & 1 & 0 & 0 & 0 & 0 & 0 & 0 \end{bmatrix} \begin{bmatrix} x_1 \\ y_1 \\ z_1 \\ x_2 \\ y_2 \\ z_2 \\ x_3 \\ y_3 \\ z_3 \end{bmatrix}$$

$$\chi(\sigma_v(xz) = 1 + (-1) + 1 = 1$$

σ_v (yz) operation

In σ_{yz} operation, all the atoms remained on same position but all their *x* vectors are reversed.

Therefore, the matrix can be represented as:

$$\sigma_v(yz).\begin{bmatrix} x_1' \\ y_1' \\ z_1' \\ x_2' \\ y_2' \\ z_2' \\ x_3' \\ y_3' \\ z_3' \end{bmatrix} = \begin{bmatrix} -1 & 0 & 0 & 0 & 0 & 0 & 0 & 0 & 0 \\ 0 & 1 & 0 & 0 & 0 & 0 & 0 & 0 & 0 \\ 0 & 0 & 1 & 0 & 0 & 0 & 0 & 0 & 0 \\ 0 & 0 & 0 & -1 & 0 & 0 & 0 & 0 & 0 \\ 0 & 0 & 0 & 0 & 1 & 0 & 0 & 0 & 0 \\ 0 & 0 & 0 & 0 & 0 & 1 & 0 & 0 & 0 \\ 0 & 0 & 0 & 0 & 0 & 0 & -1 & 0 & 0 \\ 0 & 0 & 0 & 0 & 0 & 0 & 0 & 1 & 0 \\ 0 & 0 & 0 & 0 & 0 & 0 & 0 & 0 & 1 \end{bmatrix} \begin{bmatrix} x_1 \\ y_1 \\ z_1 \\ x_2 \\ y_2 \\ z_2 \\ x_3 \\ y_3 \\ z_3 \end{bmatrix}$$

$$\chi(\sigma_v(yz) = (-1) + 1 + 1 + (-1) + 1 + 1 + (-1) + 1 + 1 + 1 = 3$$

The set of characters obtained is a 9-dimensional representation and is reducible. Hence, Γ_{3N} in C_{2v} (H_2O) is:

C_{2v}	E	C_2	$\sigma_v(xz)$	$\sigma_v(yz)$
Γ_{3N}	9	−1	1	3

The character table for C_{2v} is

C_{2v}	E	C_2	$\sigma_v(xz)$	$\sigma_v(yz)$		
A_1	1	1	1	1	z	x^2, y^2, z^2
A_2	1	1	1	1	R_z	xy
A_3	1	−1	1	−1	x, R_y	xz
A_4	1	−1	−1	1	y, R_x	yz

Using the reduction formula:

$$a_i = \frac{1}{h} \Sigma n_R\, \chi_i(R)\, \chi_j(R)$$

$$a_{A1} = \frac{1}{4}\,[1 \times 1 \times 9 + 1 \times 1 \times (-1) + 1 \times 1 \times 1 + 1 \times 1 \times 3] = 3$$

$$a_{A2} = \frac{1}{4}\,[1 \times 1 \times 9 + 1 \times 1 \times (-1) + 1 \times (-1) \times 1 + 1 \times (-1) \times 3] = 1$$

$$a_{B_1} = \frac{1}{4}\,[1 \times 1 \times 9 + 1 \times (-1) \times (-1) + 1 \times 1 \times 1 + 1 \times (-1) \times 3] = 2$$

$$a_{B_2} = \frac{1}{4}\,[1 \times 1 \times 9 + 1 \times (-1) \times (-1) + 1 \times (-1) \times 1 + 1 \times 1 \times 3] = 3$$

Thus $\Gamma_{3N} = 3\,A_1 + A_2 + 2\,B_1 + 3\,B_2$

The irreducible representation for translational and rotational representations can be known using third column of character table of C_{2v}. The Cartesian vectors *x*, *y*, and *z* transfer as B_1, B_2 and A_1, respectively and R_x, R_y and R_z rotational vectors transfer as B_2, B_1 and A_2, respectively. Therefore,

$$\Gamma_{Trans.} = A_1 + B_1 + B_2$$
$$\Gamma_{Rot.} = A_2 + B_1 + B_2$$

Thus, the contribution due to vibrational motion is given as:

$$\Gamma_{vib} = \Gamma_{3N} - (\Gamma_{Trans} + \Gamma_{Rot})$$
$$= 3\,A_1 + A_2 + 2\,B_1 + 3\,B_2 - (A_1 + B_1 + B_2 + A_2 + B_1 + B_2)$$
$$\Gamma_{vib} = 2\,A_1 + B_2$$

Total degrees of freedom will be 3 N – 6 = (3 *x* 3) – 6 = 3. Here, out of 3 modes of vibration, two belong to A_1 irreducible representation while one belongs to B_1 irreducible representation. Each vibrational mode form the basis of irreducible representation of the point group of the molecule and by performing the symmetry operation of the point group on each atom vibrational mode, the symmetry of each mode can be determined.

Vibration mode	E	C_2	$\sigma_v(xz)$	$\sigma_v(yz)$	
ν_1	1	1	1	1	A_1
ν_2	1	1	1	1	A_1
ν_3	1	–1	–1	1	B_2

$\chi(C_2) = -1$

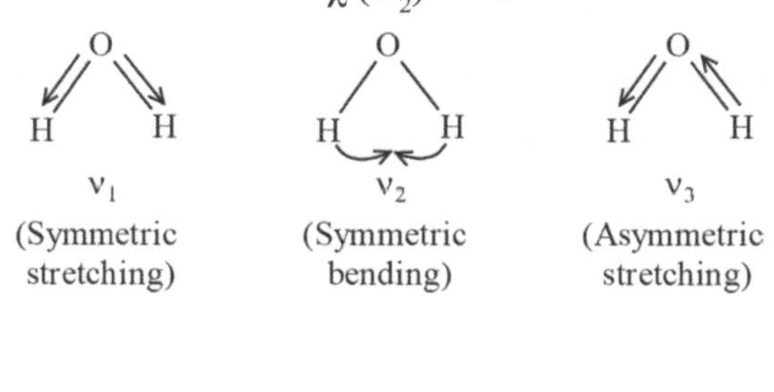

Simply by direction of the arrow, the character of each operation is determined. From the table, it can be easily found out that ν_1, ν_2, and ν_3 vibration mode belongs to symmetry A_1, A_1 and B_2, respectively and therefore, once the symmetry of normal mode of vibration is known, one can find out whether the vibration mode is IR or Raman active or not by using the selection rule.

Let us understand it by using example of H_2O molecule. The integral $\int \psi_0\, \mu\, \psi_1\, d\tau$ can be found out in terms of symmetry by direct product of ψ_0, μ and ψ_1.

ψ_0 belongs to A_1 (totally symmetric representation), μ belongs to irreducible representation to which *x*, *y*, *z* belongs. ψ_1 belongs to symmetry of any vibration mode for example ν_3 (asymmetric stretching), which belongs to B_2 symmetry.

$$\underset{\text{(Integrand)}}{\psi_o\ \mu\ \psi_1} = \underset{A_1}{\underset{\downarrow}{\psi_o}}\ \underset{\begin{bmatrix} B_1\,(x) \\ B_2\,(y) \\ A_1\,(z) \end{bmatrix}}{\underset{\downarrow}{\mu}}\ \underset{B_2}{\underset{\downarrow}{\psi_1}}$$

$$\begin{bmatrix} A_1 . B_1 . B_2 \\ A_1 . B_2 . B_2 \\ A_1 . A_1 . B_2 \end{bmatrix} = \begin{bmatrix} B_1 . B_2 \\ B_2 . B_2 \\ A_1 . B_2 \end{bmatrix} = \begin{bmatrix} A_2 \\ A_1 \\ B_2 \end{bmatrix}$$

ν_3 vibration mode is IR active, because the direct product is totally symmetric irreducible representation. In the same way, ν_1, and ν_2 are also IR active. Thus, three absorption bands can be observed in H_2O molecule.

This method is quite cumbersome and time consuming as the matrices are 3 N × 3 N dimensional. However, there is a simple way also to obtain the characters without going through the details of applying each of the group operations to each of the 3 N Cartesian coordinates. Now, the matrix contributing to the overall characters lies on the diagonal only, when the operation leaves atom unchanged or unshifted. Therefore, the problem of finding the character for each class of operation is simplified to the level of simply determining the number of atoms unshifted (NUA) during the operation and it is multiplied by character of the unshifted atom. Thus, the overall character is then equal to the character of number of such unshifted atoms multiplied by the per unshifted atom (UA). The character of unshifted atom depends on change in *X*-, *Y*-, *Z*-axes on performing symmetry operation. Its value is + 1, when axis remains unchanged and –1, when direction of axis is reversed.

Now we know that matrices for various elements of symmetry are as:

$$\overset{E}{\begin{bmatrix} 1 & 0 & 0 \\ 1 & 0 & 0 \\ 1 & 0 & 0 \end{bmatrix}} \overset{C_2(z)}{\begin{bmatrix} \cos\theta & \sin\theta & 0 \\ -\sin\theta & \cos\theta & 0 \\ 0 & 0 & 1 \end{bmatrix}}$$

$$\overset{S_n(z)}{\begin{bmatrix} \cos\theta & \sin\theta & 0 \\ -\sin\theta & \cos\theta & 0 \\ 0 & 0 & -1 \end{bmatrix}} \overset{\sigma_v(xz)}{\begin{bmatrix} 1 & 0 & 0 \\ 0 & 1 & 0 \\ 0 & 0 & -1 \end{bmatrix}}$$

$$\overset{i}{\begin{bmatrix} -1 & 0 & 0 \\ 0 & -1 & 0 \\ 0 & 0 & -1 \end{bmatrix}} \quad \overset{\sigma_v(yz)}{\begin{bmatrix} -1 & 0 & 0 \\ 0 & 1 & 0 \\ 0 & 0 & 1 \end{bmatrix}} \quad \overset{\sigma_v(xz)}{\begin{bmatrix} 1 & 0 & 0 \\ 0 & -1 & 0 \\ 0 & 0 & 1 \end{bmatrix}}$$

The characters of all these basic matrices can be tabulated as:

Symmetry element	E	$C_n(z)$	$S_n(z)$	σ	i
χ (R)/UA	3	$1+2\cos\theta$	$-1+2\cos\theta$	1	–3

The total character for each class of operation, χ(R), can be easily obtained by multiplying χ (R)/UA with the number of unshifted atom (NUA). Thus.

$$\chi(R) = (NUA)\, x\, (\chi(R)/UA)$$

For water molecule:

	E	$C_2(z)$	$\sigma_v(xz)$	$\sigma_v(yz)$
χ (R)/UA	3	–1	1	1

	E	$C_2(z)$	$\sigma_v(xz)$	$\sigma_v(yz)$
NUA	3	1	1	3

There Γ_{3N} (Total) = NUA x χ (R)/UA

C_{2v}	E	$C_2(z)$	$\sigma_v(xz)$	$\sigma_v(yz)$
Γ_{3N}	9	–1	1	3

8.6.2 INTERNAL COORDINATE METHOD

The possible internal coordinate or internal displacement coordinates can be classified as bond vectors: r_1 and r_2 (O–H bonds) = 2 and bond angles: α (∠ H_1OH_2) = 1 and hence, the total number of internal coordinates = 2 (r_1 and r_2) + 1 (α) = 3.

For symmetry identification of the normal modes in terms of the these internal coordinates, a series of matrix representations for each of the C_{2v} class will have to be worked out. If the changed coordinates after the operation can be represented as $r_1 \rightarrow r_{1'}$, $r_2 \rightarrow r_{2'}$ and $\alpha \rightarrow \alpha'$. Then,

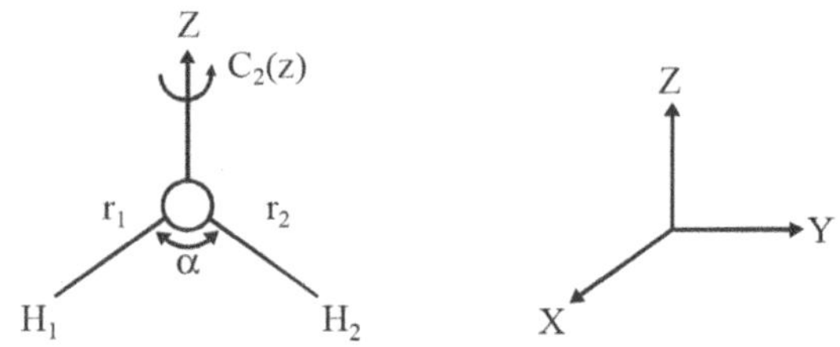

Then, one can write:

$$\begin{bmatrix} r_1' \\ r_2' \\ \alpha' \end{bmatrix} = \begin{bmatrix} \text{Transformation} \\ \text{matrix of} \\ \text{coefficients} \end{bmatrix} \begin{bmatrix} r_1 \\ r_2 \\ \alpha \end{bmatrix}$$

Thus,

$$\begin{bmatrix} r_1 \\ r_2 \\ \alpha \end{bmatrix} = \overset{E}{\begin{bmatrix} 1 & 0 & 0 \\ 0 & 1 & 0 \\ 0 & 0 & 1 \end{bmatrix}} \begin{bmatrix} r_1 \\ r_2 \\ \alpha \end{bmatrix}$$

$$\chi(E) = 3$$

$$\begin{bmatrix} r_1' \\ r_2' \\ \alpha' \end{bmatrix} = \overset{C_2}{\begin{bmatrix} 0 & 1 & 0 \\ 1 & 0 & 0 \\ 0 & 0 & 1 \end{bmatrix}} \begin{bmatrix} r_1 \\ r_2 \\ \alpha \end{bmatrix}$$

$$\chi(C_2) = 1$$

$$\sigma_v(xz)$$

$$\begin{bmatrix} r_1' \\ r_2' \\ \alpha' \end{bmatrix} = \begin{bmatrix} 0 & 1 & 0 \\ 1 & 0 & 0 \\ 0 & 0 & 1 \end{bmatrix} \begin{bmatrix} r_1 \\ r_2 \\ \alpha \end{bmatrix}$$

$$\chi(\sigma_v(xz)) = 1$$

$$\sigma_v(yz)$$

$$\begin{bmatrix} r_1' \\ r_2' \\ \alpha' \end{bmatrix} = \begin{bmatrix} 1 & 0 & 0 \\ 0 & 1 & 0 \\ 0 & 0 & 1 \end{bmatrix} \begin{bmatrix} r_1 \\ r_2 \\ \alpha \end{bmatrix}$$

$$\chi(\sigma_v(yz)) = 3$$

The character of the symmetry operation thus obtained are:

C_{2v}	E	$C_2(z)$	$\sigma_v(xz)$	$\sigma_v(yz)$
Γ_{Int}	3	1	1	3

Using the standard reduction formula and the character table of C_{2v}, Γ_{Int} can be calculated as:

$$\Gamma_{Int} = 2\,A_1 + B_2$$

The result is same as that obtained in the case of Γ_{Vib} from Cartesian coordinate method. Since no symmetry operation of this group interchanges bond vectors (r) with bond angle (α), the two vectors r_1 and r_2 form an independent basis for 2 x 2 representation, Γ_r, and Γ_α form a basis for representation

Hence, $$\Gamma_{Int} = \Gamma_\alpha + \Gamma_r$$

The matrix can also be represented in form of 2 x 2 matrices for bond vector as:

$$\text{E}\quad \begin{bmatrix} r_1' \\ r_2' \end{bmatrix} = \begin{bmatrix} 1 & 0 \\ 0 & 1 \end{bmatrix} \begin{bmatrix} r_1 \\ r_2 \end{bmatrix}$$

$$\chi(R) = 1 + 1 = 2$$

$$C_2\quad \begin{bmatrix} r_1' \\ r_2' \end{bmatrix} = \begin{bmatrix} 0 & 1 \\ 1 & 0 \end{bmatrix} \begin{bmatrix} r_1 \\ r_2 \end{bmatrix}$$

$$\chi(R) = 0 + 0 = 0$$

$$\sigma_v(xz)\quad \begin{bmatrix} r_1' \\ r_2' \end{bmatrix} = \begin{bmatrix} 0 & 1 \\ 1 & 0 \end{bmatrix} \begin{bmatrix} r_1 \\ r_2 \end{bmatrix}$$

$$\chi(R) = 0 + 0 = 0$$

$$\sigma_v(yz)\quad \begin{bmatrix} r_1' \\ r_2' \end{bmatrix} = \begin{bmatrix} 1 & 0 \\ 0 & 1 \end{bmatrix} \begin{bmatrix} r_1 \\ r_2 \end{bmatrix}$$

$$\chi(R) = 1 + 1 = 2$$

and matrices of 1 x 1 for bond angle

	E	C_2	σ_{xz}	σ_{yz}
$[\alpha]$	$[1][\alpha]$	[1]	[1]	[1]
$\chi(R)$	1	1	1	1

Finally, the result can be represented as:

C_{2v}	E	$C_2(z)$	$\sigma_v(xz)$	$\sigma_v(yz)$
Γ_r	2	0	0	2
Γ_α	1	0	1	1

Using the reduction formula or by inspecting the character table, $\Gamma_\alpha = A_1$ and $\Gamma_r = A_1 + B_2$.

Internal coordinate method has advantageously classified the nature of fundamentals into bond stretching and bending types.

(i) Bond stretching mode: $\Gamma_r = A_1$ and B_2;
(ii) Bending mode: $\Gamma_\alpha = A_1$.

The three pure vibrational modes (3 N – 6 = 3) for H_2O are:

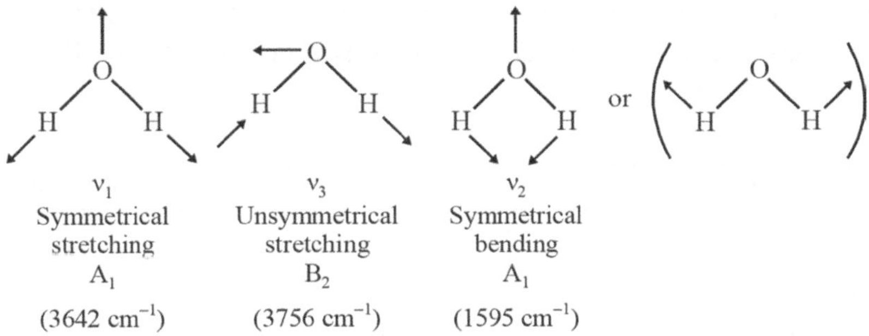

(i) One pure bond stretch, i.e., B_2 of O-H bonds (v_3)
(ii) Two of A_1 symmetry – Combination of O-H bond stretch and H-O-H bond angle deformation. with v_1 and v_2 mode (A_1) have intermixed character. It means that one mode has some character of other. The asymmetric stretching with v_3 (B_2) is only pure mode, which is important for determination of structure of molecule.

It can be assumed that v_1 (364 cm^{-1}) and v_2 (1595 cm^{-1}) are different modes. But difference in observation of calculated and experimental value shows that there is interaction between two normal modes because they have same symmetry.

Once the fundamental mode of vibration is determined. It is now easy to predict, which of them will be active in the infra-red and which one in Raman spectra. The functions *x*, *y* and *z* in the character tables represent the irreducible representations or symmetry species to which the corresponding components μ_x, μ_y and μ_z of the dipole moment belong. Infra-red transition transitions originate from the dipole moment operator, whose symmetry is same as the vector along X-, Y- and Z-axes. These are generally asymmetric vibrations from character table of C_{2v}, *z* transforms as A_1 species. Therefore, the fundamentals belonging to A_1 species (v_1 and v_2) is infra-red active. Similarly, fundamentals belonging to B_1 and B_2 should also be infra-red active as they correspond to *x* and *y* functions, respectively. However, in H_2O molecule A_1 and B_2 type of fundamentals occur and both of them are infra-red active.

Since Raman spectral absorption does not depend on the dipole moment operator, but on the polarizability operator (α), which contains binary and quadratic Cartesian functions such as x^2, y^2, z^2, xz, yz, xy and their combinations. The normal mode species will have to be checked against these functions. If any of these functions are found against the irreducible representation representing the fundamental, then the fundamental is said to be Raman active. From C_{2v} character table, it is found that A_1 and B_2 both are Raman active. Thus,

Mode	Infra-red	Raman	Nature of vibration
2 A_1	(+) Active	(+) Active	Mixed
B_2	(+) Active	(+) Active	Pure

The number of coincidences = 3 (Number of vibrations common in both)

Thus, there are three irreducible representations and three Raman (coincident with the infra-red) bands.

8.7 AB_3 PYRAMIDAL MOLECULES (C_{3v} GROUP)

Example of this type of molecules/ions are NH_3, $POCl_3$, PH_3, CH_3Cl, $CHCl_3$, etc.

8.7.1 CARTESIAN COORDINATE METHOD

The motion of atoms in NH_3 molecules are represented by a set of Cartesian coordinates. For convenience and clarity, the pyramidal NH_3 is depicted as lying in the plane of paper. But it should always be assumed that N atom is at the apex of the trigonal pyramid and the Z-axis is perpendicular to the plane of the paper and the basal triangle of three H atoms, passes through N atom and the centroid of this triangle. The number of atoms unshifted (NUA), their characters and resulting characters are as follows:

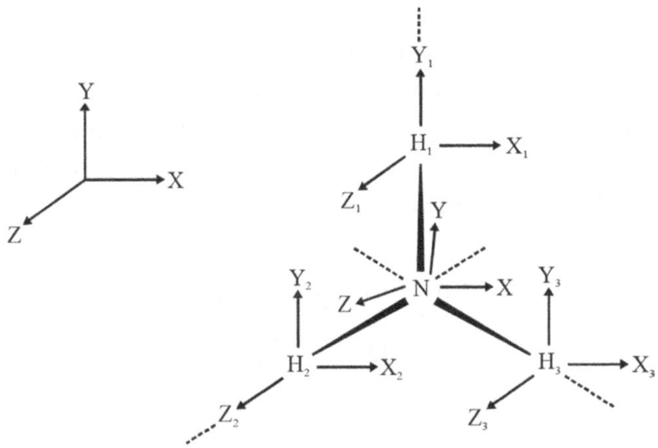

C_{3v}	E	$2C_3$	$3\sigma_v$
NUA	4	1	2
χ (R) UA	3	0	1
Γ_{3N}	12	0	2

Using the standard reduction formula, we get Γ_{3N} (Total) = $3A_1 + A_2 + 4E$

This representation includes all matrices of the molecules. Now

$$\Gamma_{Trans} = A_1(z) + E(x, y) \text{ and } \Gamma_{Rot} = A_2(R_z) + E(R_x, R_y)$$

$$\Gamma_{Trans.} = A_1 + E$$

and $\Gamma_{Rot.} = A_2 + E$

Hence $\Gamma_{Vib.} = \Gamma_{Total} - (\Gamma_{Tans.} + \Gamma_{Rot.})$

$$\Gamma_{Vib.} = 3A_1 + A_2 + 4E - (A_1 + E + A_2 + E)$$

$$\Gamma_{Vib} = 2A_1 + 2E$$

Thus, out of six modes, two vibrational modes belong to A_1 symmetry and rest four vibration modes are two pair of doubly degenerate mode, i.e., E symmetry.

8.7.2 INTERNAL COORDINATE METHOD

The internal coordinates of NH_3 consist of three bond stretch vectors and three bond angles.

Bond vectors: $r_1 - r_3$ (N–H bond) = 3

Bond angles: $\alpha_1 - \alpha_3$ (H–N–H angle) = 3

Total number of internal coordinates = 3 + 3 = 6

Since bond vectors do not exchange symmetry operations of this group, the transformation matrices corresponding to each category of internal coordinates can be separated and written as:

$$\overset{E}{\begin{bmatrix} r_1' \\ r_2' \\ r_3' \end{bmatrix} = \begin{bmatrix} 1 & 0 & 0 \\ 0 & 1 & 0 \\ 0 & 0 & 1 \end{bmatrix} \begin{bmatrix} r_1 \\ r_2 \\ \alpha \end{bmatrix}}$$

$$\chi(R) = 3$$

$$\overset{C_3}{\begin{bmatrix} r_1' \\ r_2' \\ r_3' \end{bmatrix} = \begin{bmatrix} 0 & 1 & 0 \\ 0 & 0 & 1 \\ 1 & 0 & 0 \end{bmatrix} \begin{bmatrix} r_1 \\ r_2 \\ r_3 \end{bmatrix}}$$

$$\chi(R) = 3$$

$$\overset{\sigma_v}{\begin{bmatrix} r_1' \\ r_2' \\ r_3' \end{bmatrix} = \begin{bmatrix} 1 & 0 & 0 \\ 0 & 0 & 1 \\ 0 & 1 & 0 \end{bmatrix} \begin{bmatrix} r_1 \\ r_2 \\ r_3 \end{bmatrix}}$$

$$\chi(\sigma_v) = 3$$

$$\overset{E}{\begin{bmatrix} \alpha_1' \\ \alpha_2' \\ \alpha_3' \end{bmatrix} = \begin{bmatrix} 1 & 0 & 0 \\ 0 & 1 & 0 \\ 0 & 0 & 1 \end{bmatrix} \begin{bmatrix} \alpha_1 \\ \alpha_2 \\ \alpha_3 \end{bmatrix}}$$

$$\chi(E) = 3$$

$$\overset{C_3}{\begin{bmatrix} r_1' \\ r_2' \\ r_3' \end{bmatrix} = \begin{bmatrix} 0 & 1 & 0 \\ 0 & 0 & 1 \\ 1 & 0 & 0 \end{bmatrix} \begin{bmatrix} r_1 \\ r_2 \\ r_3 \end{bmatrix}}$$

$$\chi(C_3) = 0$$

$$\sigma_v$$

$$\begin{bmatrix} r_1' \\ r_2' \\ \alpha_3' \end{bmatrix} = \begin{bmatrix} 1 & 0 & 0 \\ 0 & 0 & 1 \\ 0 & 1 & 0 \end{bmatrix} \begin{bmatrix} r_1 \\ r_2 \\ \alpha_3 \end{bmatrix}$$

$$\chi(\sigma_v) = 1$$

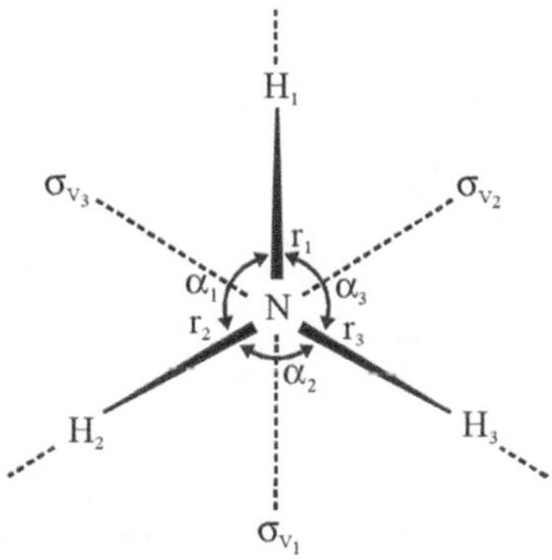

Thus, the result are:

C_{3v}	E	$2C_3$	$3\sigma_v$
Γ_r	3	0	1
Γ_α	3	0	1

The two representations can be reduced using reduction formula.

$$\Gamma_r = A_1 + E$$

$$\Gamma_\alpha = A_1 + E$$

Hence

$$\Gamma_{Int} = \Gamma_r + \Gamma_\alpha$$

$$\Gamma_{Int} = 2\,A_1 + 2\,E$$

Results obtained from internal and Cartesian coordinate methods are same.

Out of the six normal modes of vibration (3 N – 6), two are totally symmetric (A_1 type) and the other four are two pairs of double degenerate (E type) modes. Two symmetrical modes will consist of a simultaneous symmetrical stretching of all N-H bonds and a symmetry movement of all

H-N-H angles in an umbrella like manner. On the other hand, the double degenerate E modes cannot be easily understood. Further from character table, it is found that A_1 and E modes are both infra-red and Raman active. (E is combination of bond stretching and bending modes).

Mode	Infrared	Raman	Nature of vibration
2 A_1	(+) Active	(+) Active	Mixed
2 E	(+) Active	(+) Active	Mixed

The number of coincidences = 4

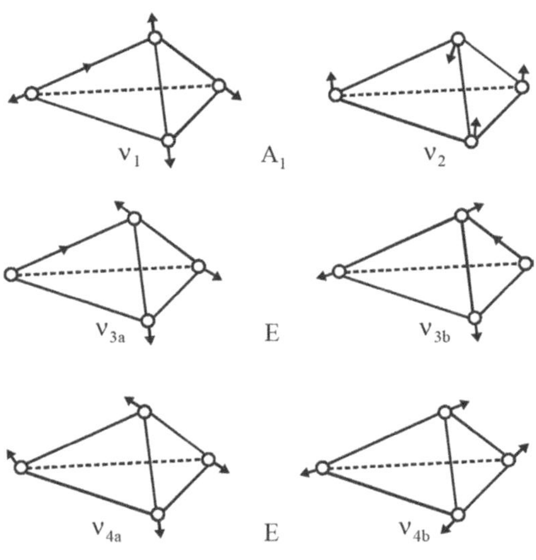

8.8 AB_4 MOLECULES (T_d POINT GROUP)

Examples of this type of molecules/ions are CH_4, BH_4^+, SiH_4, ClO_4^- or SO_4^{2-}.

Such molecules have T_d symmetry. As the molecule is non-linear, five atomic species will have 3 x 5–6 = 9 vibrational modes (degrees of internal freedom).

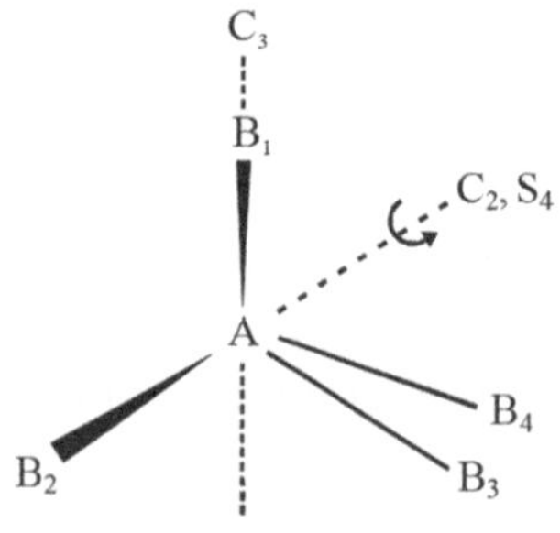

8.8.1 CARTESIAN COORDINATE METHOD

The set of 15 Cartesian displacement vectors along the Cartesian coordinate axes X-, Y-, and Z- on each atom of the molecule forms a basis for the following representation.

T_d	E	$8C_3$	$3C_2$	$6S_4$	$6\sigma_d$
NUA	5	2	1	1	3
χ (R) UA	3	0	−1	−1	1
Γ_{3N} (Total)	15	0	−1	−1	3

Using standard reduction formula, this can reduced as Γ_{3N} (Total) = $A_1 + E + T_1 + 3\,T_2$

From character table, it follows that $\Gamma_{Rot.} = T_1\ (R_x, R_y, R_z)$ and $\Gamma_{Trans} = T_2\ (x, y, z)$

Hence, the vibration modes are

$\Gamma_{Vib} = \Gamma_{3N}$ (Total) $- (\Gamma_{Rot.} + \Gamma_{Trans.}) = A_1 + E + T_1 + 3\,T_2 - (T_1 + T_2) = A_1 + E + 2\,T_2$

8.8.2 INTERNAL COORDINATE METHOD

A-B bond lengths and bond angles ∠ B-A-B form the basis for a representation to find the contribution of the internal coordinates.

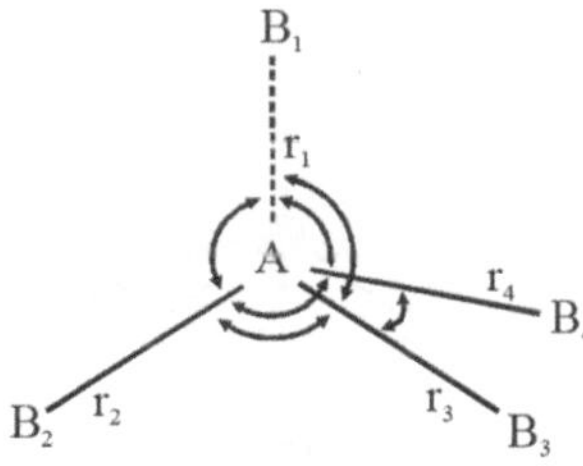

Bond Angles: As a_1 to α_6 = 6; Bond vectors: r_1 to r_4 = 4, therefore, total number of internal coordinates will be 6 + 4 = 10 (one more than 3 N – 6)

T_d	E	$8C_3$	$3C_2$	$6S_4$	6σ
Γ_r	4	1	0	0	2
Γ_α	6	0	2	0	2

Using standard reduction formula and character table of T_d, it is found that $\Gamma_r = A_1 + T_2$ and $\Gamma_\alpha = A_1 + E + T_2$ and hence $\Gamma_{Int} = \Gamma_\alpha + \Gamma_r = 2\,A_1 + E + 2\,T_2$.

It will be seen that the total dimensionality of these two representations is ten (one is excess of the correct number calculated by 3 N – 6 formula). Specifically, there is an extra A_1 representation. It is easy to determine that the representation is the one in Γ_α, although it is possible for all the four of A-B (C-H) distances (bond lengths) to change independently, it is not possible for all the six bond angles to change independently. If any five are arbitrarily changed, then the change of the sixth one is automatically fixed. For A_1 vibration, all the six angles would have to change in the same way at the same time (i.e., all increase or all decrease), and it is clearly impossible. Hence, we obtain the results that the A_1 vibration of CH_4 consists purely of C-H stretching, and the E vibration is purely HCH angle deformation, while both; bond stretching and angle bending, contribute to each of the normal vibrations of T_2 symmetry. Thus,

$$\Gamma_\alpha = E + A_2$$

Here, A_1 is not included.

IR and Raman active vibrations can be determined as:

Mode	Infrared	Raman	Nature of vibration
A_1	(–) Inactive	(+) Active	Pure stretching
E	(–) Inactive	(+) Active	Pure bending
2 T_2	(+) Active	(+) Active	Mixed

The character table also shows activities of these fundamentals.

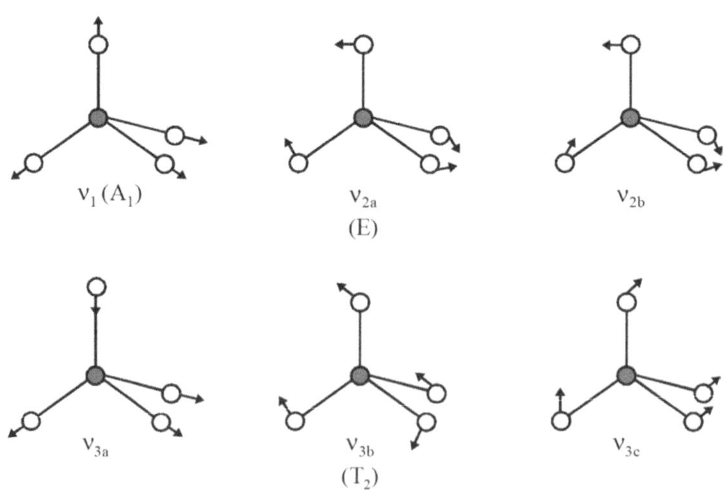

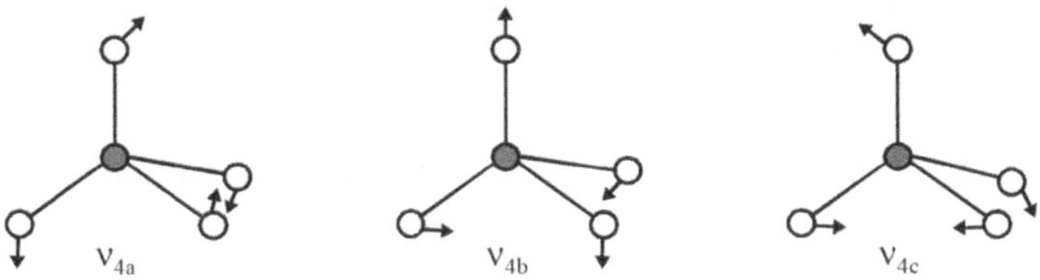

(T_2)

The spectral activity of vibration for molecules of T_d point group are represented below:

Point group	IR Active	Raman active	Polarized	Number of coincidences
T_d	2	4	1	2
	$2\,T_2$	A_1, E, $2\,T_2$	A_1	$2\,T_2$
	ν_3, ν_4	1 + 1 + 2	ν_1	ν_3, ν_4
		$\nu_1 + \nu_2 + \nu_3 + \nu_4$		

8.9 AB_6 MOLECULES (O_h POINT GROUP)

Example of this type of molecule is SF_6.

8.9.1 CARTESIAN COORDINATE METHOD

This molecule belongs to O_h point group. It has 3 × 7–6 = 15 degrees of internal freedom. The 21 Cartesian displacement vectors along *X*-, *Y*-, and *Z*-axes in each of molecule generates the representation Γ_{3N} as:

O_h	E	$8C_3$	$6C_2'$	$6S_4$	$3C_2 (=C_4^2)$	*i*	$6S_4$	$8S_6$	$3\sigma_h$	$6\sigma_d$
NUA	7	1	1	3	3	1	1	1	5	3
χ(R) / UA	3	0	–1	1	–1	–3	–1	0	1	1
Γ_{3N}(Total)	21	0	–1	3	–3	–3	–1	0	5	3

Using reduction formula, the representation Γ_{Total} reduced as follows:

$$\Gamma_{3N} = A_{1g} + E_g + T_{1g} + 3\,T_{1u} + T_{2g} + T_{2u} = 21 \text{ mode}$$

The character table of the point group shows that the rotation and translation belong, respectively to the T_{1g} and T_{1u} representation. Hence, the vibrational modes are:

$\Gamma_{Trans.} = T_{1u}\ (x, y, z)$ and $\Gamma_{Rot.} = T_{1g}\ (R_x, R_y, R_z)$

$$\begin{aligned}\Gamma_{Vib.} &= \Gamma_{Total} - (\Gamma_{Rot} + \Gamma_{Trans.}) \\ &= A_{1g} + E_g + T_{1g} + 3\,T_{1u} + T_{2g} + T_{2u} - (T_{1g} + T_{1u}) \\ &= A_{1g} + E_g + T_{2g} + 2\,T_{1u} + T_{2u} = 15 \text{ vibrational mode}\end{aligned}$$

8.9.2 INTERNAL COORDINATE METHOD

The set of six S-F bonds and the set of 12 FSF angles also forms the representation. Thus,

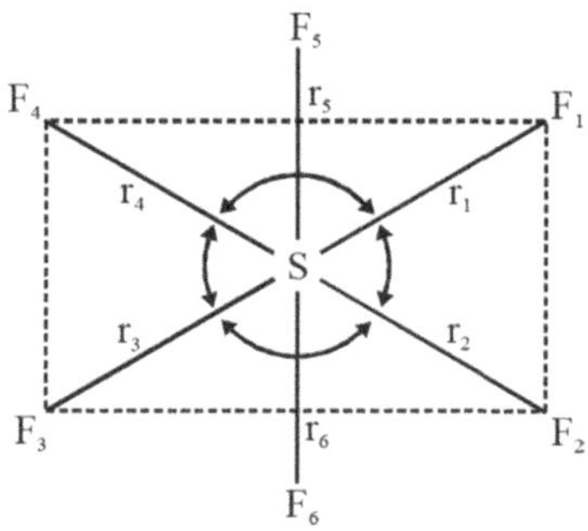

Bond vectors are r_1 to $r_6 = 6$
Bond angles are α_1 to $\alpha_{12} = 12$
Total number of internal coordinates = 6 + 12 = 18
These are 3 more than calculated; (3 N – 6) = 3 x 7–6 = 21–6 = 15
The results can be expressed in tabular form as:

O_h	E	$8C_3$	$6C_2'$	$6C_4$	$3C_2$	i	$6S_4$	$8S_6$	$3\sigma_h$	$6\sigma_d$
Γ_r	6	0	0	2	2	0	0	0	4	2
Γ_α	12	0	2	0	0	0	0	0	4	2

Using standard reduction formula and character table of point group O_h, Γ_r and Γ_α can be reduced as:

$$\Gamma_r = A_{1g} + E_g + T_{1u}$$
$$\Gamma_\alpha = A_{1g} + E_g + T_{2g} + T_{1u} + T_{2u}$$

Therefore, $\Gamma_{Int.}\ \Gamma_r + \Gamma_\alpha = 2\,A_{1g} + 2\,E_g + T_{2g} + 2\,T_{1u} + T_{2u}$

So total degrees of freedom = 18, which exceeds 15 by 3. Obviously, there is some redundancy here. Since the S-F coordinates (Γ_r) are completely independent and the before, redundancy must be entirely in Γ_{FSF} (Γ_α).

By comparing Γ_{Int}. with genuine internal modes $\Gamma_{Vib.}$, we see that the A_{1g} and E_g occurring in Γ_α are the spurious one. Therefore, $\Gamma_\alpha = T_{2g} + T_{1u} + T_{2u}$

Hence, $\Gamma_{Int.} = A_{1g} + E_g + T_{2g} + 2\,T_{1u} + T_{2u}$

Thus, we conclude that each of the two T_{1u} modes will involve a combination of bond stretching and angle deformation will be IR active (ν_3 and ν_4). The A_{1g} and E_g modes will involve only bond stretching and will be Raman active (ν_1 and ν_2), while T_{2g} (ν_5) and T_{2u} (ν_6) modes will involve only angle deformation mode. Hence, T_{2g} will be Raman active and T_{2u} mode will be inactive in IR and Raman spectra, both.

The character table also shows activities of these fundamentals.

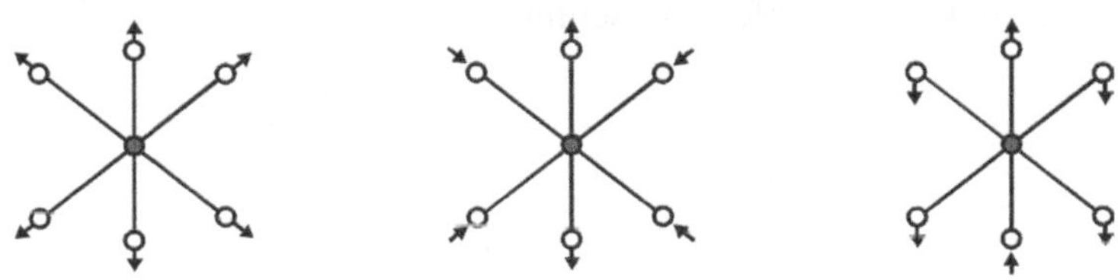

$\nu_1\,(A_{1g})$ $\nu_2\,(E_g)$ $\nu_3\,(T_{1u})$

Stretching modes

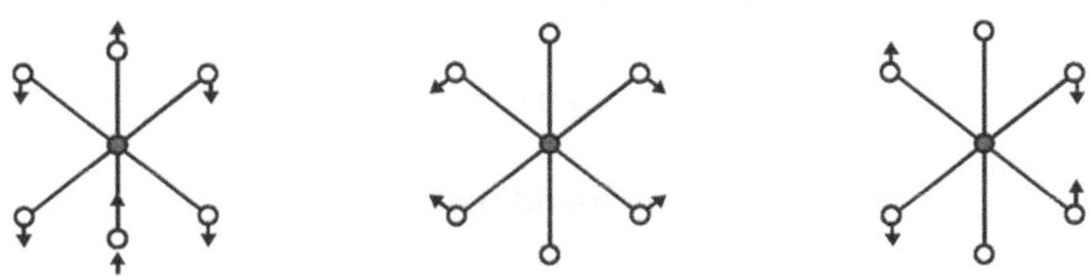

$\nu_4\,(T_{1u})$ $\nu_5\,(T_{2g})$ $\nu_6\,(T_{2u})$

Bending modes

Using character table of O_h point group, we can identify the IR and Raman active vibrations:

Mode	Infrared	Raman	Nature of vibration
A_{1g}	(–) Inactive	(+) Active	Bond stretching (pure)
E_g	(–) Inactive	(+) Active	Bond stretching (pure)
T_{2g}	(–) Inactive	(+) Active	Angle deformation (pure)
$2\,T_{1u}$	(+) Active	(–) Inactive	Mixed, i.e., stretching + bonding
T_{2u}	(+) Active	(–) Inactive	Angle deformation (pure)

From this table, we can see that mutual exclusion rule is satisfied, which means molecules have center of symmetry.

8.10 A_2B_2 MOLECULES (C_{2h} GROUP)

Example of this type of molecule is trans – N_2F_2.

This molecule has plane or but non-linear structure. It belongs to the point group C_{2h}. It is a non-linear four atomic molecule and it has $3 \times 4–6 =$ 6 degrees of internal freedom.

8.10.1 CARTESIAN COORDINATE METHOD

The set of 12 Cartesian displacement vectors for the entire molecule generates the following reducible representation:

C_{2h}	E	C_2	i	σ_h
NUA	4	0	0	4
χ (R) UA	3	−1	−3	1
Γ_{3N} (Total)	12	0	0	4

This can be reduced as follows:

$$\Gamma_{3N} = 4\,A_g + 2\,B_g + 2\,A_u + 4\,B_u$$

From the character table, it is found that

$$\Gamma_{Rot.} = A_g + B_g + B_g = A_g + 2\,B_g$$

$$\Gamma_{Trans.} = A_u + B_u + B_u = A_u + 2\,B_u$$

Thus, the genuine normal vibrations for this molecule is

$$\Gamma_{Vib.} = \Gamma_{3N}\,(\text{Total}) - (\Gamma_{Rot.} + \Gamma_{Trans.})$$

$$= 4\,A_g + 2\,B_g + 2\,A_u + 4\,B_u - (A_g + 2\,B_g + B_g + A_u + 2\,B_u)$$

$$= 3\,A_g + A_u + 2\,B_u = 6\ (\text{Normal modes})$$

8.10.2 INTERNAL COORDINATE METHOD

The nature of these six vibrations may be further specified in terms of the contribution made to each one of them by the various internal coordinates.

A_g and B_u vibrations must involve only motions within the molecular plane, since the characters of the representation A_g and B_u with respect to σ_h are positive. The A_u vibration will however, involve out-of-plane deformation, since the character of A_u with respect to σ_h is negative. Thus, we may describe the normal mode of A_u symmetry as out-of-plane (∞p) deformation. The symmetry table for this can be worked out as:

C_{2h}	E	C_2	i	σ_h	IR
$\Gamma_{\infty p}$	1	1	−1	−1	A_u

In order to treat the remaining five in-plane vibration, we need a set of five internal coordinates such that changes in them may occur entirely in the molecular plane. A suitable set, related to the bonding in the molecule, consists of two N-F distances, the two NNF angles, and the N = N distance. It is found that two N-F distances form the basis for the representation Γ_{r_2}, v_3 the two angles NNF for Γ_α and the N=N distance for Γ_{r_1}.

Thus, Bond vectors: r_1 to $r_3 = 3$

Bond angles: α_1 and $\alpha_2 = 2$

Total number of internal coordinates = 5 (One less than the 3 N – 6) mode

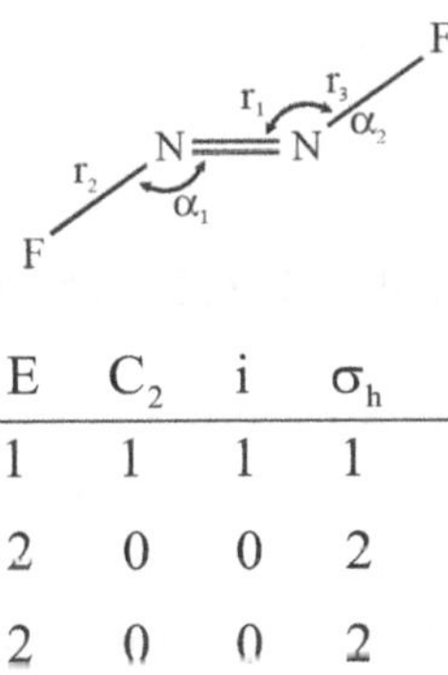

C_{2h}	E	C_2	i	σ_h	IR
Γ_{r_1}	1	1	1	1	A_g
Γ_{r_2}, r_3	2	0	0	2	$A_g + B_u$
Γ_α	2	0	0	2	$A_g + B_u$

From character table, it is found that:

$$\Gamma_{r_1} = A_g;\ \Gamma_{r_2}, r_3 = A_g + B_u \text{ and } \Gamma_\alpha = A_g + B_u$$

Thus $$\Gamma_{Int.} = \Gamma_{r_1} + \Gamma_{r_2}, r_3 + \Gamma_\alpha = 3\,A_g + 2\,B_u = 5$$

The missing sixth mode is obviously an out-of-plane mode. With the addition of $\Gamma_{\infty p}$, $\Gamma_{Int.}$ now becomes $\Gamma_{Int.} = 3\,A_g + 2\,B_u + A_u$

Therefore, it follows that the three Raman active vibrations (A_g) will be compounded to symmetric N-F stretching, symmetric NNF bending and N = N stretching, the relative amount of each involved in each normal mode depend, of course, on the actual value of the force constants and atomic masses. Similarly, the two N-F stretching and NNF angle bending.

Mode	Infrared	Raman	Nature of vibration
3 A_g	(–) Inactive	(+) Active	Mixed
2 B_u	(+) Active	(–) Inactive	Mixed
A_u	(+) Active	(–) Inactive	Pure, out-of-plane deformation

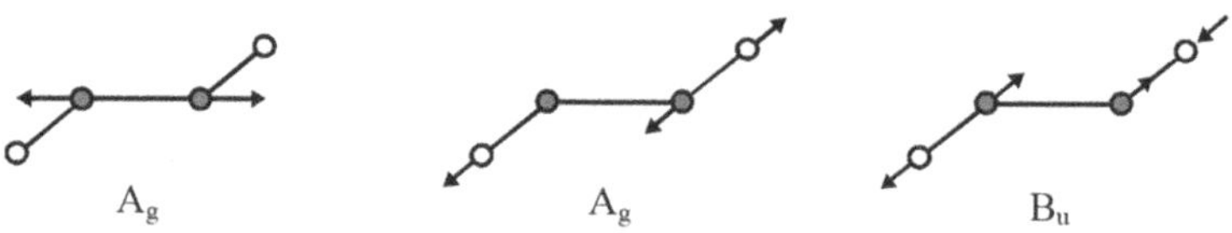

In-plane stretching modes (2 A_g + B_u)

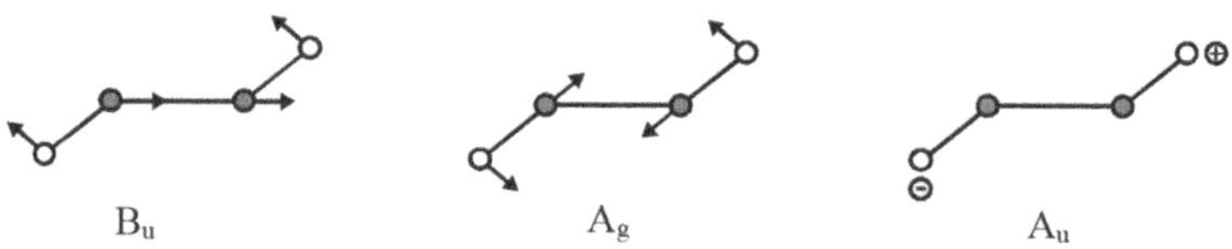

In-plane bending modes (A_g + B_u)

Out-of-plane deformation (angle bending) mode

8.11 OVERTONES AND BINARY/TERNARY COMBINATION BANDS

The number and spectral activities of the fundamental transition of the normal mode of a polyatomic molecule can be determined by normal mode analysis. But changes (less or more) in band are often observed, when actual spectra are studied. This variation in band in comparison to the results of normal mode analysis may be because of other bands called overtones, combinations and Fermi resonance bands.

8.11.1 OVERTONE BANDS

When a molecule absorbs radiation of appropriate energy, then transition occur from the ground state to its first vibrationally excited state and in one vibration mode, $\Delta v = \pm 1$. Such transition gives a fundamental band.

But overtone band occurs due to excitation to the second, third or even fourth vibrational excited state, i.e., ($v_0 \rightarrow v_2$, $v_0 \rightarrow v_3$, $v_0 \rightarrow v_4$). It means overtone band appears because of the excitation beyond v = 1 level by single photon. Therefore,

Band		Vibrational mode	
$2\nu_1$	$\rightarrow$	± 2	(First overtone)
$3\nu_1$	$\rightarrow$	± 3	(Second overtone)
$4\nu_1$	$\rightarrow$	± 4	(Third overtone)
$n\nu_1$	$\rightarrow$	$\pm n$	

Overtones have low intensity band than fundamental band and therefore, they are very weak. When molecule in first vibrational state is excited to third vibrational level, then energy required for this transition is not exactly twice of that energy, which is required for excitation to the second vibrational level. It is so, because higher levels lie relatively closer together than lower levels.

8.11.2 COMBINATION BANDS

When single photon has exact (precise) energy to excite two vibrations at the same time, then another type of overtone like band is created, which is known as combination band. For this, the energy of the combination band must be exactly the sum of two independent frequencies. Suppose ν_i and ν_j are two different fundamental bands, then sum of ν_i and ν_j ($\nu_i + \nu_j$) gives a combination band.

8.11.3 DIFFERENCE BANDS

There is another type of band, called difference band, which occurs rarely but these are reported at higher temperature. Difference bands have $\nu_i - \nu_j$ frequency. Again like combination band, difference band has low intensity than fundamental band.

Higher order combination bands are seldom observed.

$$\left.\begin{matrix} 2\nu_1 + \nu_2 \\ \nu_1 + \nu_2 + \nu_3 \end{matrix}\right] \text{Ternary combination band}$$

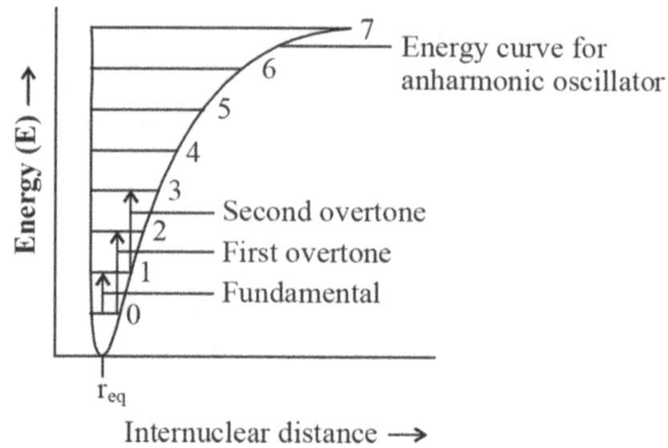

Vibration level diagram

The symmetry of overtone and combination species and their spectroscopic activity (Infrared and Raman) can be easily determined. When two or more non-degenerate vibrations combine, the symmetry of that level is given by the direct product of the representations to which that individual vibration belongs. The fundamentals may combine to give binary, ternary, etc., combination bands, which may be of overtone or combination (sum or difference) band type.

For example, consider a combination band arising out of Γ_i (ν_1) and Γ_j (ν_3) fundamentals. A direct product of these two is taken, $\Gamma_i \times \Gamma_j$, which is quite often reducible to a combination of some irreducible representations. If some of these irreducible representations correspond to x, y, z or α_{ij} functions in the character table of that molecular point group, then that combination band may occur in either infrared or Raman spectrum. Similar considerations also apply to overtones, except that the direct product has to be taken of the same irreducible representations ($\Gamma_i \times \Gamma_j$). It is enough, if one of the species of irreducible representations of the product satisfies the requirement.

Let us try to illustrate it by taking BF_3 molecule as an example, which belongs to D_{3h} point group. The normal mode analysis has shown that the normal modes of this molecule belong to A_1,″ 2 E′ and A_2'' species.

D_{3h} character table indicates that A_1' is infrared inactive, whereas A_2'' and E′ are infra-red active. If we consider a combination band of A'_1 (ν_1) and (ν_3) modes, which can be written as ($\nu_1 \pm \nu_3$), then the direct product of A_1' and E′ gives $A_{1'}$ x E′ = E′ (Infra-red active).

Consider the first overtone of A_2'' ($\nu_2 + \nu_2 = 2\ \nu_2$), whose direct product gives the species as:

$$A_2'' \times A_2'' = A_1' \text{ (Infra-red inactive)}$$

While the fundamental A_2'' being infrared active, its first overtone becomes inactive in infrared. Similarly, the second overtone of A_2'' ($3\nu_2$) is

$$A_2'' \times A_2'' \times A_2'' \text{ (Infra-red active)}$$

The second overtone, though active in the infrared, occurs as a very weak band. The infra-red and Raman activity of other combinations can be similarly worked out.

Consider another example of SO_2 molecule, which belongs to C_{2v} point group. SO_2 is predicted to have three normal modes as $(3N - 6) = 3$, but the spectral data show the presence of more than three bands. The bands at 1361, 1151 and 519 cm^{-1} are the three fundamentals designated as ν_3, ν_1 and ν_2, respectively. They can be easily identified based on their intense character.

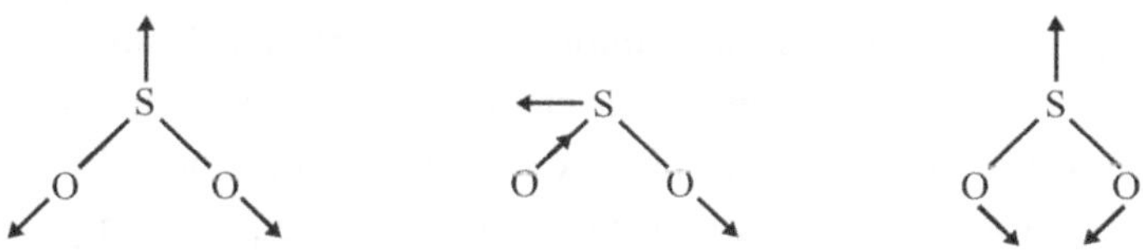

$\nu_1(A_1)$ Symmetrical stretching — $\nu_3(B_2)$ Unsymmetric stretching — $\nu_2(A_1)$ Symmetric stretching

Since all these fundamentals are also active in Raman spectrum, some of these binary combinations can also be obtained in Raman with decreased intensity. Coupling of group vibrations occurs, if these are of the same symmetry type. For example, in case of acetylene, the symmetric C-H stretching vibration and C-C stretching vibration are of the same symmetry type, and these are highly coupled. The determination of the species contained in an overtone of a degenerate fundamental is difficult and we have to consider, what are called symmetrical products. A variant of the direct product is the symmetrical products, which must be used in ascertaining the symmetry of the overtones of degenerate fundamentals.

The IR and Raman spectra of a molecule generally exhibit a number of strong bands due to fundamentals and a rather large number of weaker bands, which correspond to the overtone and combination bands of the molecule.

The completely assigned infra-red bands of SO_2 molecule is:

ν (cm^{-1})	**Frequency**	**Mode symmetry**	**Assignment**
519	ν_2	A_1	Fundamental band
606	$\nu_1 - \nu_2$	A_1	Difference band

ν (cm^{-1})	Frequency	Mode symmetry	Assignment
1151	ν_1	A_1	Fundamental band
1361	ν_3	B_2	Fundamental band
1871	$\nu_2 + \nu_3$	B_2	Combination (sum) band
2305	ν_1	A_1	Overtone of ν_1
2499	$\nu_1 + \nu_3$	B_2	Combination (sum) band

Therefore, it can be concluded that diatomic molecules, may at the most have some overtones in addition to their fundamentals, whereas in polyatomic molecules, overtones as well as some combination bands will occur, in addition to the number of fundamentals permitted by 3 N – 6 (non-linear) or 3 N – 5 (linear) formula. Overtones may occur at frequencies approximately twice those of the corresponding fundamental vibration. Combination (sum or difference) tones/bands may occur at frequencies approximately equal to the sum or difference of the frequencies of any two or more fundamentals. Thus, the overtones and combination bands are called as binary, ternary or quaternary combinations depending on whether two, three or four fundamentals are involved. Fundamentals occur generally as intense bands whereas the binary and other combination appear as weak intensity band in the spectrum.

The activity or allowedness of an overtone or a combination band in infra-red and Raman spectra depends on the overall symmetry of the combination. The allowed combinations appear as weak bands in the infra-red, but are usually too weak to be observed in Raman. Higher order combinations are still even weaker.

If a binary or any other combination and a fundamental vibration have the same symmetry and approximately the same frequency, then the two may interact to give rise to a pair of bands of comparable intensity. This interaction is termed as 'Fermi resonance.' ψ_1 and ψ_2 with energies E_1 and E_2 belong to the same irreducible representation, and they interact by a typical quantum mechanical resonance. Then the energy of interaction, E′, is given by

$$E' = \int \psi_1 H \psi_2 d\tau \tag{8.10}$$

where H is the Hamiltonian for the interaction. As a result of this interaction, the states ψ_1 and ψ_2 mix to give rise to two new states ψ'_1 and ψ'_2 having energies E'_1 and E'_2, such that $E'_1 = E_1 + E'$ and $E'_2 = E_2 - E'$ (if $E_1 > E_2$).

Considering the example of CO_2 molecule, bands at 2349, 1340 (which is infact a doublet at 1286 and 1388 cm^{-1}) and 667 cm^{-1} have been assigned to ν_3, ν_1 and ν_2, respectively. The band at 1340 cm^{-1} (ν_1) belongs to symmetry species whereas the one at 667 cm^{-1} (ν_2) belongs to π_u species. The assignment of the band at 1340 cm^{-1} has actually been simplified, which is an intense doublet. The resonance interaction between the fundamental mode ν_1 (at 1340 cm^{-1}) and the first overtone of ν_2 [i.e., 2 ν_2, $\Sigma_8^+ = 2 \times 667 =$ 1334 cm^{-1}], both being active in Raman, takes place as they occur almost at the same frequency. 2 ν_2 has the same symmetry (Σ_8^+) and same frequency (1334 cm^{-1}) as that of the fundamental v_1. Here, ν_1 is expected to be much more intense than 2 ν_2. In fact, after the interaction, these were found to be of same intensity. The weak overtone band (2 ν_2) is said to have borrowed intensity from the fundamental band (ν_1). In this process, ν_1 is raised from 1340 to 1388 cm^{-1} and 2 ν_2 is depressed from 1334 to 1286 cm^{-1}. Thus, the energy of interaction, E′ corresponds to 48 cm^{-1} and the new bands are separated by 96 cm^{-1} (2 E′).

8.12 FERMI RESONANCE

When two atoms attached to a common atom vibrates with similar frequency, then coupling vibration may occur by interaction of fundamental vibration with the overtone of some other vibration (i.e., overtone or combination) of same energy and symmetry. Such coupling is called Fermi resonance, Fermi resonance has high intensity because overtone or combination band borrows intensity from the fundamental band of the same symmetry. Basically, in Fermi resonance, two bands (fundamental band and the overtone or combination band) interact and split, losing their individual identity and form a pair of bands of similar energy.

Fermi resonance is named after Enrico Fermi, who discovered it. The mixing of two bands shift both the energy levels away, which leads to shift of higher energy band to higher energy and lower energy band shift to lower energy side and both have approximately equal intensity. The interacted band can be said as accidentally degenerated.

For example, if fundamental vibration (totally symmetric) has same frequency as that of first overtone (totally symmetric) of non-degenerated fundamental, then there will be an accidental degeneracy, i.e., double degeneracy for non-degenerated totally symmetric species.

Thus, the possibility of Fermi resonance should be considered, whenever the spectrum shows a doublet of bands while only one band is expected.

Fermi resonance may even occur between any two combination bands, when both belong to the same irreducible representation and have approximately the same frequency. One may be an overtone and the other may be a suitable combination band of any order. However, since overtones and combination bands occur as very weak bands, the resonance interaction that might take place between them may not be often noticeable.

8.13 SOLID STATE EFFECTS

The vibrations of an individual molecule in the gas phase are subject only to the symmetry restrictions based on its own intrinsic point symmetry, but when the molecule resides in a crystal, it is in principle, subject only to the symmetry restrictions arising out of its crystalline environment.

To be completely rigorous, the molecule cannot even be treated as a discrete entity; instead the entire array of molecules must be analyzed. However, such a completely rigorous approach is essentially impossible for practical reasons and unnecessary for most of the purposes, and therefore, approximations have been made. Two levels of approximation have frequently been used.

- The site symmetry approximation, and
- The correction field (sometimes called factor group) approximation.

The first is conceptually very simple and very often, it is entirely adequate. However, sometimes it fails, and then the more abstruse correlation field treatment must be employed. These approximations are:

8.14 SITE SYMMETRY APPROXIMATION

The number of vibrations or bands depends on the symmetry of the molecule. When symmetry changes from higher to lower side due to change in the state of the compound, then the number of bands increases in number. Increase in band number may the due to splitting or formation of new band.

Site of symmetry is the main criteria for analysis of any spectra of a crystalline sample. It is the symmetry of the environment of a molecule in the crystal. In gaseous or liquid state, the symmetry may be higher, whereas, in solid state, the symmetry is lowered. This mechanism is called site symmetry lowering. The number of vibration changes in the different state of molecule. Lowering of symmetry will lead to splitting of degenerate vibrations. It shows transitions, which were forbidden in gas liquid state.

This phenomenon occurs due to site symmetry lowering, strong intermolecular force, lattice vibration, and intermolecular vibration coupling in condense phase. When unit cell consists of more than one chemically equivalent molecule, then vibration in the individual molecule may couple with each other and increases complication in the spectrum.

Effects of site symmetry lowering

- Change in selection rule
- Splitting of degeneracies

(i) Change in selection rule

Selection rules used for IR and Raman spectra are valid for gaseous and then for liquid state. But, for solid state, new selection rule is required.

(ii) Splitting of degeneracies

IR spectrum of liquid/gas state gives one band, but by site symmetry lowering, number of bands increases due to splitting. Therefore, it has been observed that band, which is forbidden in gas and liquid state, appears in the solid state. For differentiating crystal structure, infra-red spectrum are more sensitive in comparison to X-ray diffraction method.

Therefore, in solid state, the selection rule for vibrational spectrum is governed by site symmetry.

A non-degenerate vibration may be inactive in the high symmetry of the free molecule but active in the symmetry of one or more subgroups of the same molecule. For example, the A_1' mode of the carbonate ion (totally symmetric C-O stretching) is not infra-red active under the full D_{3h} symmetry of CO_3^{2-}. The compound $CaCO_3$ occurs in two crystallographically different forms, calcite and aragonite. In the former, the site symmetry of the CO_3^{2-} ion is D_3, while in the latter, C_s totally symmetric vibrations are infra-red active. In agreement with these expectations, the symmetric C-O stretching mode of CO_3^{2-} (known from the Raman spectrum of solutions of carbonates) is not observed in calcite but it appears weakly in aragonite.

The effect of low site symmetry in splitting degeneracy can also be demonstrated in different forms of $CaCO_3$. The E′ representation of the group D_{3h} correlates with the E representation in D_3. Hence, in calcite, both v_3 and v_4 are observed as single peaks. In the group C_s, there are no representations of order greater than 1, which means that the degenerate vibrations of CO_3^{2-} must be split by the C_s site symmetry of aragonite. Actually v_3 is still observed as a single peak, indicating that the magnitude of the splitting is too small to permit resolution or that one component has very low intensity, but ν_4 is distinctly split into two peaks separated by 14 cm^{-1}.

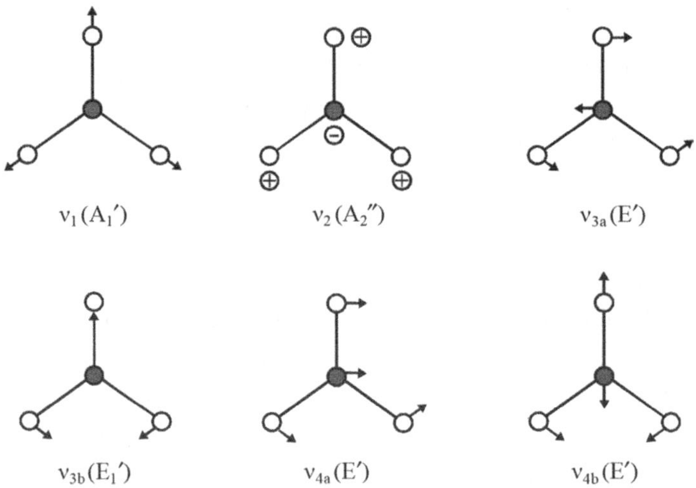

Total six normal modes
of vibrations in CO_3^{2-} ion $= 3\text{ N} - 6$
$= (3 \times 4) - 6 = 6$

KEYWORDS

- **Fermi resonance**
- **Infra-red**
- **Mutual exclusion rule**
- **Overtone**
- **Raman**

CHAPTER 9

COORDINATION COMPOUNDS AND OTHERS

CONTENTS

9.1 SPLITTING OF LEVELS AND TERMS IN A CHEMICAL ENVIRONMENT

In order to determine representation of point group of a particular environment (like O_h, T_d), we use wave function as basis. Firstly, the elements of the matrix were determined and then the sum of diagonal elements of matrix gives $\chi(\alpha)$. Character of each symmetry operation can be known with help of this formula. Character of these symmetry operations will then be character of representation of that particular environment.

As wave function $\psi(r, \theta, \phi)$ is equal to $r.\theta.\phi.\ \Psi_s$, it is assumed that spin function r is invariant to all operations in a point group. The function θ depends only on angle θ. Therefore, if all the rotations are carried out about an axis, from which θ is measured, θ will be invariant. Thus, by always choosing the axis of rotation in this way, only the function ϕ will be altered by rotation. The explicit form of ϕ function, aside form a normalizing constant is:

$$\phi = e^{im\varphi}$$

If the function $e^{im\varphi}$ is taken and rotated by an angle α, the set of ϕ wave function I becomes wave function II by α rotation.

$$\underset{\text{I}}{\begin{bmatrix} e^{li\varphi} \\ e^{(l-1)i\varphi} \\ \vdots \\ e^{(1-l)i\varphi} \\ e^{-li\varphi} \end{bmatrix}} \xrightarrow[\text{angle } \alpha]{\text{Rotated by}} \underset{\text{II}}{\begin{bmatrix} e^{li(\varphi+\alpha)} \\ e^{(l-1)i(\varphi+\alpha)} \\ \vdots \\ e^{(1-l)i(\varphi+\alpha)} \\ e^{-li(\varphi+\alpha)} \end{bmatrix}}$$

The matrix necessary to produce this transformation is:

$$\begin{bmatrix} e^{li\alpha} & 0 & 0 & 0 \\ 0 & e^{(l-1)i\alpha} & 0 & 0 \\ 0 & 0 & e^{(1-l)i\alpha} & 0 \\ 0 & 0 & 0 & e^{-li\alpha} \end{bmatrix}$$

and the character of this representation becomes:

$$\chi(\alpha) = e^{li\alpha} + e^{(l-1)i\alpha} + \ldots + e^{(1-l)i\alpha} + e^{-li\alpha}$$

$$\chi(\alpha) = \frac{\sin(l + 1/2)\,\alpha}{\sin \alpha/2} \qquad (\alpha \neq 0)$$

The formula is also valid for the case, when $\alpha = 0$

$$\chi(E) = 2l + 1$$

This result can also be obtained as:

$$\chi = \lim_{\alpha \to 0} \frac{\sin(l + 1/2)\,\alpha}{\sin \alpha/2} = \frac{(l + 1/2)\,\alpha}{\alpha/2}$$

$$= 2(l + 1/2) = 2l + 1$$

Let us proceed with set of five d orbitals, having m value l, $l - 1$, …., 0, ….$1 - l$, $-l$, namely 2, 1, 0, – 1, – 2. The matrix after rotation of angle ϕ by an angle α will be:

$$\begin{bmatrix} e^{2i\alpha} & 0 & 0 & 0 & 0 \\ 0 & e^{i\alpha} & 0 & 0 & 0 \\ 0 & 0 & e^{0} & 0 & 0 \\ 0 & 0 & 0 & e^{-i\alpha} & 0 \\ 0 & 0 & 0 & 0 & e^{-2i\alpha} \end{bmatrix}$$

This five dimensional matrix is only a special case for a set of d function.

Now keeping $l = 2$ and $\alpha = \pi$ (i.e., 180°) the character $\chi(\alpha)$ of the representation is determined.

For two-fold rotation (C_2)

$$\chi(C_2) = \frac{\sin(2 + 1/2)\pi}{\sin \pi/2} = \frac{\sin 5\pi/2}{\sin \pi/2}$$

$$= \frac{\sin 450^\circ}{\sin 90^\circ} = \frac{1}{1} = 1$$

For three-fold rotation (C_3)

$$\chi(C_3) = \frac{\sin 5\pi/3}{\sin \pi/3} = \frac{\sin 300^\circ}{\sin 60^\circ} = \frac{-0.86}{0.86} = -1$$

For four-fold rotation (C_4)

$$\chi(C_4) = \frac{\sin 5\pi/4}{\sin \pi/4} = \frac{\sin 225^\circ}{\sin 45^\circ} = \frac{-0.70}{0.70} = -1$$

$$\chi(E) = 2l + 1 = 2 \times 2 + 1 = 5$$

For $l = 1$ (p level)

For two-fold rotation (C_2)

$$\chi(C_2) = \frac{\sin(1 + 1/2)\pi}{\sin \pi/2} = \frac{\sin 3\pi/2}{\sin \pi/2} = \frac{\sin 270^\circ}{\sin 90^\circ} = \frac{-1}{1} = -1$$

For three-fold rotation (C_3)

$$\chi(C_3) = \frac{\sin 3\pi/3}{\sin \pi/3} = \frac{\sin 120^\circ}{\sin 60^\circ} = \frac{0}{0.70} = 0$$

For four-fold rotation (C_4)

$$\chi(C_4) = \frac{\sin 3\pi/4}{\sin \pi/4} = \frac{\sin 135^\circ}{\sin 45^\circ} = \frac{0.70}{0.70} = 1$$

$$\chi(E) = 2l + 1 = 2 \times 3 + 1 = 7$$

For l = 3 (f level)

$$\chi(C_2) = \frac{\sin(3+1/2)\pi}{\sin \pi\ /2} = \frac{\sin 7\pi\ 2}{\sin \pi\ 2} = \frac{\sin 630^{\circ}}{\sin 90^{\circ}} = \frac{-1}{1} = -1$$

$$\chi(C_3) = \frac{\sin 7\pi/3}{\sin \pi/3} = \frac{\sin 420^{\circ}}{\sin 60^{\circ}} = \frac{0.86}{0.80} = 1$$

$$\chi(C_4) = \frac{\sin 7\pi/4}{\sin \pi/4} = \frac{\sin 315^{\circ}}{\sin 45^{\circ}} = \frac{-0.70}{0.70} = -1$$

$$\chi(E) = 2l + 1 = 2 \times 3 + 1 = 7$$

and for $l = 4$ (g level)

$$\chi(C_2) = \frac{\sin(4+1/2)\pi}{\sin \pi/2} = \frac{\sin 9\pi/2}{\sin \pi/2} = \frac{\sin 810^{\circ}}{\sin 90^{\circ}} = \frac{-1}{1} = 1$$

$$\chi(C_3) = \frac{\sin 9\pi/3}{\sin \pi/3} = \frac{\sin 540^{\circ}}{\sin 60^{\circ}} = \frac{0}{0.70} = 0$$

$$\chi(C_4) = \frac{\sin 9\pi/4}{\sin \pi/4} = \frac{\sin 405^{\circ}}{\sin 45^{\circ}} = \frac{-0.70}{0.70} = \frac{-1}{1} = 1$$

$$\chi(E) = 2l + 1 = 2 \times 4 + 1 = 9$$

Character can be determined in a similar manner for h, i level and so on. All these results are summarized in the following table:

Type of level	l	$\chi(E)$	$\chi(C_2)$	$\chi(C_3)$	$\chi(C_4)$
s	0	1	1	1	1
p	1	3	–1	0	1
d	2	5	1	–1	–1
f	3	7	–1	1	–1
g	4	9	1	0	1
h	5	11	–1	–1	1
i	6	13	1	1	–1

The orbitals, which are degenerate in the free atom or ion, do not remain degenerate, when an atom or ion is placed in an environment with O_h, T_d or any other symmetry. So before finding representations or splitting in different environment, one should know that small letter is used to represent the state for a single electron in the environment of various symmetries, corresponding with the use of small letters, s, p, d, f, ..., to represent their state in the free atom while capital letter is used to represent the state, after splitting terms of the free ion in a specific environment For example, an f state of a free ion will be split into the state A_2, T_1 and T_2, when ion is placed in the center of a tetrahedral environment. Now, the splitting or representations in various symmetry of environment can be the find out.

Type of level	*l*	χ (E)	χ (C_2)	χ (C_3)	χ (C_4)	Irreducible representations spanned
s	0	1	1	1	1	A_{1g}
p	1	3	–1	0	1	T_{1u}
d	2	5	1	–1	–1	$E_g + T_{2g}$
f	3	7	–1	1	–1	$A_{2u} + T_{1u} + T_{2u}$
g	4	9	1	0	1	$A_{1g} + E_g + T_{1g} + T_{2g}$
h	5	11	–1	–1	1	$E_u + 2\,T_{1u} + T_{2u}$
i	6	13	1	1	–1	$A_{1g} + A_{2g} + E_g + T_{1g} + 2\,T_{2g}$

The proof of this splitting is that sum of the character of irreducible representation taken from character table is equal to the character of reducible representation Γ_d.

9.1.1 SPLITTING OF d LEVELS IN O_h SYMMETRY ENVIRONMENT

O_h	Ç(E)	Ç(C_2)	Ç(C_3)	Ç(C_4)
E_g	2	0	–1	0
T_{2g}	3	1	0	–1
"$_d$	5	1	–1	–1

When characters of irreducible representation (E_g and T_{2g}), i.e., 2 and 3, respectively (taken from O_h character table), are added, then the result is equal to Γ_d, i.e., 5. Therefore, it proof that the *d* orbital (5 set of *d* orbitals) split into $E_g + T_{2g}$ in O_h symmetry.

Another example is for f level in O_h symmetry:

O_h	Ç(E)	Ç(C_2)	Ç(C_3)	Ç(C_4)
A_{2u}	1	–1	1	–1
T_{1u}	3	–1	0	1
T_{2u}	3	1	0	–1
"$_d$	7	–1	1	–1

Similarly, it can be determined for s, p, g, h, i levels and so on.

g and u subscripts are used. g is used, when environment has center of symmetry and u is used, when environment is antisymmetric to inversion. All AOs are centrosymmetric and if the *l* is even (s, d, g…) to inversion then these will be of g character, while antisymmetric AOs, for which *l* is add (p, f, h….) to inversion, will be of u character.

9.1.2 SPLITTING IN T_d SYMMETRY ENVIRONMENT

Type of level	Symmetry of environment T_d
s	a_1
p	t_2
d	$e + t_2$
f	$a_2 + t_1 + t_2$
g	$a_1 + e + t_1 + t_2$
h	$e + t_1 + 2\,t_2$
i	$a_1 + a_2 + e + t_1 + 2\,t_2$

As there is no subscript, it only means that this environment has no center of symmetry. Five set of *d* orbitals split into e + t_2 in T_d symmetry environment.

T_d	Ç(E)	Ç(C_2)	Ç(C_3)
E	2	2	–1
T_2	3	–1	0
"$_d$	5	1	0

Taking the characters of irreducible representations from T_d character table, sum of character of irreducible representations e + t_2 gives the character of Γ_d.

For f levels in T_d symmetry environment.

T_d	Ç(E)	Ç(C_2)	Ç(C_3)
A_1	1	1	1
T_1	3	–1	0
T_2	3	–1	0
“$_d$	7	–1	1

Similarly, it can be determined for s, p, g, h, i levels and so on.

9.1.3 SPLITTING IN D_{4h} SYMMETRY ENVIRONMENT

Type of level	**Symmetry of environment T_d**
s	a_{1g}
p	$a_{2u} + e_u$
d	$a_{1g} + b_{1g} + b_{2g} + e_g$
f	$a_{2u} + b_{1u} + b_{2u} + 2\ e_u$
g	$2\ a_{1g} + a_{2g} + b_{1g} + b_{2g} + 2\ e_g$
h	$a_{1u} + 2\ a_{2u} + b_{1u} + b_{2u} + 3\ e_u$
i	$2\ a_{1g} + a_{2g} + 2\ b_{1g} + 2\ b_{2g} + 3\ e_g$

where, *a*, or *b* is one set of orbital, while *e* is two set of orbitals.

Five set of *d* orbitals split into $a_{1g} + b_{1g} + b_{2g}$ +, i.e., in D_{4h} environment (Lower symmetry environment).

D_{4h}	Ç(E)	Ç(C_2)	Ç(C_4)
A_{1g}	1	1	1
B_{1g}	1	1	–1
B_{2g}	1	1	–1
E_g	2	–2	0
“$_d$	5	1	–1

Taking characters of irreducible representations from D_{4h} character table, sum of character of irreducible representations of $a_{1g} + b_{1g} + b_{2g}$ +, i.e., is the character of Γ_d.

Similarly, it can be determined for s, p, *d*, g, h, i, levels and so on.

In a similar manner, one can determine splitting of various set of complexes belonging to D_{2d}, C_{2v}, D_3, etc. symmetry. Splitting of one electron for D_3 and D_{2d} symmetry is given here.

Type of level	Symmetry of environment	
	D_3	D_{2d}
s	a_1	a_1
p	$a_2 + e$	$b_2 + e$
d	$a_1 + 2\,e$	$a_1 + b_1 + b_2 + e$
f	$a_1 + 2\,a_2 + 2\,e$	$a_1 + a_2 + b_1 + 2\,e$
g	$2\,a_1 + a_2 + 3\,e$	$2\,a_1 + a_2 + b_1 + b_2 + 2\,e$
h	$a_1 + 2\,a_2 + 4\,e$	$a_1 + 2\,a_2 + b_1 + 2\,b_2 + 3\,e$
i	$3\,a_1 + 2\,a_2 + 4\,e$	$2\,a_1 + a_2 + 2\,b_1 + 2\,b_2 + 3\,e$

Another point is to be mentioned here regarding the splitting of terms of the free ion in chemical environment, and it is concerned with the spin multiplicity. The chemical environment doesn't interact directly with the electron spin; thus, all of the states, into which a particular term is split, have the spin multiplicity as the parent term.

In order to illustrate the splitting of terms of a d^2 configuration in different environments, the states for a d^2 ion in several point group are:

Free ion term	State of point groups		
	O_h	T_d	D_{4h}
^1S	$^1A_{1g}$	$^1A_1$	$^1A_{1g}$
^1G	$^1A_{1g}\ ^1T_{2g}$	$^1A_{1-}\,^1T_2$	$2^1A_{1g}\ ^1B_{2g}$
	$^1E_g\ ^1T_{1g}$	$^1E\ ^1T_1$	$^1A_{2g}\ 2^1E_g\ ^1B_{1g}$
^3P	$^3T_{1g}$	$^3T_1$	$^3A_{2g}\ ^3E_g$
^1D	$^1E_g\ ^1T_{2g}$	$^1E\ ^1T_2$	$^1A_{1g}\ ^1E_g\ ^1B_{1g}\ ^1B_{2g}$
^3F	$^3A_{2g}\ ^3T_{1g}\ ^3T_{2g}$	$^3A_{2-}\,^3T_{2-}\,^3T_1$	$^3A_{2g}\ 2^3E_g$
			$^3B_{1g}\ ^3B_{2g}$

9.2 SPLITTING OF *d* ORBITALS

According to quantum mechanics, maximum number of *d* orbital is five.

$$\text{Number of orbitals} = 2\,l + 1 \ (l = 2 \text{ for } d \text{ orbital})$$

$$= 2 \times 2 + 1$$

$$= 5 \text{ orbitals}$$

These are d_{xy}, d_{yz}, d_{xz}, $d_{x^2-y^2}$ and d_{z^2} orbitals.

There is a misconception about the shape of *d*-orbital that they are all having dumb-bell shapes. The four *d*-orbitals are dumbbell shaped except d_{z^2}. The shape of d_{z^2} orbital is different from other four *d*-orbitals. There is a possibility of six *d*-orbitals, i.e., d_{xy}, d_{yz}, d_{xz}, $d_{x^2-y^2}$, $d_{y^2-z^2}$, and $d_{z^2-x^2}$. Three of them are in between the axes and remaining three are along axes. As six *d*-orbital are not possible, a linear combination of $d_{y^2-z^2}$ and $d_{z^2-x^2}$ was taken to form d_{z^2} orbital. Linear combination of these two orbitals does not take place in same way as presently they are. The lobes along Z-axis in $d_{y^2-z^2}$ and $d_{z^2-x^2}$ have different orientations. So, when these are to be combined, $d_{y^2-z^2}$ has to be rotated by 90° and it results in $d_{z^2-y^2}$.

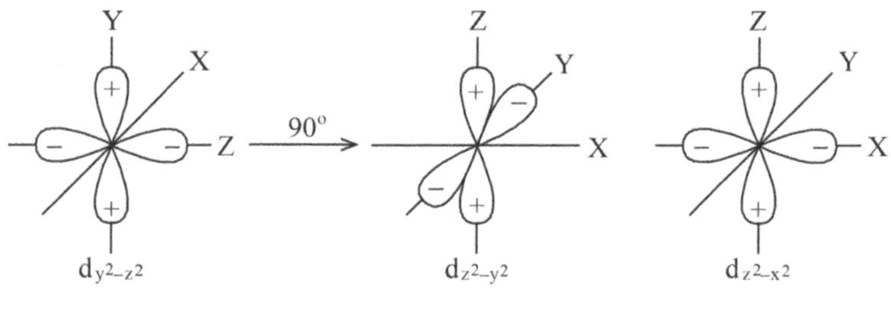

Now, this $d_{z^2-y^2}$ may combine with $d_{z^2-x^2}$ as:

$$d_{z^2-y^2} + d_{z^2-x^2} = d_{2z^2-x^2-y^2}$$

$d_{2z^2-x^2-y^2}$ is written in abbreviated form as d_{z^2}. Therefore, lobes along Z-axis are larger in size than the lobes of other four *d*-orbitals.

The $d_{2z^2-x^2-y^2}$ (= d_{z^2}), orbital consists of four lobes in xy plane, which on overlap look like a ring while two lobes are along Z-axis and these are larger in size in comparison to other four lobes.

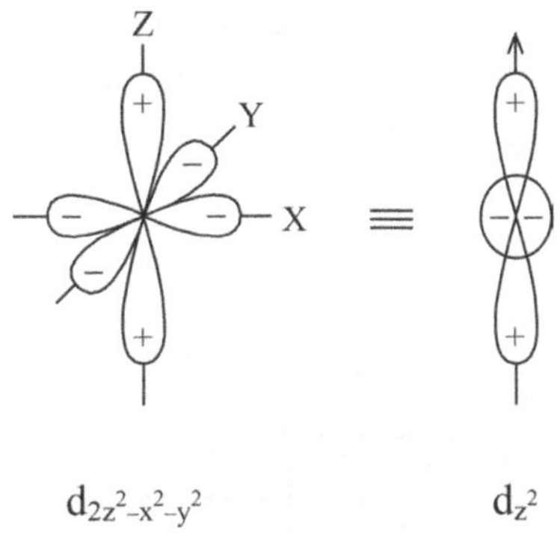

Crystal field theory considers that the ligand ions create an electrical field around the metal ion and thus, they perturb the energies of the metal orbitals. If the electrical field is spherical, it raises the energies of s, p or *d* orbitals uniformly. In other words, due to the presence of the negatively charged ligand field, the electrons in the metal orbitals in the vicinity feel repulsion and hence, the energy of metal orbitals is raised. But in such circumstances, spherical field still retains the triple degeneracy of p orbitals or penta-degeneracy of *d* orbitals.

However, in case of an octahedral complex, the ligands are present at the corners (apices) of an octahedron, they still affect the p orbitals equally, because p_x, p_y and p_z are along the axes, Thus, the triple degeneracy to p orbitals is retained.

Five *d* orbitals are oriented in different ways. $d_{x^2-y^2}$ and d_{z^2} orbitals are along the axes while d_{xy}, d_{xz} and d_{yz} orbitals are in between the axes. Thus, in case of a non-spherical field, the effect of the ligand field is different on the different *d* orbitals and their degeneracy is resolved. In such a case, five *d* energy level are split depending upon the environment of the ligands, i.e., tetrahedral, square planar, octahedral, etc.

The splitting of these *d* orbitals can be understood in terms of group theory.

Any mathematical function (wave function) can be a basis of representation. The symmetries of atomic orbitals (AOs) in various geometries (point groups) have also been summarized. It is also known that subscripts of the orbitals indicate its transformation properties in point groups. Thus, in cubic symmetry (O_h or T_d), the five *d* orbitals are split into t_{2g} and, e_g and t_2 and e in O_h and T_d, respectively. Using pictorial description of *d* atomic orbitals wave function (or vectorial representations along these lobes) and considering the cubic symmetry (Table 9.1). Embodying octahedron and tetrahedron, we can tabulate the results of carrying out only rotation operations of O_h, i.e., O sub group and corresponding operations of T_d.

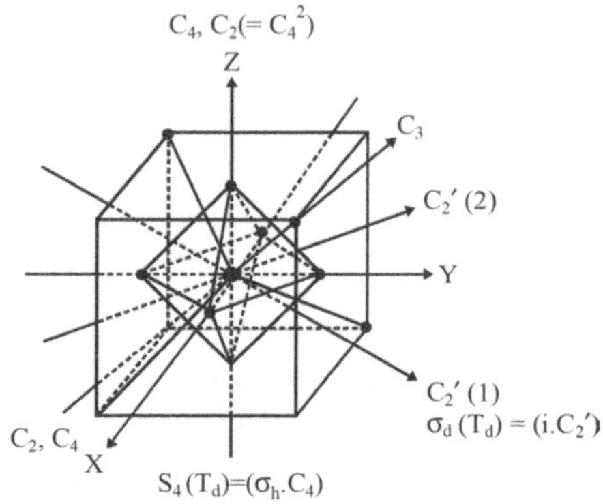

Cubic symmetry

TABLE 9.1 Transformation Properties of *d* Orbitals in Cubic Symmetry (O and Td)

T_d	E	$8\,C_3$	$3\,C_2$	$(\sigma_h.C_4) = S_4$	$(i.C^1_2) = \sigma_d$
O	E	$8\,C_3$	$3\,C_2$	$6\,C_4$	C^1_2
d_{xy}	d_{xy} (1)	d_{xz} (0)	d_{xy} (+1)	$-d_{xy}$ (−1)	d_{xy} (+1)
d_{yz}	d_{yz} (1)	d_{xy} (0)	d_{yz} (−1)	d_{xz} (0)	$-d_{xz}$ (0)
d_{xz}	d_{xz}(1)	d_{yz}(0)	$-d_{xz}$ (−1)	$-d_{yz}$ (0)	$-d_{yz}$ (0)
$d_{x^2-y^2}$	$d_{x^2-y^2}$(1)	$-d_{x^2-z^2}{}^*$ (−1)	$d_{x^2-y^2}$(+1)	$-d_{x^2-y^2}$(−1)	$-d_{x^2-y^2}$(−1)
d_{z^2}	d_{z^2} (1)	$+d_{2y^2-x^2-z^2}{}^*$ (0)	d_{z^2}(+1)	d_{z^2}(+1)	d_{z^2}(+1)
G_d	5	−1	1	−1	1

*In terms of original $d_{x^2-y^2}$ and d_{z^2}, where $d_{x^2-y^2} = \frac{\sqrt{3}}{2}(x^2 - y^2)$ and $d_{z^2} = \frac{1}{2}(2z^2 - x^2 - y^2)$, the two orbitals transform as the linear combination of the two. Thus,

$-d_{x^2-z^2} = -\frac{1}{2} d_{x^2-y^2} + \frac{\sqrt{3}}{2} d_{z^2}$ and $d_{z^2} = -d_{x^2-2} - \frac{1}{2} d_{z^2}$ and in matrix form:

$$C_3 \begin{bmatrix} d_x 2_{-y} 2 \\ d_z 2 \end{bmatrix} = \begin{bmatrix} -\frac{1}{2} + \frac{\sqrt{3}}{2} \\ -\frac{\sqrt{3}}{2} - \frac{1}{2} \end{bmatrix} \begin{bmatrix} d_x 2_{-y} 2 \\ d_z 2 \end{bmatrix}$$

$$\chi(C_3) = -\frac{1}{2} - \frac{1}{2} = +1$$

Overall Γ_d values are obtained by utilizing the well known deduction that unshifted functions (vectors) contribute + 1 to the character, and those that shifted to their negative forms (+ lobe replacing – lobe in pictorial description or + vector replacing – vector in vectorial representation) contribute – 1. But when it is totally shifted to a new wave function (orbital designation, i.e., $d_{xy} \rightarrow d_{xz}$, etc.), then they contribute zero to the character of matrices of representation. All these χ_S are indicated in brackets in each class of operation.

In order to arrive at the resolution of *d* orbital degeneracy in tetrahedral field, operations of T_d point group are performed on the *d* orbitals and the total character of the reducible representation is obtained. This is reduced to the irreducible representation.

$$\Gamma_d = E + T_2$$

Similarly for octahedral (O_h) and square planer complexes, (D_{4h}) the irreducible representation Γ_d can be reduced to:

$$\Gamma_d (O_h) = T_{2g} + E_g$$

$$\Gamma_d (D_{4h}) = A_{1g} + B_{1g} + B_{2g} + E_g$$

Though the group theory shows that the two sets have different energies, it does not indicate the order.

The tetrahedral point group has no inversion operation (center of symmetry) and hence, there will be no subscript g or u in the symbols of the irreducible representations.

The orbitals corresponding to the irreducible representations are as follows:

$$E = d_{x^2-y^2}, d_{z^2}; T_2 = d_{xy}, d_{xz}, d_{yz}$$

Thus, the tetrahedral field also splits the penta-degenerate *d* orbitals into two sets, a higher energy triply degenerate orbitals and a lower energy doubly degenerate orbitals, t_2 and e, respectively. The energies of the two sets can be worked out by considering tetrahedral perturbation over the *d* orbital wave functions.

$$d_{xy}, d_{yz}, d_{xz} (t_2) = +4\,Dq = 0.4\,\Delta_t$$

$$d_{x^2-y^2}, d_{z^2} (e) = -6\,Dq = -0.6\,\Delta_t$$

It can be seen that there is a reversal in the order of energies of the two sets of tetrahedral as compared to the octahedral field. If the orientation of the ligands is considered in the tetrahedral field, they are at the opposite corners of the two opposite faces of a cube and the metal ion at the center. In other words, the ligands are located between two axes. Hence, the electrons going to the orbitals in between the axes (d_{xy}, d_{yz} and d_{xz}) face greater repulsion from the ligand field, than those along the axis, ($d_{x^2-y^2}$ and d_{z^2}). Thus, t_2 set of *d*-orbitals is greater in energy than e set of orbitals.

It is further expected that in case of tetrahedral structure, the field is created by the charges on the four ligands, and hence, Δ_t should be 2/3 Δ_0, where there are six charged ligands. Δ_t is further reduced to 4/9 Δ_t because the ligands at the opposite corners of two opposite faces in tetrahedral complexes are not exactly in between the two axes. Hence, the relative order of stabilization and destabilization of e and t_2 is not same as in the octahedral complexes. These two factors result in $\Delta t = 4/9\ \Delta_0$.

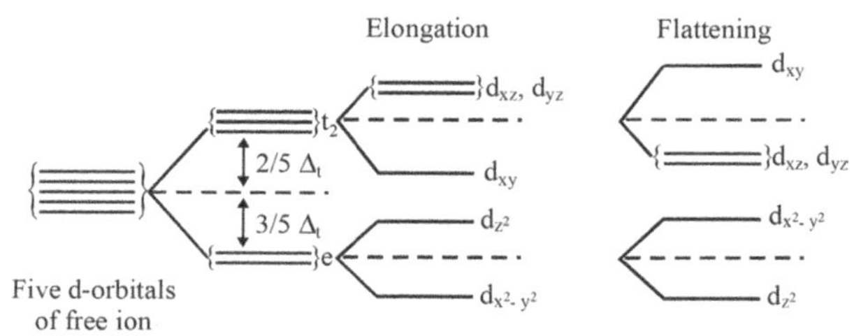

Effect of distortion

There can be distortion in tetrahedral complexes also. In the case of elongation (distortion) (C_{2v}), the angle L-M-L between two pairs of ligands becomes less than the tetrahedral angle, and consequently, the field along the *Z*- axis is more than in the equatorial plane.

In case of flattening (distortion) (C_{2v}), the angle L-M-L increases and hence, field along *Z*- axis is less than in the equatorial plane. The splitting of e and t_2 sets is reverse to that in case of elongation. Crystal field theory can explain the spectra, magnetic properties, thermodynamic and kinetic properties of complexes very well but it has a serious limitation.

Crystal field theory considers the metal ion and ligand ion or ligand dipole as a point charge and the metal ligand interaction is considered to be purely electrostatic.

In octahedral complexes, the ligands are oriented at the corners of the axes, the *d* orbitals directed towards the axes, i.e., $d_{x^2-y^2}$ and d_{z^2} experience greater repulsion from the negative ligand field, the electron is raised to higher energy in populating these orbitals and therefore, i.e., orbitals are higher in energy. The d_{xy}, d_{xz} and d_{yz} orbitals, being in between the axes, feel less repulsion and hence, these are lowered in energy. The lowering and raising of the energy is with respect to the energies of the *d* orbitals in the spherical field.

The separation in the energy of the two sets is called crystal field splitting energy and symbolized as Δ_o (o is subscript signifying octahedral). Since there is no external source of energy, the quantum mechanics requires that the total energy of the *d* orbitals should be same. In other words, the increase in the energy of the e_g, orbitals should be equal to the lowering in the energy of the t_{2g} orbitals. It can thus be shown that t_{2g} orbitals are lowered down by – 2/5 Δ_o and, i.e., orbitals are raised by 3/5 Δ_o.

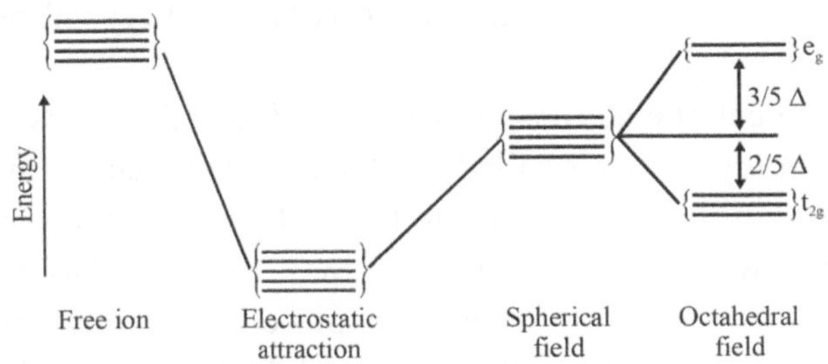

Splitting of d-orbitals

The value of Δ_o depends on different parameters. An octahedral perturbation over *d* orbitals can be calculated and this gives the energy of the *d* orbitals quantitatively.

$$E_g = d_{x^2-y^2}, d_{z^2} \quad e_g = 6\ Dq = 0.6\ \Delta_o$$

$$T_{2g} = d_{xy}, d_{xz}, d_{yz} \quad t_{2g} = -4\ Dq = -0.4\ \Delta_o$$

$$\Delta E = E(e_g) + E(t_{2g})$$

$$= 2(+6\ Dq) + 3(-4\ Dq)$$

$$= 0$$

It mans splitting occur in such a way that no net change in energy take place. The difference of energy Δ_o is fixed, i.e., 10 Dq, where D and q are dependent on some parameters of the complex.

$$D = \frac{35\,Ze^2}{4\,a^5} \quad \text{and } q = \frac{2\,(\overline{r_2})^4}{105}$$

$$Dq = \frac{1}{6}\;\frac{Z\,e^2\,(\overline{r_2})^4}{a^5}$$

where e = electronic charge; z = charge on ligand; r_2 = radius of *d* orbital stationary; $(\overline{r_2})^4$ = fourth power of mean average radius of *d*-orbitals; a = distance between metal and ligand.

Thus, depending on the nature of the metal ion and the ligand, the value of crystal field splitting (D_o) changes. If the value of D_o is high, the ligand is said to create a strong field and if D_o is less, the crystal field is said to be weak.

Rearrangement of electrons in these split *d* orbitals results in lowering of total orbital energy, which is called crystal field stabilization energy (CFSE).

Depending on the strength of the field, D_o may be greater than pairing energy (P) or lower. This does not affect the CFSE calculation upto d^3 case. From d^4 to d^7, pairing of electrons takes place in t_{2g} orbitals in strong field ligand ($\Delta_o > P$), whereas in weak field ligand, high spin complex are formed (i.e., $\Delta_o < P$)

The difference in energy between t_2 and e set of orbitals is denoted by Δ_t (= 10 Dq). The relationship between crystal field splitting energy of O_h and T_d is:

$$\Delta_t = \frac{4}{9}\Delta_o \text{ or } 0.45\,\Delta_o$$

Distorted octahedral complexes of the type ML_6 can be formed due to Jahn-Teller effect. The two ligands in the axial direction are at a greater or smaller distance from the metal ion than the remaining four in the equatorial plane.

Distorted octahedral field also exists in trans complexes of the type $[ML_4X_2]$, where two X ligands along the axial direction create a different field than the four ligands (L) present in the equatorial plane.

Thus, the symmetry of the ligand field is reduced to D_{4h} (square planer complex). In order to work out the splitting of the *d* orbitals in D_{4h} field, the operations of D_{4h} point group are performed, and the total character, and the reducible representation are obtained. The total character can be reduced to following irreducible representations:

$$\Gamma_d = A_{1g} + B_{1g} + B_{2g} + E_g$$

It can, therefore, be concluded that in a D_{4h} field, the *d* orbitals split up into four sets, three are non-degenerate and one is doubly degenerate. The *d* orbitals corresponding to the irreducible representations are as follows:

	Orbitals
a_{1g}	d_{z^2}
b_{1g}	$d_{x^2-y^2}$
b_{2g}	d_{xy}
e_g	d_{yz}, d_{xz}

Thus, the doubly degenerate set, i.e., in octahedral field splits up into two levels, an upper b_{1g} ($d_{x^2-y^2}$) and a lower a_{1g} (d_{z^2}) in D_{4h} field. Similarly, triply degenerate t_{2g} set splits up into one non-degenerate set b_{2g} (d_{xy}) and one doubly degenerate set e_g (d_{xz}, d_{yz}).

The order of energies can be determined quantitatively, one has to make use to quantum mechanics. However, qualitatively, the order can be arrived at by simple method also.

In tetragonally distorted octahedral field, the field along the Z-axis being less than in the xy plane, d_{z^2} (a_{1g}) orbital has lower energy than $d_{x^2-y^2}$ (b_{1g}). Similarly, electrons in d_{xz} and d_{yz} orbitals feel less repulsion form the ligands than one in present d_{xy} orbital. Thus, d_{xz} and d_{yz} orbitals are lowered in energy than d_{xy} orbital.

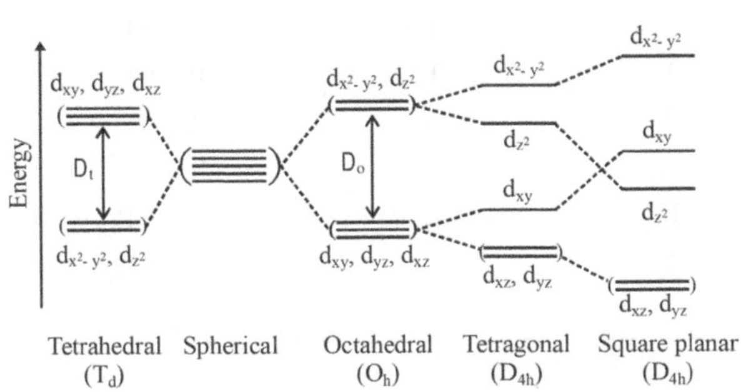

Splitting of d-*orbitals in different environments*

In square planar complexes also, the field symmetry is D_{4h} and hence, the splitting pattern remains same as that of tetragonal complex. However, there is a change in the order of their energy levels. Since, there is no ligand along *Z*-axis, the repulsion felt by the electron in d_{z^2} orbitals is very small and d_{z^2} orbital is even lowered than d_{xy}. One will expect d_{z^2} orbital to be even lower in energy than d_{xz} and d_{yz}, as d_{z^2} orbital is oriented along *Z*-axis and is farther from the ligands situated along *X* and *Y*-axes than d_{xz} and d_{yz} orbitals. However, the electrons in the d_{z^2} orbitals have 1/3 probability of occurring along the collar in the xy plane and hence, feel the repulsion due to equatorial ligands. d_{xz} and d_{yz} have nodes along xy plane. The energy spacing between b_{2g} (d_{xy}) and b_{1g} ($d_{x^2-y^2}$) level is designated as Δ.

If the square planar complexes are of $[ML_2X_2]$ type (cis- or trans-), the field symmetry is reduced to C_{2v} or D_{2h}, respectively.

The doubly degenerate set (e_g) gets split into two non-degenerate sets b_{2g} (d_{xz}) and b_{3g} (d_{yz}) in D_{2h} and b_1 (d_{xz}) and b_2 (d_{yz}) in C_{2v} point group.

The square planar geometry is favored by d^8 configuration in presence of strong field. In this condition, Δ is so large that all the eight electrons are paired in d_{xz}, d_{yz} and d_{xy} orbitals, while $d_{x^2-y^2}$ remains unoccupied and thus, form low spin complexes.

It can be concluded that this distortion results in descending symmetry from O_h to D_{4h}, which leads to loss of degeneracy. The correlation table of O_h with other point group is given in Tables 9.2 and 9.3.

TABLE 9.2 Correlation Table

O_h	O	T_d	D_{4h}	C_{2v}
A_{1g}	A_1	A_1	A_{1g}	A_1
A_{2g}	A_2	A_2	B_{1g}	A_2
E_g	E	E	$A_{1g} + B_{1g}$	$A_1 + A_2$
T_{1g}	T_1	T_1	$A_{2g} + E_g$	$A_2 + B_1 + B_2$
T_{2g}	T_2	T_2	$B_{2g} + E_g$	$A_1 + B_1 + B_2$
A_{1u}	A_1	A_1	A_{1u}	A_2
A_{2u}	A_2	A_2	B_{1u}	A_1
E_u	E	E	$A_{1u} + B_{1u}$	$A_1 + A_2$
T_{1u}	T_1	T_1	$A_{2u} + E_u$	$A_1 + B_1 + B_2$
T_{2u}	T_2	T_2	$B_{2u} + E_u$	$A_2 + B_1 + B_2$

TABLE 9.3 Splitting of Different Orbitals in Various Fields

Orbitals	O_h	T_d	D_{4h}	D_{2d}
s	a_{1g}	a_1	a_{1g}	a_1
p	t_{1u}	t_1	$a_{2g} + e_u$	$b_2 + e$
d	$e_g + t_{2g}$	$e + t_2$	$a_1 + b_{1g} + b_{2g} + e_g$	$a_1 + b_1 + b_2 + e$
f	$a_{2u} + t_{1u} + t_{2u}$	$a_2 + t_1 + t_2$	$a_{2u} + b_{1u} + b_{2u} + 2\,e_u$	$a_1 + a_2 + b_2 + 2\,e$

The splitting in type of orbitals in various fields are represented in the table.

Similarly, splitting pattern can be obtained for g, h, and i, orbitals also.

9.3 ELECTRONIC SPECTRA OF COMPLEXES

A spectrum arises because the electrons may be promoted from one energy level to another. Such electronic transitions are of high energy, and in addition, much lower energy vibrational and rotational transitions always occur. The vibrational and rotational levels are too close in energy to be resolved into separate absorption bands, but they result in considerable broadening of the electronic absorption bands in *d–d* spectra. Bandwidths are commonly found to be of the order of 1000–3000 cm^{-1}.

All the theoretically possible electronic transitions are not actually observed. The position is formalized into a set of selection rules, which distinguish between allowed and forbidden transitions. Allowed transitions occur quite commonly. While forbidden transitions do occur, but less frequently, and they are consequently of much lower intensity.

9.3.1 LAPORTE ORBITAL SELECTION RULE

According to Laporte rule, only those transitions are possible, in which change in parity occurs. That is:

$$\underset{(g)}{\text{Gerade}} \rightleftharpoons \underset{(u)}{\text{Ungerade}} \quad \text{(Allowed transition)}$$

Orbitals with centrosymmetry are represented by gerade (g) and without centrosymmetry by ungerade (u).

s p d f ....so on

g u g u

According to this rule, *d–d* transitions are formally forbidden. But UV/vis spectroscopy and optical spectroscopy for complexes involves *d–d* transition. This is due to relaxation in selection rule. Due to this relaxation, *d–d* transitions can occur, but only at low intensities.

The selection rule can be relaxed, when unsymmetrical vibrations of complexes temporarily destroy its centrosymmetry, and allowed transition, which would otherwise be Laporte forbidden.

In tetrahedral complex, there is no center of symmetry and therefore, orbitals have no *g* or *u*. On splitting *d* orbitals in T_d orientation, they will form *e* and t_2 orbitals. Among these two, *e* is pure form of atomic *d* orbitals and thus, their *g* character is maintained even in the complex. On the other hand, t_2 molecular orbitals are formed from atomic *d* (gerade) and *p* (ungerade) orbitals, i.e., by *d–p* mixing, which give *u* character to the t_2 level in the complex. Thus, Laporte rule is relaxed.

9.3.2 SPIN SELECTION RULE

An electron does not change its spin during transitions between energy levels, that is DS = 0. In d^2 configuration in an octahedral field, the ground state (T_{1g}) has a multiplicity of 3 and that three are three excited states with the same multiplicity ($^3T_{2g}$, $^3A_{2g}$, and $^3T_{1g}$). Thus, spin allowed transition is:

$$^3T_{1g} \longrightarrow {}^3T_{2g}$$

$$^3T_{1g} \longrightarrow {}^3A_{2g}$$

$$^3T_{1g} \longrightarrow {}^3T_{1g}(P)$$

Transitions from triplet ground state to singlet excited are spin forbidden.

Thus, in the case of Mn^{2+} in a weak octahedral field, such as $[Mn(H_2O_6]^{2+}$, the *d–d* transitions are spin forbidden because each of the *d* orbitals is singly occupied. Many Mn^{2+} compounds are off white or pale flesh colored, but the intensity is only about one hundredth of that for a spin allowed transition.

Since the spin forbidden transitions ($\Delta S = 0$) are very weak, analysis of the spectra of transition metal complexes can be greatly simplified by ignoring all such spin forbidden transitions and considering only those excited states, which have the same multiplicity as the ground state.

There are incomplete *d* orbitals in transition metals. The penta-degenerate *d* orbitals get split up into t_{2g} and e_g sets in an octahedral field. Rearrangement of electrons takes place in such a way that the energy is least (minimum). In case of d^1 metal ion, the arrangement is $t_{2g}^{\ 1}\ e_g^{\ 0}$ in an octahedral field. On being excited, the electron in the t_{2g} orbitals absorbs energy equal to the crystal field splitting and moves to the e_g orbitals. Since the value of Δ_o is low, absorption takes place in the visible region and the transition metal complexes are colored. This $d \rightarrow d$ transition appears to be the simple explanation for the color in the transition metal salts and complexes. However, the absorption spectra of the octahedral complexes show that the molar absorbance of such $d \rightarrow d$ transition bands are low. This is because of the selection rules.

(i) In case of octahedral complexes, $d \rightarrow d$ transitions are $t_{2g} \rightarrow e_g$ transition, i.e., $g \rightarrow g$ transitions, which can not cause any change in the dipole moment i.e. $\left| \int \psi\, \mu_M\, \psi^* d\tau \right|^2 = 0$ for the octahedral complexes with center of symmetry, $d \rightarrow d$ transitions should be Laporte forbidden.

9.3.3 RELAXATION IN SELECTION RULES

However, such forbidden transitions do become allowed in complex compounds due to the following reasons:

(a) Octahedral symmetry may get distorted during the vibration of the molecule and the center of symmetry is lost. $d \rightarrow d$ transitions become allowed in octahedral complexes and low intensity bands are observed because of coupling of the electronic and vibrational wave functions, i.e., vibronic coupling.
(b) There may be some mixing of *d* and p orbitals in the complex and thus, $t_{2g} \rightarrow e_g$, i.e., transitions are not purely $d \rightarrow d$ transitions.
(c) The intensity of the bands in some complexes is much greater than it can be expected from these two reasons. This can be explained by considering that the metal *d* orbitals overlap with the ligand orbitals

and as a result, the pure *d* orbital character is lost. This is an evidence for metal and ligand orbital overlap in complex compounds and is also in support of ligand field theory.

(ii) Number of unpaired spins or the multiplicity should not change during $d \rightarrow d$ transitions. However, in $[Mn(H_2O)_6]^{2+}$ complex with d^5 configuration, the ground state has multiplicity six but the excited state will have lower multiplicity 4. Thus, the electronic transition in $[Mn(H_2O)_6]^{2+}$ is doubly forbidden (Laporte and spin) and the intensity of the bands is very low. This is the reason, why bivalent manganese salts or complexes are very light pink in color.

Spin forbidden transitions, though less intense, are relatively sharper than the spin allowed transitions. This is because of the fact that there is change in the position of the ligands during the vibration of the molecule. In this case, ligand field undergoes change and as a result, there is change in the extent of splitting of the *d* orbitals. The spin allowed electronic transitions, which are dependent on Δ, differ in energy for different states of vibration in different molecules and hence, these give broad band. The spin forbidden transitions are not dependent on the value of Δ and therefore, there is no such significant broadening due to vibrational change.

(iii) Simultaneous excitation of more than one electron does not take place. However, low intensity bands, corresponding to two electron transitions, are observed in some complexes.

Hund's rules are used in order to find out the ground state of a free metal ion.

- The magnetic quantum number is maximized. The highest magnetic quantum number corresponds to the total angular momentum quantum number L. The state with highest L has the lowest energy.
- The state with highest spin has the lowest energy.
- For a multiple state with different possible J values, $(L + S \rightarrow L - S)$, the state with the lowest J value is lowest in energy, provided that the atomic orbitals are less than half-filled. But in the case, where the atomic orbitals are more than half filled, the state with the highest value of J has the lowest energy.

However, Δ is normally greater than spin orbit coupling, i.e., separation between J values, so J states are not distinguished in considering the splitting of the ground state.

In d^1 case, the highest value of $m_L = +2$ and hence, $L = 2$ and $S = ½$. The ground state is $^{(2S+1)}L_{J_{L+S}}$ and $^{(2S+1)}L_{J_{L-S}}$, i.e., $^2D_{5/2}$ and $^2D_{3/2}$.

In case of multi-electronic atoms, the ground state and higher energy states may be S, P, D, F, etc. The spectral states split up on application of the ligand field.

The terms S, P, D, F, etc. have same symmetry corresponding to s, p, d, f, etc., orbitals. It means D term is split by an octahedral field in exactly same pattern as a set of *d* orbitals and F term is split in same manner as a set of f orbitals in O_h and so on.

9.4 ORGEL DIAGRAMS FOR TETRAHEDRAL COMPLEXES

It has been observed that the splitting pattern of the orbitals in the tetrahedral field are same as in octahedral field and spitting of free ion states in tetrahedral complexes are also same as in octahedral complexes, but the order of the split up states are, however, reversed.

d^1 case

In d^1 case, the ground state ^2D splits up into ^2E and $^2T_2$ states. The ground state electronic configuration is $e^1t_2^0$ and hence, the ground state is ^2E ($d_{x^2-y^2}^{1}$ $d_{z^2}^{0}$ t_2^0, $d_{x^2-y^2}^{0}$ $d_{z^2}^{1}$ t_2^0). The electronic configuration in the excited state is $e^0 t_2^1$ and therefore, it is a T_2 state (e^0 d_{xy}^1 d_{xz}^0 d_{yz}^0, e^0 d_{xy}^0 d_{xz}^1 d_{xz}^0, e^0 d_{xy}^0 d_{xz}^0 d_{xz}^1). The spectral transition can be shown as $E \rightarrow T_2$, and this corresponds to Δ. Since the tetrahedral complex has no center of symmetry, subscript g has not been used.

d^9 case

In case of d^9 the ground state electronic configuration is e^4 t_2^5 with one unpaired electron in the t_2 orbital. Here, the state is T_2 (e^4 t_2^5, i.e., e^4 d_{xy}^2 d_{xz}^2 d_{yz}^1, e^4 d_{xy}^1 d_{xz}^2 d_{yz}^2, e^4 d_{xy}^2 d_{xz}^1 d_{yz}^2). In the excited state, the electron moves to the t_2 orbital resulting in the configurtion e^3 t_2^6. The unpaired electron is now in the e orbital, so the state is ^2E. ($d_{x^2-y^2}^{2}$ $d_{z^2}^{1}$ t_{2g}^6, $d_{x^2-y^2}^{1}$ $d_{z^2}^{2}$ t_{2g}^6. The transition can be shown as $^2E \leftarrow {}^2T_2$.

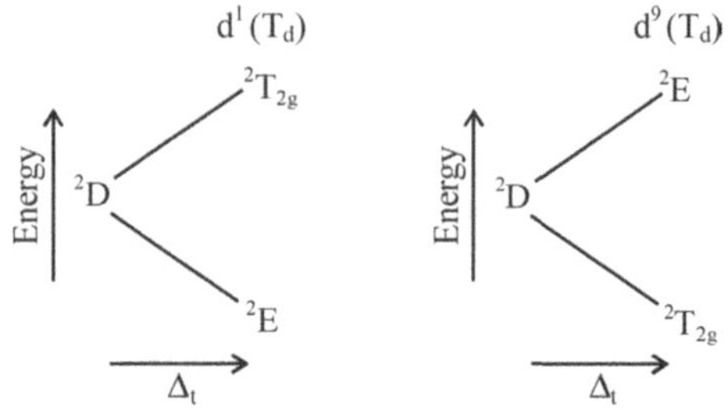

d^6 and d^4 are similar to that in cases of d^1 and d^9.

d^4 case

d^4 ground state is $^5T_2$ ($e^3\,t_2^2$, i.e., $e^3\,d_{xy}^{\,1}, d_{xz}^{\,1}\,d_{yz}^{\,0}\;e^3\,d_{xy}^{\,0}\,d_{xz}^{\,1}\,d_{yz}^{\,1}, e^2\,d_{xy}^{\,1}\,d_{xz}^{\,0}\,d_{yz}^{\,1}$). The excited state configuration is $e^1\,t_2^3$ and it corresponds to ^5E state ($d_{x^2-y^2}^{\,1}\,d_{z^2}^{\,0}\,t_2^3, d_{x^2-y^2}^{\,0}\,d_{z^2}^{\,1}\,t_2^3$).

d^6 case

d^6 ground state is $^5E_2$, $e^3\,t_2^3$, i.e., ($d_{x^2-y^2}^{\,2}\,d_{z^2}^{\,1}\,t_{2g}^3, d_{x^2-y^2}^{\,1}\,d_{z^2}^{\,2}\,t_2^3$) and excited state is $^5T_2$ ($e^2\,t_2^4$, i.e., $e^2\,d_{xy}^{\,2}\,d_{xz}^{\,1}\,d_{yz}^{\,1}, e^2\,d_{xy}^{\,1}\,d_{xz}^{\,2}\,d_{yz}^{\,1}, e^2\,d_{xy}^{\,1}\,d_{xz}^{\,1}\,d_{yz}^{\,2}$).

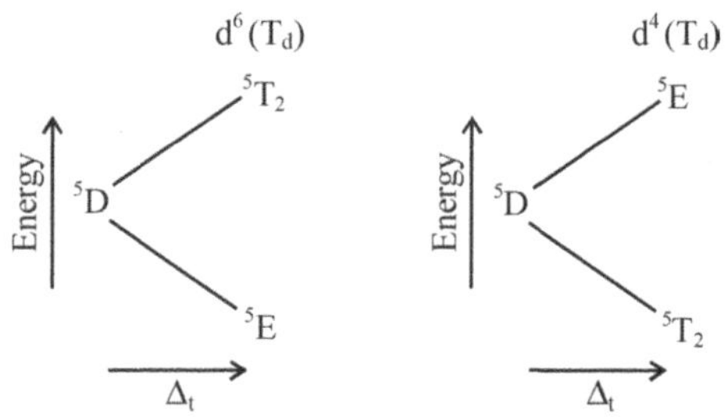

In all these cases, the transition energy corresponds to Δ. It has been known that for the same ligands, Δ_t is 4/9 of D_o. Thus, the transition energy in the tetrahedral complexes is less and the band occurs in the higher wavelength region.

Further, the tetrahedral complexes have no center of symmetry and therefore, $d \rightarrow d$ transitions are not Laporte forbidden. $d \rightarrow d$ transitions will result in change in the dipole moment and the intensity of the transitions in tetrahedral complexes is high. Thus, a high intensity transition in the higher wavelength region indicates that the structure is tetrahedral.

In all these cases, it has been presumed that the field is weak. Pairing of the electrons does not take place in d^4 and d^6 cases, i.e., the separation between the free ion ground state and higher energy state of lower multiplicity is more and hence, the crystal field splitting cannot mix the ground state with the higher state.

Orgel has suggested diagrams for such weak field cases. He has plotted the splitting of the ground spectral state as a function of crystal field splitting. One diagram can represent the splitting pattern in d^1, d^4, d^6 and d^9 metal ions in octahedral and tetrahedral fields.

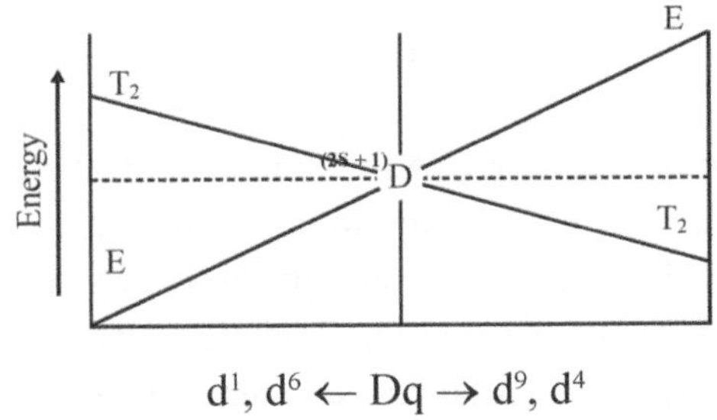

In d^2 configuration, levels split into ^1S, ^3P, ^1D, ^3F and ^1G terms because of interelectronic repulsion. ^3F is the ground with lowest energy and ^3P, ^1G, ^1D and ^1S are excited states. The transition between $^3F \rightarrow {}^3P$ is allowed, whereas transition from $^3F \rightarrow {}^1S$, ^1D, and ^1G are forbidden according to spin selection rule. F state breaks up into A_2, T_2 and T_1, while P state gets converted to T_1.

The *d* orbitals get split up into lower energy e and higher energy t_2 orbitals in the tetrahedral field. The ground electronic arrangement is $e^2\, t_2^0$ and corresponds to non-degenerate A_2 ($d_{x^2-y^2}^{1}\, d_{z^2}^{1}\, t_2^0$).

On excitation of one electron, the arrangement is $e^1\, t_2^1$ and it corresponds to two triply degenerate states $T_2(F)$ and $T_1(P)$. Two electron excitation results in the electronic arrangement $e^0\, t_2^2$ and this corresponds to triply degenerate $T^1(F)$ state ($e^0\, d_{xy}^1\, d_{xz}^1 d_{yz}^0$, $e^0\, d_{xy}^0\, d_{xz}^1\, d_{yz}^1$, $e^0\, d_{xy}^1\, d_{xz}^0\, d_{yz}^1$).

Similarly, it can be seen that splitting pattern of d^2 is same as d^7, whereas d^3 and d^8 cases in tetrahedral field are inverse of d^2 and d^7.

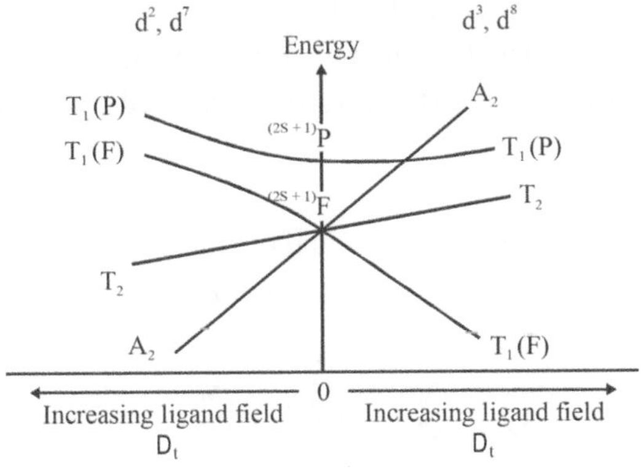

9.5 ORGEL DIAGRAMS FOR OCTAHEDRAL COMPLEXES

The spectral states get split up in octahedral field. The combination of the irreducible representations in O_h character table represents the considered atomic orbitals with a particular spectral terms. These are:

$$S - A_{1g}$$
$$P - T_{1g}$$
$$D - E_g, T_{2g}$$
$$F - A_{1g}, T_{1g}, T_{2g}$$

This means a non-degenerate (only one symmetric electron arrangement) spectral state remains nondegenerate in the O_h field also. A triply degenerate D state remains triply degenerate, though the energy is affected. However, a penta-degenerate D state gets split up into a doubly degenerate state E_g and a triply degenerate state T_{2g}. A hepta-degenerate F state splits up into a non-degenerate A_{2g}, and two triply degenerate states, T_{1g} and T_{2g}. Let us consider different cases now.

d^1 Case

In this case, the ground state is ^2D. It splits up into T_{2g} and E_g states in the octahedral field. Group theory does not tell us, which state will be of the lower energy and which one will be the higher energy state? This can be understood by seeing the electronic arrangement. The *d* orbitals split up into t_{2g} orbitals and e_g orbitals in the octahedral field. In the lower energy arrangement ($t_{2g}^1\, e_g^0$), the electron in the t_{2g} orbital will correspond to the irreducible representation T_{2g} and hence, the lower energy state is T_{2g}. There are three electronic arrangements with the same energy $d_{xy}^1\, d_{xz}^0\, d_{yz}^0$, $d_{xy}^0\, d_{xz}^1\, d_{yz}^0$ or $d_{xy}^0\, d_{xz}^0\, d_{yz}^1$, making the state triply degenerate. In the excited state, $t_{2g}^0\, e_g^1$, the electron is in E_g state. Doubly degenerate arrangements are $d_{x^2-y^2}^1\, d_{z^2}^0$ or $d_{x^2-y^2}^0\, d_{z^2}^1$. Thus, the transition $t_{2g}^1\, e_g^0 \rightarrow t_{2g}^0\, e_g^1$ is transfer of electron from T_{2g} state to E_g state and it is written as $T_{2g} \rightarrow E_g$ transition. The band corresponding to this transition will involve energy equal to Δ. It should be kept in mind that the atomic orbitals are always represented by small letters, whereas spectral states (terms) are represented by capital letters.

d^9 Case

In the arrangement of nine electrons in *d* orbitals, the highest magnetic quantum number works out to be + 2 and hence, L = 2.

$$L = 2 \quad \text{i.e. D}$$

$$S = \frac{1}{2}$$

$$2S + 1 = 2$$

Therefore, spectral state (term) is ^2D.

The magnetic quantum numbers of all the orbitals with electrons are added. If two electrons are present in an orbital, then its magnetic quantum number is added twice. There is one unpaired electron and hence, the free ion ground state is ^2D. It is same as in case of d^1, because in d^9, there is a hole (vacancy) instead of an electron. There can be ten possible arrangements.

In d^9 configuration, ^2D splits into T_{2g} and E_g. The ground state electronic configuration $t_{2g}^{6}\,e_g^{3}$ has one unpaired electron in the e_g orbital. The symmetry of the state is determined by the unpaired electron only. The paired electrons correspond to a totally symmetrical irreducible representation. Hence, the ground state is E_g. The two electronic arrangements of same energy are $t_{2g}^{6}\,d_{x^2-y^2}^{2}\,d_{z^2}^{1}$ or $t_{2g}^{6}\,d_{x^2-y^2}^{1}\,d_{z^2}^{2}$.

The excited state configuration is $t_{2g}^{5}\,e_g^{4}$, and here, the unpaired electron is in the t_{2g} orbital. Now the state is T_{2g} ($d_{xy}^{1}\,d_{xz}^{2}\,d_{yz}^{2}\,e_g^{4}$, $d_{xy}^{2}\,d_{xz}^{1}\,d_{yz}^{2}\,e_g^{4}$, $d_{xy}^{2}\,d_{xz}^{2}\,d_{yz}^{1}\,e_g^{4}$). The electronic transition is represented as $^2E_g \rightarrow {}^2T_{2g}$. Thus, the splitting pattern in d^9 case is opposite of the d^1 case.

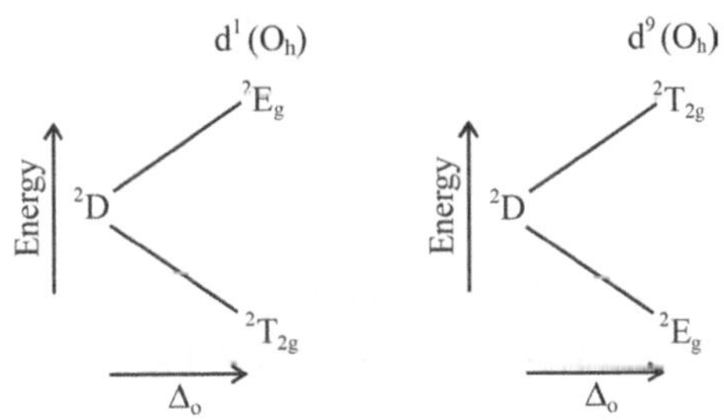

d^6 Case

In this case, there are half filled *d* orbitals and an extra electron. Following the same method of finding highest magnetic quantum number, one observes that L = 2 and S = 2,

$$L = 2 \quad \text{(i.e., D)}$$

$$S = 2$$

$$2S + 1 = 5$$

Therefore, term symbol is ^5D.

The free ion ground state is ^5D. In an octahedral field, this ^5D state gets split up into T_{2g} of lower energy ($t_{2g}^4 e_g^2$, i.e., $d_{xy}^2 d_{xz}^1 d_{yz}^1 e_g^2$, $d_{xy}^1 d_{xz}^2 d_{yz}^1 e_g^2$ and $d_{xy}^1 d_{xz}^1 d_{yz}^2 e_g^2$) and E_g of higher energy ($t_{2g}^3 e_g^3$, i.e., $t_{2g}^2 d_{x^2-y^2}^2 d_{z^2}^1$ and $t_{2g}^2 d_{x^2-y^2}^1 d_{z^2}^2$). Thus, the transition can be shown as $^5T_{2g} \rightarrow {}^5E_g$. This is similar to that in case of d^1 metal ion.

d^4 Case

Here, one electron is less than the half filled *d* orbital. The ground state has $L = 2$ and $S = 2$ and hence, it is also a ^5D state. This gets split up in an octahedral field into ^5E_g state of lower energy ($t_{2g}^3 e_g^1$, i.e., $t_{2g}^3 d_{x^2-y^2}^1 d_{z^2}^0$ or $t_{2g}^3 d_{x^2-y^2}^0 d_{z^2}^1$) and excited state $^5T_{2g}$ ($t_{2g}^2 e_g^2$, i.e., $d_{xy}^1 d_{xz}^1 d_{yz}^0 e_g^2$, $d_{xy}^0 d_{xz}^1 d_{yz}^1 e_g^2$, and $d_{xy}^1 d_{xz}^0 d_{yz}^1 e_g^2$). Thus, the transition can be shown as $^5T_{2g} \rightarrow {}^5E_g$.

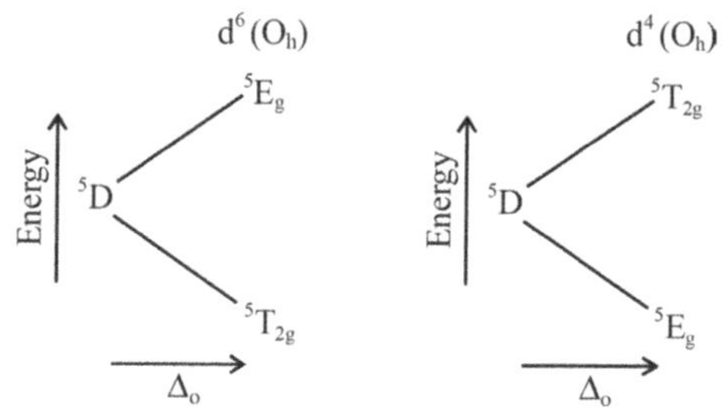

On imposing a field due to the ligand, the spectral states get split up. Two cases may arise, which are weak field and strong field.

(i) Weak field

d^2 Case

The octahedral perturbation is less than the inter-electronic repulsion in such cases. Thus, the splitting of the individual spectral states is less than the separation between them and there cannot be any mixing of the spectral states under the influence of the crystal field. The splitting of the individual spectral states is only considered. The spectral states of d^2 system also split up in the same way as corresponding atomic orbitals as per group theoretical considerations:

State	**Irreducible representation**
^1S	$^1A_{1g}$ (No splitting)
^1G	^1E_g, $^1T_{1g}$, $^1T_{2g}$, $^1A_{1g}$
^3P	$^3T_{1g}$ (No splitting)
^1D	^1E_g, $^1T_{2g}$
^3F	$^3T_{1g}$, $^3T_{2g}$, $^3A_{2g}$

In weak field case only, the splitting of the free ion ground spectral state ^3F and the higher state of some multiplicity ^3P has to be considered. Transitions to split up states from ^1D are possible but they will be both Laporte and spin forbidden and hence, their intensity is very low.

Group theory does not tell about the order of the energies of the split up states of ^3F. This can be understood by seeing the electronic arrangements. ^3F and ^3P states correspond to the microstates with parallel spins. The *d* orbitals get split up into t_{2g} and e_g orbitals in an octahedral field. The lowest energy arrangement is $t_{2g}^{2}\, e_g^{0}$. It has triple degeneracy ($d_{xy}^{1}\, d_{xz}^{1}\, d_{yz}^{0}\, e_g^{0}$, $d_{xy}^{1}\, d_{xz}^{0}\, d_{yz}^{1}\, e_g^{0}$ and $d_{xy}^{0}\, d_{xz}^{1}\, d_{yz}^{1}\, e_g^{0}$ and is termed $^2T_{1g}$. On excitation of electron, the arrangement $t_{2g}^{1}\, e_g^{1}$ is obtained. This results in six possibilities. ($d_{xy}^{1}\, d_{xz}^{0}\, d_{xz}^{0}\, d_{x^2-y^2}^{1}\, d_{z^2}^{0}$, $d_{xy}^{0}\, d_{xz}^{1}\, d_{yz}^{0}\, d_{x^2-y^2}^{2}$, $d_{z^2}^{0}$, $d_{xy}^{0}\, d_{xz}^{0}\, d_{yz}^{1}\, d_{x^2-y^2}^{1}\, d_{z^2}^{0}$, $d_{xy}^{1}\, d_{xz}^{0}\, d_{yz}^{0}\, d_{x^2-y^2}^{0}\, d_{z^2}^{1}$, $d_{xy}^{1}\, d_{xz}^{0}\, d_{yz}^{0}\, d_{x^2-y^2}\, d_{z^2}^{1}\, d_{xy}^{0}\, d_{xz}^{0}\, d_{yz}^{1}\, d_{x^2-y^2}\, d_{z^2}^{1}$).

So $^3T_{1g}$ is the ground state and it is triply degenerate. Next higher term is $^3T_{2g}$ and highest term is $^3A_{2g}$. In $^3T_{1g}$, subscript 1 denotes symmetric with respect to rotation axis other than principal axis of symmetry and in $^3T_{2g}$, the subscript 2 shows that it is antisymmetric with respect to the other rotational axis.

Thus three electronic transitions are possible, $t_{2g}^{2}\, e_g^{0} \rightarrow t_{2g}^{1}\, e_g^{1}$ corresponding to two transitions $T_{2g}\,(F) \leftarrow T_{1g}\,(F)$ and $T_{1g}\,(P) \leftarrow T_{1g}\,(F)$. The second transition $t_{2g}^{2}\, e_g^{0} \rightarrow t_{2g}^{0}\, e_g^{2}$ can be represented in spectral terms as $^3A_{2g}\,(F) \leftarrow {}^3T_{1g}\,(F)$.

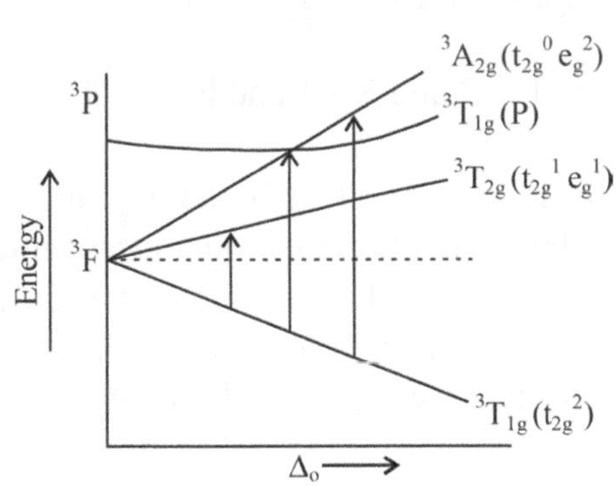

Thus, three bands can be expected in the spectrum of the octahedral complexes of d^2 ions (^3F spectral state). The lowest energy first band $^3T_{2g}$ (F) $\rightarrow$ $^3T_{2g}$ (F) corresponds to transfer of one electron from t_{2g} to e_g orbital and hence, the band energy is equal to the difference in energies of e_g and t_{2g} orbitals, i.e., equal to Δ.

The splitting pattern can be shown as follows:

Δ depends on ligand field. When Δ_o is small then $^3T_{1g}$ (P) and $^3A_{2g}$ (F) tend to cross over, and therefore, only two transitions take place. As a result, 2 bands may be observed.

$$^3T_{1g}\ (F) \longrightarrow\ ^3T_{2g}\ (F)$$

$$^3T_{1g}\ (F) \longrightarrow\ ^3A_{2g}\ (F)\ (=\ ^3T_{1g}\ (P)$$

e. g. in $(V\ (H_2O)_6)^{3+}$ 2 Bands (2 Transitions)
but in $(V\ (NH_3)_6)^{3+}$ 3 Bands (3 Transitions)

d^8 Case

The possible microstates can be worked out in the cases of ions with more than two electrons in *d* orbitals also and the ground state can be obtained. In d^8 case, there are two vacancies (hole) in *d* orbitals. Their positions can be altered in the same way as those of the two electrons and 45 arrangements can be obtained. This corresponds to ^3F, ^3P, ^1D, ^1G, and ^1S, out of which the ground state is ^3F. Ground state can also be calculated by finding the highest values of magnetic quantum number.

L = 3 and S = 1 and hence, ^8F

^3F state gets split up into A_{2g}, T_{1g} (F) and T_{2g} and ^3P gets converted to T_{1g} (P) in the octahedral field. The electronic arrangements corresponding to these states can be understood. In an octahedral field, the ground state arrangement is $t_{2g}^{\ 6}\ e_g^{\ 2}$. This is non-degenerate ($t_{2g}^{\ 6}\ d_{x^2-y^2}^{\ 1}\ d_{z^2}^{\ 1}$) and hence, it is $^3A_{2g}$. On excitation of one electron, $t_{2g}^{\ 5}\ e_g^{\ 3}$ arrangement is obtained. There are six electronic arrangements possible corresponding to T_{2g} (F) and T_{2g} (P). With two-electron excitation, the possible corresponding arrangement is t_{2g}^{4} e_g^{4}. This is triply degenerate ($d_{x^2-y^2}^{2}\, d_{xz}^{1}\, d_{yz}^{1}\, e_g^{4}$, $d_{xy}^{1}\, d_{xz}^{2}\, d_{yz}^{1}\, e_g^{4}$ and $d_{xy}^{1}\, d_{xz}^{1}\, d_{yz}^{2}\, e_g^{4}$). This forms $^3T_{1g}$ (F) state.

Three electronic transitions are also possible. $t_{2g}^{6}\, e_g^{2} \rightarrow t_{2g}^{5}\, e_g^{3}$, which corresponds to $^3T_{2g}$ (F) $\leftarrow$ $^3A_{2g}$ and $^3T_{1g}$ (P) $\leftarrow$ $^3A_{2g}$ transitions while $t_{2g}^{6}\, e_g^{2} \rightarrow t_{2g}^{4}\, e_g^{4}$ corresponds to $^3T_{1g}$ (P) $\leftarrow$ $^3A_{2g}$ transition. Thus, three bands can be expected. The lowest energy band $^3T_{1g} \leftarrow {}^3A_{2g}$ corresponds to Δ. Thus, there is a reversal in the energies of the spectral states in d^8 ions as compared to d^2 case.

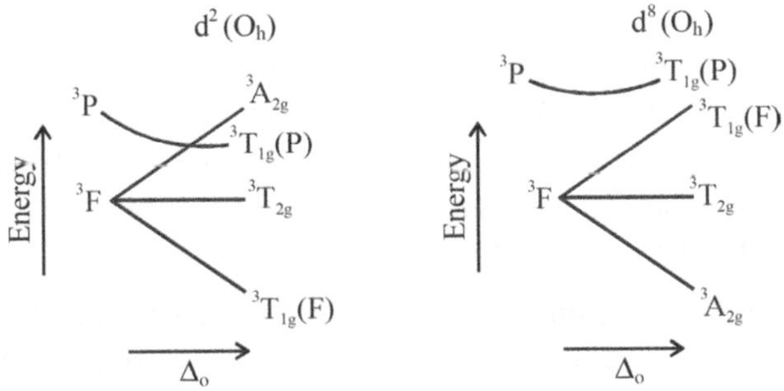

d^7 Case

The free ion ground state is ^{4}F, as highest magnetic quantum number is 3, i.e., L = 3 and S = 3/2.

Another state of same multiplicity is ^{4}P. ^{4}F and ^{4}P states split up into $^4T_{1g}$ (F), $^4T_{2g}$, $^4T_{1g}$ (P), and $^4A_{2g}$, in an octahedral field. The ground state electronic arrangement $t_{2g}^{5}\, e_g^{2}$ corresponds to triply degenerate $^4T_{1g}$. Excitation of one electron gives rise to electronic arrangement $t_{2g}^{4}\, e_g^{3}$ corresponding to T_{2g} (F) and T_{2g} (F). Two-electron transition gives rise to $t_{2g}^{3}\, e_g^{4}$, a non-degenerate A_{2g} state. Thus, the spectral states and their energies are same as in d^2 case and three transitions are possible giving rise to three bands.

$$^4T_{1g} \longrightarrow {}^4T_{2g}$$

$$^{4}T_{1g} \longrightarrow {}^{4}A_{2g}$$

$$^{4}T_{1g}\ (F) \longrightarrow {}^{4}T_{2g}\ (P)$$

d^3 Case

d^3 case has the following electronic arrangement in the free ion.

The highest magnetic quantum number is 3 and S = 3/2 and hence, the ground state is ^4F. This is a case similar to d^8 with two holes. Another spectral state of same multiplicity is ^4P. The splitting pattern of ^4F and ^4P is same as in case of d^8. The ground state is $^4A_{2g}$ and three transitions are possible $^4T_{2g} \leftarrow {}^4A_{2g}$, $^4T_{1g}(F) \leftarrow {}^4A_{2g}$, $^4T_{1g}(P) \leftarrow {}^4A_{2g}$.

Thus, d^2 and d^7 configuration of O_h have same splitting as in case of d^3 and d^8 configuration of T_d while d^3 and d^8 configuration of O_h have same splitting as that of d^2 and d^7 configuration of T_d.

General Orgel diagram:

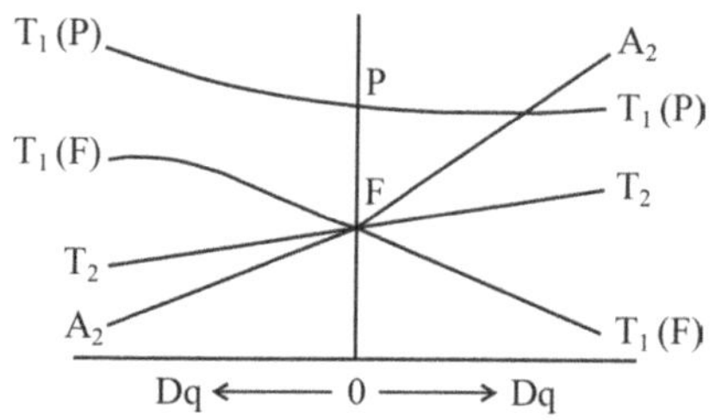

Thus, it can be concluded that

(i) $d^n (T_d) = d^{10-n} (O_h)$

(Same splitting and multiplicity)

(ii) $d^n (T_d)$ inverse of $d^{10-n} (T_d)$

$d^n (O_h)$ inverse of $d^{10-n} (O_h)$

(Inverse of splitting but same multiplicity)

(iii) $d^5 (T_d) \equiv d^5 (O_h)$

Case/Configuration		Ground	Excited term (same multiplicity)
T_d	O_h		
d^1	d^9	$^2E_2$	$^2T_2$
d^2	d^8	$^3A_2$	$^3T_2$, $^3T_1$ (F), $^3T_1$ (P)
d^3	d^7	$^4T_1$ (F)	$^4T_2$, $^4A_2$, $^4T_1$ (P)
d^4	d^6	$^5T_2$	$^5E_2$
d^5	d^5	$^6A_1$	None
d^6	d^4	$^5E_2$	$^2T_2$
d^7	d^3	$^4A_2$	$^4T_2$, $^4T_1$ (F), $^4T_1$ (P)
d^8	d^2	$^3T_1$ (F)	$^3T_2$, $^3A_2$, $^3T_1$ (P)
d^9	d^1	$^2T_2$	$^2E_2$

This table also gives information about number of allowed transition in T_d and O_h.

For example, d^1, d^4, d^6 and d^9 configurations with weak field ligands in O_h give one allowed transition, whereas d^2, d^3, d^7, and d^8 give three spin allowed transitions. But in d^5 case, the spin allowed transition occurs.

(ii) Strong Field

In these complexes, crystal field splitting of a free ion spectral state may be more than the separation between the two free ion states.

Individual splitting of the free ion states are not considered in such cases. Spitting of the *d* orbitals by the ligand field is first considered. The electrons are arranged in the split up *d* orbitals and their interactions are considered to arrive at the probable spectral states resulting on imposing the crystal field.

Let us consider the d^2 case. The free ion states are ^3F, ^3P, ^3D, ^1G and ^1S. Their splitting in a weak O_h field has already been considered. In a strong field also, *d* orbitals get split up into e_g and t_{2g} sets. The possible electronic configurations are $t_{2g}^{\ 2}\ e_g^{\ 0}$, $t_{2g}^{\ 1}\ e_g^{\ 1}$ and $t_{2g}^{\ 0}\ e_g^{\ 2}$. The electrons in the different configurations undergo orbital and spin interactions resulting in microstates.

For $t_{2g}^{\ 2}$ case, orbital degeneracy is three and spin degeneracy is two. There are two unpaired electrons and hence, the number of microstates is equal to $6 \times 5/2 = 15$.

d^5 Ions

The d^5 configuration occurs with Mn(II) and Fe(III) ions. In high spin octahedral complexes formed with weak ligands, for example $[MnF_6]^{4-}$, $[Mn(H_2O)_6]^{2+}$ and

$[FeF_6]^{3-}$, there are five unpaired electrons with parallel spins. Any electronic transition within the *d* level must involve a reversal of spins, and in common with all other 'spin forbidden' transitions, any absorption bands will be extremely weak. This accounts for the very pale pink color of most Mn(II) salts, and the pale violet color of iron(III) alum. The ground state term is ^6S. None of the 11 excited states can be attained without reversing the spin of an electron, and hence, the probability of such transitions is extremely low. Of the 11 excited states, the four quartets ^4G, ^4F, ^4D and ^4P involve the reversal of only one spin. The other seven states are doublets, and these are doubly spin forbidden, and therefore, these are unlikely to be observed. In an octahedral field, these four states split into ten states, and hence, up to ten extremely weak absorption bands may be observed. Several features in the spectrum of $[Mn(H_2O)_6]^{2+}$ are unusual.

The Orgel energy level diagram for octahedral Mn^{2+} is:

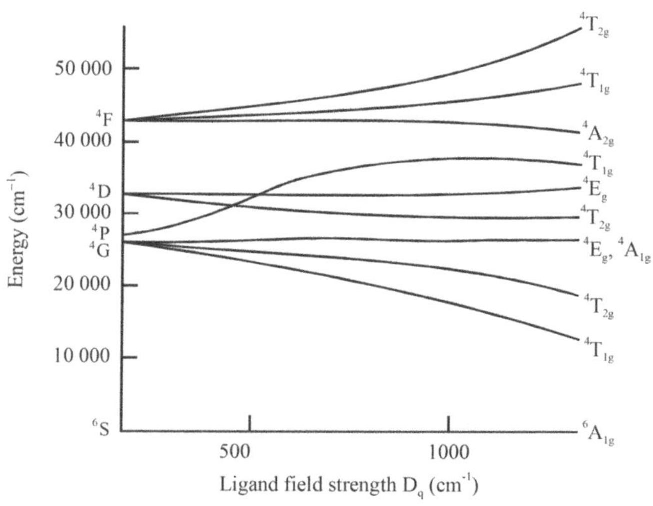

Orgel energy level diagram for Mn^{2+} (d^5) in octahedral field.

Only the quarter terms have been included because transitions to the others are doubly spin forbidden.

It is to be noted that the ground state ^6S does not split, and transforms to the $^6A_{1g}$ state, as shown along the horizontal axis. It is to be noted also that the ^4E_g(G). $^4A_{1g}$, ^4E_g(D), and $^4A_{2g}$(F) terms are also horizontal lines on the diagram, so their energies are independent of the crystal field. The ligands in a complex vibrate about mean positions, so the crystal field strength 10 Dq varies about a mean value. Thus, the energy for a particular transition varies about a mean value, and hence, the absorption peaks are broad. The degree of broadening of the peaks is related to the slope of the lines on the Orgel diagram. Since the slope of the ground state term $^6A_{1g}$ is zero, and the slopes of the ^4E_g (G), $^4A_{1g}$, ^4E_g (D), and $^4A_{2g}$ (F) terms are also zero, transitions from the ground state to these four states should give rise to sharp peaks. By the same reasoning, transitions to states with appreciable slope such as $^4T_{1g}$(G) and $^4T_{2g}$(G) give broader bands.

The bands are assigned as follows:

$^6A_{1g} \rightarrow {}^4T_{1g}$	18,900 cm^{-1}
$^6A_{1g} \rightarrow {}^4T_{2g}$(G)	23,100 cm^{-1}
$^6A_{1g} \rightarrow {}^4E_g$	24,970 and 25,300 cm^{-1}
$^6A_{1g}$ (S) $\rightarrow {}^4A_{1g}$ (G)	
$^6A_{1g} \rightarrow {}^4T_{2g}$ (D)	28,000 cm^{-1}
$^6A_{1g} \rightarrow {}^4E_g$(D)	29,700 cm^{-1}

The same diagram applies to tetrahedral d^5 complexes, if the *g* subscripts are omitted.

9.6 TANABE–SUGANO DIAGRAMS

The simple Orgel energy level diagrams are useful for interpreting spectra, but they have two important limitations:

(i) They treat only the high-spin (weak field) case.

(ii) They are only useful for spin allowed transitions, when the number of observed peaks is greater than or equal to the number of empirical parameters: crystal field splitting Dq, modified Racah parameter B′ and bending constant X.

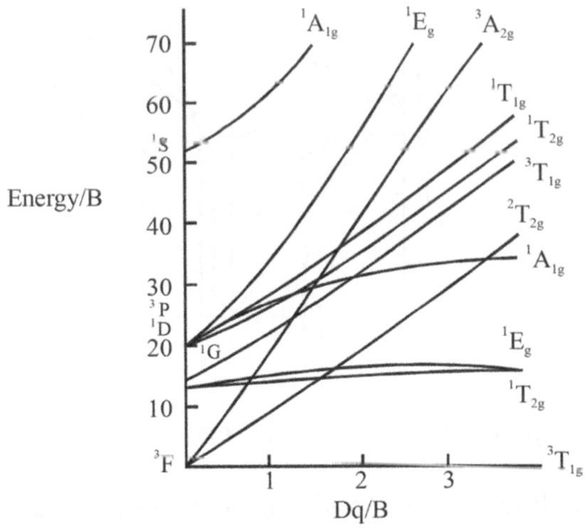

Tanabe–Sugano diagram for d^2 case, i.e., V^{3+}

Though it is possible to add low-spin states to an Orgel diagram, Tanabe–Sugano diagrams are commonly used instead for the interpretation of spectra including both—weak and strong fields. Tanabe–Sugano diagrams are similar to Orgel diagrams as they also show, how the energy levels change with Dq. but they differ in several ways:

(i) The ground state is always taken as the abscissa (horizontal axis) and provides a constant reference point. The other energy states are plotted relative to this.
(ii) Low-spin terms, i.e., states, where the spin multiplicity is lower than the ground state, are included.
(iii) In order to make the diagrams general for different metal ions with the same electronic configuration, and to allow for different ligands both of which affect Dq and B (or B′), the axes are plotted in units of energy/B and Dq/B.

A different diagram is required for each electronic arrangement. Only two examples are given here. The Tanabe–Sugano diagram for a d^2 case such as V^{3+}. It is to be noted that in this case, there is no fundamental difference between strong and weak fields.

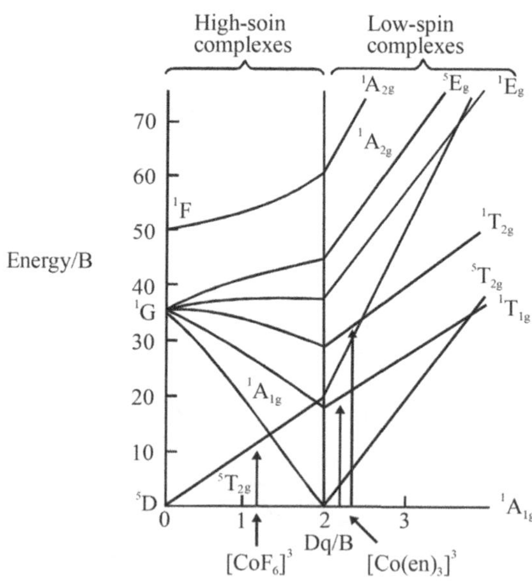

Tanabe–Sugano diagram for d^6 case, i.e., Co^{3+}

This is a simplified version and only the singlet and quintet terms are shown. There is a discontinuity at 10 Dq/B = 20, and this is shown by a vertical line. At this point, spin pairing of electrons occurs. On left of this line, we have high spin complexes (weak ligand field) and on right, we have low spin complexes (strong ligand field). The free ion ground state is ^5D. This is split by an octahedral field into the $^5T_{2g}$ ground state and the ^5E_g excited state. The singlet state in the free ion is of high energy. This is split by the octahedral field into five different states, out of which the $^1A_{1g}$ is important. This state is greatly stabilized by the ligand and drops rapidly in energy as the ligand field strength increases. At the point, where 10 Dq/B = 20, the $^1A_{1g}$ line crosses the horizontal line for the $^5T_{2g}$ state (which is the ground state). At still higher field strengths, the $^1A_{1g}$ state is the lowest in energy, and becomes the ground state. Since the ground state is taken as the horizontal axis, the right hand part of the diagram must be redrawn.

Since the fluoride ion is a weak field ligand, the complex $[CoF_6]^{3-}$ is high spin. The complex is blue in color, and a single peak occurs at 13000 cm^{-1}. This is explained by the transition $^5T_{2g} \rightarrow {}^5E_g$ shown as an arrow in the left hand part of the diagram. The spectrum of a low spin complex such as $[Co(ethylenediamine)_3]^{3+}$ should show the transitions $^1A_{1g} \rightarrow {}^1T_{1g}$ and $^1A_{1g} \rightarrow {}^1T_{2g}$ (shown as two arrows in the right hand part of the diagram).

9.7 ORGEL AND TANABE–SUGANO DIAGRAMS

In the Orgel diagram, the splitting of the free ion spectral state is considered in the weak field, and the energies of the split up spectral states (in cm^{-1}) is plotted as a function of Dq (also in cm^{-1}). Ground state of the complex is derived from the ground spectral state of the free ion and hence, the multiplicity of the ground state in the complex remains same as of the free ion ground state. The splitting pattern of the higher energy free ion states is not of much importance in the Orgel diagram.

However, it has been seen that in a strong field, there is greater interelectronic repulsion and hence, the ground state of the complex may be derived from a free ion state of higher energy and lower multiplicity than the free ion ground state. Hence, the splitting pattern of the ground and also the higher spectral states of the free ion as a function of Dq in weak and strong fields has been considered by Tanabe and Sugano. The diagrams have following characteristics:

(i) In these diagrams, the energies of the split up states divided by Racah parameter B, i.e., E/B is plotted against Dq/B. (The values of E are obtained as in Orgel diagram and the value of B for a metal ion is obtained from the emission spectrum). Thus, the positions of the spectral states are shown as a function of two parameters Dq and B and hence, the Tanabe–Sugano diagrams are valid for all central ions of a particular configuration d^n. For example, d^5 diagram is valid for both; Mn^{2+} and Fe^{3+}, where for each metal ion, the corresponding B value is used.

(ii) In transition state diagrams, the ground term is made the horizontal base line so that the energy of the transition of electrons from the ground to the excited states can be calculated by the vertical distance from the base line. But in Orgel diagram, such determination becomes difficult because no splitting of terms occur as horizontal base line.

(iii) For d^1, d^2, d^3, d^8, and d^9 cases, the ground state remains the same in weak and strong fields. For d^4, d^5, d^6, and d^7 cases, there is crossing of terms for a critical value of field strength. A state with lower multiplicity derived from a higher free ion spectral state becomes ground state with increasing Δ. In case of d^4 metal ions in the weak octahedral field, ground electronic configuration is $t_{2g}^3e_g^1$ and it results in a quintet state ^3E_g derived from free ion ^5D. However, in a strong O_h field, the ground state electronic configuration is $t_{2g}^4e_g^0$ and it results in a triplet $^2T_1$ state derived from free ion ^2H state. The crossing point and the multiplicities of the ground states in other cases, can be seen in the Tanabe–Sugano diagrams. In all these cases, a higher state becomes ground state in the strong field and becomes the base line. Thus, there is a break in the original ground state and a sharp change in the slopes of all the lines. But these breaks in the lines are due to the change in the base line of the diagram and they do not represent any discontinuity in the energies of the states.

(iv) In Tanabe–Sugano diagrams, some spectral states are shown by curved rather than straight lines. This is because of the fact that there is interaction between states of same symmetry and same multiplicity derived from different free ion spectral states.

9.8 CONSTRUCTION OF ENERGY LEVEL DIAGRAMS

Orbital is one electron wave function, and if more than one electron is present in systems, then interelectronic repulsion should be taken into consideration. The electron-electron repulsion gives rise to the energy state called the term. Now the question arises, what the relative energy of these state are and how these energies depend on the strength of the chemical interaction of the ion with its surroundings?

The energy level diagrams are based on the so-called one electron model, even if the atom or ion has more than one *d* electrons, i.e., the effect of interelectronic repulsion has been ignored. First, energy level diagrams of d^2 configuration are developed and then, the effect of interelectronic repulsion and the surrounding environment are added. The separation of the two sets of orbitals into which the group of five *d*-orbitals is split can be taken as a measure of this interaction. The magnitude of Δ_o or Δ_t is plotted as abscissa and energy as ordinate. In free ion term, the value of energies Δ_o or Δ_t is zero.

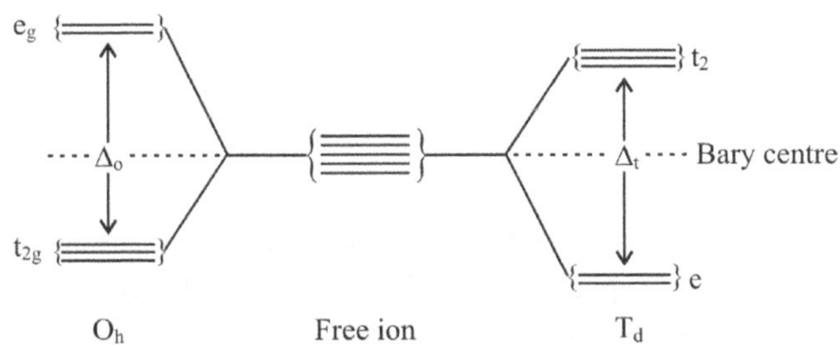

Relative energies of e and t_2 orbital resulting from the splitting of the set of *d* orbitals in O_h and T_d environment.

Now the method of constructing an energy level diagram by treating d^2 in O_h environment will be described. For d^2 configuration, electron-clectron interaction comcs in play; thus, giving rise to ground state free ion term (^3F) and numbers of excited state terms (^3P, ^1G, ^1D and ^1S). Terms in order of increasing energy are:

$$^3F < {}^1D < {}^3P < {}^1G < {}^1S$$

In limit of an extremely large splitting of *d*-orbitals, the following 3 configurations, in order of increasing energy are possible:

$$t_{2g}^{\ 2},\ t_{2g}.\ e_g,\ e_g^{\ 2}$$

Now, if the strong interaction of the environment with the ion is allowed to relax so that the electron starts feeling the presence of one another. It results in coupling, in such a way that it gives rise to a set of state of the entire configuration.

Configuration	Direct product	States
t_{2g}^2	$t_{2g} \times t_{2g}$	$A_{1g} + E_g + T_{1g} + T_{2g}$
$t_{2g}\, e_g$	$t_{2g} \times e_g$	$T_{1g} + T_{2g}$
e_g^2	$e_g \times e_g$	$A_{1g} + A_{2g} + E_g$

These are the symmetries of the orbital state produced by interaction of electrons.

Let us see, how these states can be obtained from direct product. For example, $e_g^{\,2} = e_g \times e_g$ giving A_{1g}, A_{2g}, and E_g states.

O_h	E	C_2	i	S_4
e_g	2	0	2	0
e_g	2	0	2	0
Γ_P	4	0	4	0

The direct product of degenerate representation is a reducible representation. No irreducible representation has order greater than 3; however, the direct product Γ_P must be reducible. Now Γ_P decomposes as A_{1g}, A_{2g}, and E_g.

This decomposition is that sum of the character for the irreducible representations are the character of the reducible representation, Γ_P.

Similarly, $t_{2g}^{\,2} = t_{2g} \times t_{2g}$ gives $A_{1g} + E_g + T_{1g} + T_{2g}$

O_h	E	C_2	C_4
t_{2g}	3	1	−1
t_{2g}	3	1	−1
Γ_P	9	1	1

Since product representation is greater than 3, Γ_P decomposes as discussed earlier. The sum of the character for the irreducible representation is the character of the reducible representation Γ_P. In same way, direct product of $e_g.t_{2g}$ gives $T_{1g} + T_{2g}$. This decomposition can be explained.

Thus, it can be concluded that symmetry of state can be determined by multiplying character of the operations to obtain the total character for two electrons. After then, application of reduction formula to reducible representation will reduce it into the sum of irreducible representations.

The character of the representation of a direct product is equal to the product of the character of the representation based on the individual sets of functions. It is bases of the direct product method.

$$\chi_{\text{Direct product}}(R) = \chi_a(R).\ \chi_b(R).\ \chi_c(R)\ldots\ \chi_i(R)$$

In O_h point group, total character of reducible representations, i.e., direct product representation of the irreducible representations of O_h group are:

O_h	E	$8C_3$	$3C_2\ (=C_4^2)$	$6C_2$	$6C_4$
A_{1g}	1	1	1	1	1
A_{2g}	1	1	1	–1	–1
E_g	2	–1	2	0	0
T_{1g}	3	0	–1	–1	1
T_{2g}	3	0	–1	1	–1
$T_2 \times T_2$	9	0	1	1	1
$T_2 \times E$	6	0	–2	0	0
$E \times E$	4	1	4	0	0

Hence, direct product representation of O_h group for two electrons in t_{2g}, $t_{2g}.e_g$ and e_g are:

O_h	E	$8C_3$	$6C_2$	$6C_4$	$3C_2$	i	$6S_4$	$8S_6$	$3\sigma_h$	$6\sigma_d$
$\Gamma_{T_{2g}\times T_{2g}}$	9	0	1	1	1	9	1	0	1	1
$\Gamma_{T_{2g}'\times T_{2g}'}$	6	0	0	0	–2	6	0	0	–2	0
$\Gamma_{E_g\times E_g}$	4	1	0	0	4	4	0	1	4	0

Now we have to determine multiplications of strong field state. As 2 electrons are involved, they must be either singlet or triplet according to (2 S + 1). Considering first t_{2g}^2 configuration, it may be regarded that as a set of six boxes for t_{2g} level.

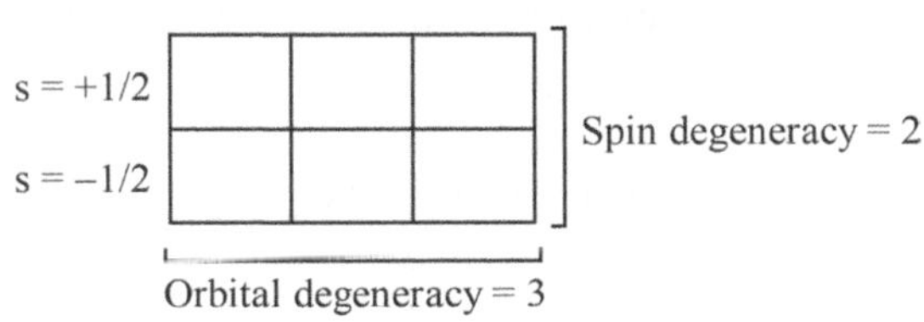

$$\text{Total degeneracy of } t_{2g}^2 = \frac{6 \times 5}{2} = 15.$$

Here, 2 in denominator stands for indistinguishability of the electrons.

It means that the number of ways, in which two electrons can occupy the 6 boxes, are 15. When strong field is decreased (relaxed), then the orbital state separate into A_{1g}, E_g, T_{1g} and T_{2g}. Total degeneracy of these states must remain 15.

$$t_{2g} \times t_{2g} = a.A_{1g} + b.E_g + c.T_{1g} + d.T_{2g}$$

$$1.a + 2.b + 3.c + 3.d = 15$$

where *a, b, c,* and *d* are either 1 or 3. Now making combination of 1 and 3 for *a, b, c,* and *d* in such a way that on putting these values in this equation should give total as 15. So such possibilities can be:

	a	b	c	d
I	1	1	1	3
II	1	1	3	1
III	3	3	1	1

I set $1 \times 1 + 2 \times 1 + 3 \times 1 + 3 \times 3 = 15$

II set $1 \times 1 + 2 \times 1 + 3 \times 3 + 3 \times 1 = 15$

III set $3 \times 1 + 2 \times 3 + 3 \times 1 + 3 \times 1 = 15$

Set second seems to be more correct, which can be proved by correlation diagram.

For $t_{2g}.e_g$ configuration, one can place electron in any of the 6 boxes and other electron in any of the 4 boxes; thus, total 24 arrangements are possible. Here, it must be noted that there is no possibility of two electrons being in the same box. It means that in all arrangements spin may be either paired or unpaired. Configuration $t_{2g}.e_g$ gives rise to T_{1g} and T_{2g} states.

These states may be singlet or triplet. Thus, $t_{2g}.e_g$ gives ${}^1T_{1g}$, ${}^3T_{1g}$, ${}^1T_{2g}$, and ${}^3T_{2g}$ states.

$$t_{2g}.e_g = {}^1T_{1g} + {}^3T_{1g} + {}^1T_{2g} + {}^3T_{2g}$$

$$3 \times 1 + 3 \times 3 + 3 \times 1 + 3 \times 3 = 24$$

Similarly, in e_g^2 configuration, two electrons can be placed in four boxes. Thus, total degeneracy will be $\frac{4 \times 3}{2} = 6$.

$$e_g \times e_g = a.A_{1g} + b.A_{2g} + c.E_g$$

$$1.a + 1.b + 2.c = 6$$

Possible combination of 1 and 3 for *a, b, c* can be

	a	b	c
I	1	3	1
II	3	1	1

I set $1 \times 1 + 1 \times 3 + 2 \times 1 = 6$

II sct $1 \times 3 + 1 \times 1 + 2 \times 1 = 6$

Set second is correct, which can also be confirmed by correlation diagram.

Order of energy in spectral state can be obtained by modified Hund's rules.

(a) Higher is the multiplicity (2S + 1), lower will be the energy.
(b) When 2S + 1 is same, then orbital degeneracy is considered. Higher is the orbital degeneracy, lower will be the energy.

e.g.,	T	〉	E	〉	A
	Triply degenerate		Doubly degenerate		Non – degenerate

Correlation Diagram of d^2 Ions

In correlation diagram, extreme left shows free ion with zero field ligand field and extreme right shows complex with strong ligand field. The increasing order of energy (in strong field) has configuration $(t_{2g})^2$ (ground state), $(t_{2g})^1$, $(e_g)^1$ and $(e_g)^2$ at highest state. The electronic states for some d orbitals are given in Table 9.4.

Construction of correlation diagram follows two principles:

(i) As one goes from weak to the strong interaction with the environment, the symmetry properties of the system is not changed. Thus, there must be same number of each kind of state through out.

TABLE 9.4 Electronic Configuration

Free ion	Ion in octahedral	Electronic state
d^1, d^9	$(e_g)^1$	^2E_g
	$(t_{2g})^1$	$^2T_{2g}$
d^2, d^8	$(e_g)^2$	$^3A_{2g} + {}^1A_{1g} + {}^1E_g$
	$(t_{2g})^1 (e_g)^1$	$^3T_{1g} + {}^3T_{2g} + {}^1T_{1g} + {}^1T_{2g}$
	$(t_{2g})^2$	$^3T_{1g} + {}^1A_{1g} + {}^1E_g + {}^1T_{2g}$

(ii) As the strength of the interaction changes, state of same spin degeneracy and symmetry cannot cross. It is called non-crossing rule.

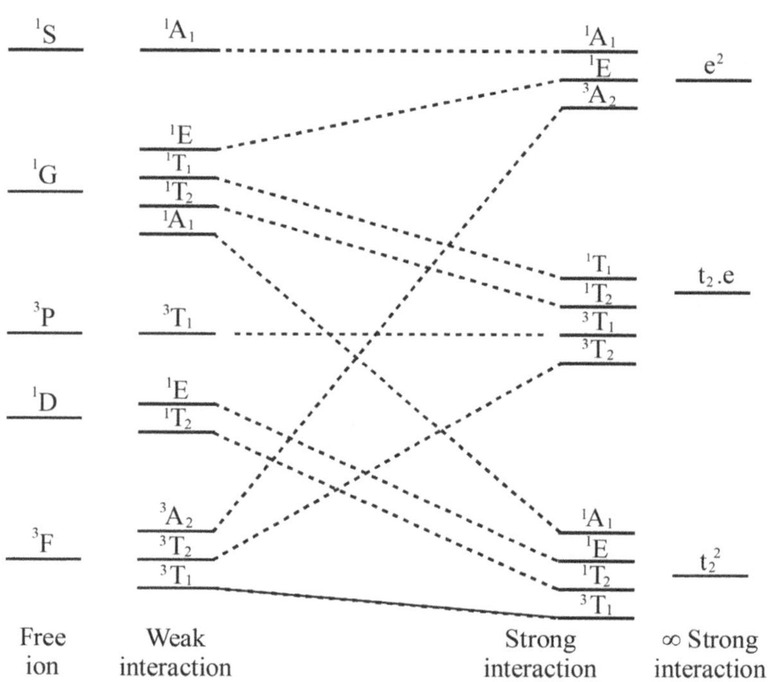

Correlation diagram for O_h

In this diagram, left side shows the state of free ion and right side shows the state, into which these free ion state split under the influence of octahedral environment.

Now, the multiplicity of all states is known. At extremely right side (hypothetical), a case of strong interaction, whereas immediately to its left side is strong but not infinitely strong. In order that each state on the left side goes over into a state of the same kind on the right without violation of the non-crossing rule.

9.9 MOLECULAR ORBITALS FOR σ-BONDING IN ML_n COMPLEXES

9.9.1 TETRAHEDRAL COMPLEXES (ML_4)

In tetrahedral complex, $[ZnCl_4]^{2-}$, the ligands L are at the corners of tetrahedron as:

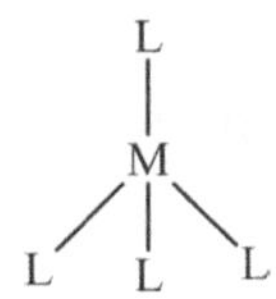

It can be represented in cube as:

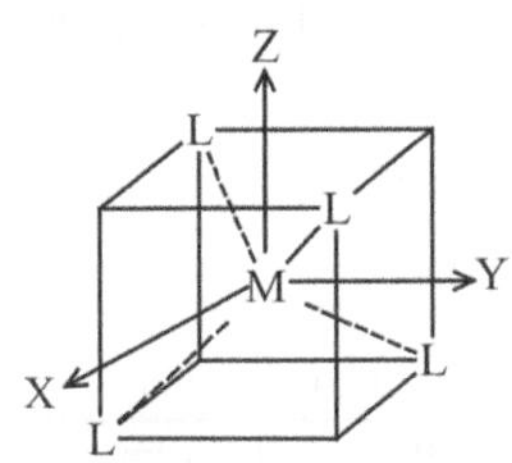

The operations of T_d point group are performed on ligand AO's to determine the symmetries of the σ MOs.

We first need to find out the reducible representation, for which entire set of σ orbitals form a basis. For this purpose, we may represent each σ orbital by a vector pointing from M to L atom, and denote these vectors as r_1, r_2, r_3, and r_4.

(i) Applying E (Identity) operation, we obtain new vector set as r_1', r_2', r_3', and r_4'.

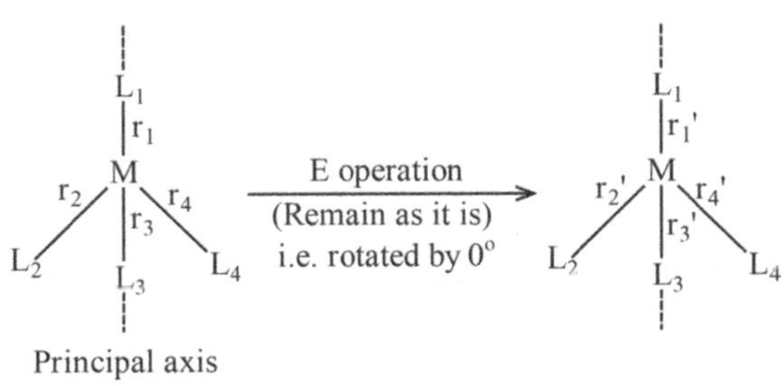

Because if the complex is rotated through 0° or 360°, all the vectors remained at their respective positions.

The new vectors are related to original vectors as:

$$r_1' = 1.r_1 + 0.r_2 + 0.r_3 + 0.r_4$$
$$r_2' = 0.r_1 + 1.r_2 + 0.r_3 + 0.r_4$$
$$r_3' = 0.r_1 + 0.r_2 + 1.r_3 + 0.r_4$$
$$r_4' = 0.r_1 + 0.r_2 + 0.r_3 + 1.r_4$$

$$\begin{bmatrix} r_1' \\ r_2' \\ r_3' \\ r_4' \end{bmatrix} = \begin{bmatrix} 1 & 0 & 0 & 0 \\ 0 & 1 & 0 & 0 \\ 0 & 0 & 1 & 0 \\ 0 & 0 & 0 & 1 \end{bmatrix} \begin{bmatrix} r_1 \\ r_2 \\ r_3 \\ r_4 \end{bmatrix}$$

Matrix representation

(ii) For C_3 operation, set of vector is rotated by $2\pi/3 = 120°$ about the C_3 axis, and we get:

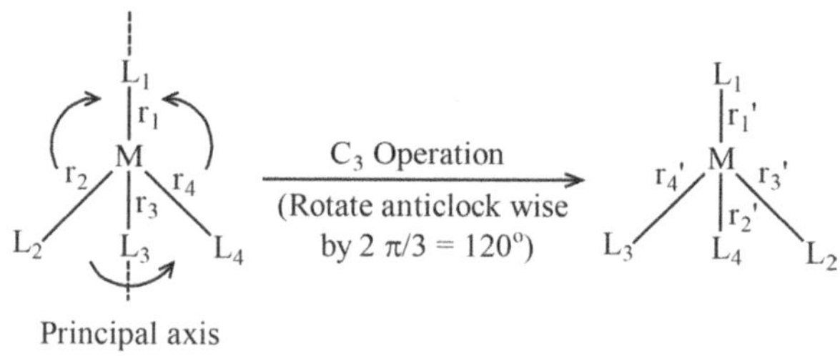

$$\begin{bmatrix} r_1' \\ r_2' \\ r_3' \\ r_4' \end{bmatrix} = \begin{bmatrix} 1 & 0 & 0 & 0 \\ 0 & 0 & 0 & 1 \\ 0 & 1 & 0 & 0 \\ 0 & 0 & 1 & 0 \end{bmatrix} \begin{bmatrix} r_1 \\ r_2 \\ r_3 \\ r_4 \end{bmatrix}$$

Matrix representation

$$\chi(C_3) = 1$$

Proceeding in the same way for C_2, S_4 and σ_d operation.

C_2 operation

$$r_1' = 0.r_1 + 0.r_2 + 0.r_3 + 1.r_4$$
$$r_2' = 0.r_1 + 0.r_2 + 1.r_3 + 0.r_4$$
$$r_3' = 0.r_1 + 1.r_2 + 0.r_3 + 0.r_4$$
$$r_4' = 1.r_1 + 0.r_2 + 0.r_3 + 0.r_4$$

$$\begin{bmatrix} r_1' \\ r_2' \\ r_3' \\ r_4' \end{bmatrix} = \begin{bmatrix} 0 & 0 & 0 & 1 \\ 0 & 0 & 1 & 0 \\ 0 & 1 & 0 & 0 \\ 1 & 0 & 0 & 0 \end{bmatrix} \begin{bmatrix} r_1 \\ r_2 \\ r_3 \\ r_4 \end{bmatrix}$$

$$\chi(C_2) = 0$$

S_4 operation

$$r_1' = 0.r_1 + 0.r_2 + 1.r_3 + 0.r_4$$
$$r_2' = 0.r_1 + 0.r_2 + 0.r_3 + 1.r_4$$
$$r_3' = 1.r_1 + 0.r_2 + 0.r_3 + 0.r_4$$
$$r_4' = 0.r_1 + 1.r_2 + 0.r_3 + 0.r_4$$

$$\begin{bmatrix} r_1' \\ r_2' \\ r_3' \\ r_4' \end{bmatrix} = \begin{bmatrix} 0 & 0 & 1 & 0 \\ 0 & 0 & 0 & 1 \\ 1 & 0 & 0 & 0 \\ 0 & 1 & 0 & 0 \end{bmatrix} \begin{bmatrix} r_1 \\ r_2 \\ r_3 \\ r_4 \end{bmatrix}$$

$$\chi(S_4) = 0$$

σ_d operation

$$r_1' = 1.r_1 + 0.r_2 + 0.r_3 + 0.r_4$$
$$r_2' = 0.r_1 + 0.r_2 + 1.r_3 + 0.r_4$$
$$r_3' = 0.r_1 + 1.r_2 + 0.r_3 + 0.r_4$$
$$r_4' = 0.r_1 + 0.r_2 + 0.r_3 + 1.r_4$$

$$\begin{bmatrix} r_1' \\ r_2' \\ r_3' \\ r_4' \end{bmatrix} = \begin{bmatrix} 1 & 0 & 0 & 0 \\ 0 & 0 & 1 & 0 \\ 0 & 1 & 0 & 0 \\ 0 & 0 & 0 & 1 \end{bmatrix} \begin{bmatrix} r_1 \\ r_2 \\ r_3 \\ r_4 \end{bmatrix}$$

$$\chi(\sigma_d) = 2$$

Finally, the following set of character for the representations is generated:

T_d	E	$8C_3$	$3C_2$	$6S_4$	$6\sigma_d$
Γ_{tetra}	4	1	0	0	2

An easier way to achieve this is also available. The character is equal to the number of vectors that are unshifted by the operation.

Using character table of T_d group, this representation can be reduced in the following way with help of reduction formula.

$$\Gamma_{Tetra} = A_1 + T_2$$

Thus, there are four MOs that will be equivalent to the set of four σ orbitals. One orbital of A_1 symmetry and three orbitals of T_2 representation. The character table shows that AOs of the M fall into these categories:

Irreducible representation	**Orbitals**
A_1	s
T_2	p_x, p_y, p_z or d_{xy}, d_{xz}, d_{yz}

The central atom M uses appropriate set of p orbitals for σ-bonding of SiF_4, $AlCl_4^-$, $ZnCl_4^{2-}$, etc.

We now know, which particular AOs of the central atom will be used to form the MOs of A_1 and T_2 symmetry?

Now SALCs are to constructed by employing the projection operation technique.

For A_1, SALC must have the same symmetry as the s orbital on atom M. s orbital is spherical and it has positive signs. Four ligand orbitals combine with s orbital. Thus, normalized A_1 SALC has to be:

$$s = \sigma_1 + \sigma_2 + \sigma_3 + \sigma_4$$

For T_2 SALCs must match the symmetries of the p-orbitals on the atom M. The combination must be as follows to match the p orbital.

$$p_z = \sigma_1 - \sigma_2 - \sigma_3 + \sigma_4$$

$$p_x = \sigma_1 - \sigma_2 + \sigma_3 - \sigma_4$$

$$p_y = \sigma_1 + \sigma_2 - \sigma_3 - \sigma_4$$

Now one can form MO's by bringing the central atom orbital and the SALCs together to give positive or negative overlap; thus, forming a bonding or an antibonding MO (Both have same symmetry, but quite different energy). The bonding combination, ψ_b, is slightly lower in energy and the antibonding, ψ_a, is high by same amount of energy.

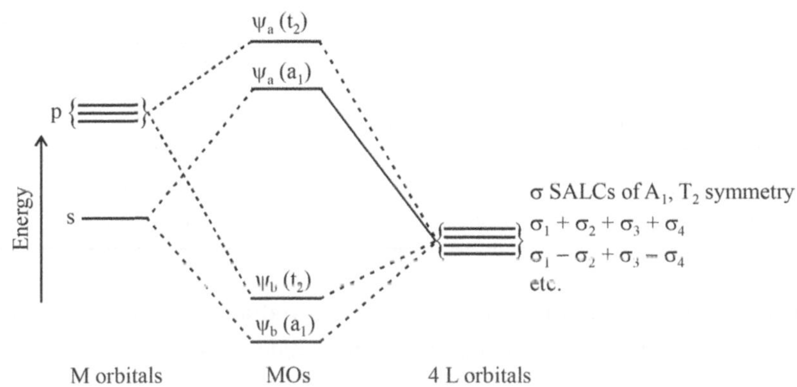

MO energy level diagram for ML_4 like tetrahedral complex showing both; the A_1 and T_2 type interactions

9.9.2 OCTAHEDRAL COMPLEXES (ML_6)

In octahedral complex, ML_6, we need a set of six σ-bonding orbitals, which will give rise to following representation:

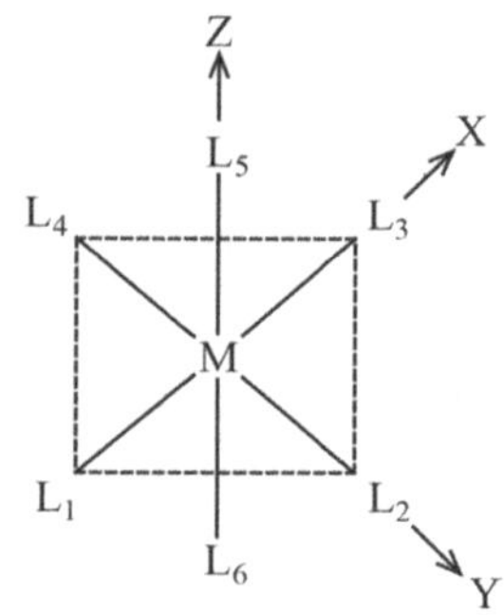

O_h	E	$8C_3$	$6C_2$	$6C_4$	$3C_2$	i	$6S_4$	$8S_6$	$3\sigma_h$	$3\sigma_d$
Γ_σ	6	0	0	2	2	0	0	0	4	2

An easier way to determine these representations is there without writing complete matrix, i.e., The character is equal to number of vectors that are unshifted by the operation.

Operation	**Unshifted vector**	χ
Identity operation	All vectors	6
Rotation operation, which are not on axes X–, Y–, and Z– (i.e., bond axis)	No vector	0
Rotation along X–, Y–, and Z– (bond axis)	Two vectors	2
σ_h operation	Four vectors	4

Thereafter, this reducible representation is reduced by reduction formula to know the contribution of irreducible representations as:

$$\Gamma_\sigma = A_{1g} + E_g + T_{1u}$$

Irreducible representation	**Orbitals**
A_{1g}	s
E_g	$d_{z^2}, d_{x^2-y^2}$
T_{1u}	p_x, p_y, p_z

Central atom M also has T_{2g} symmetry, which belongs to d_{xy}, d_{yz}, and d_{xz}. But for σ-bonding, it requires SALCs of A_{1g}, E_g and T_{1u}. Therefore,

T_{2g} remain non-bonding for σ system. It is therefore possible to make the full set of MOs, because all the necessary orbitals are available on the central atom.

Now constructing SALCs by projection operation technique assuming that the σ bonds are oriented towards the central atom with its positive lobe. For A_{1g}, SALC must match the totally symmetry atomic s orbital and it has positive sign. Six ligand orbitals (LO's) are combined. Thus, the normalized A_{1g} SALC has to be:

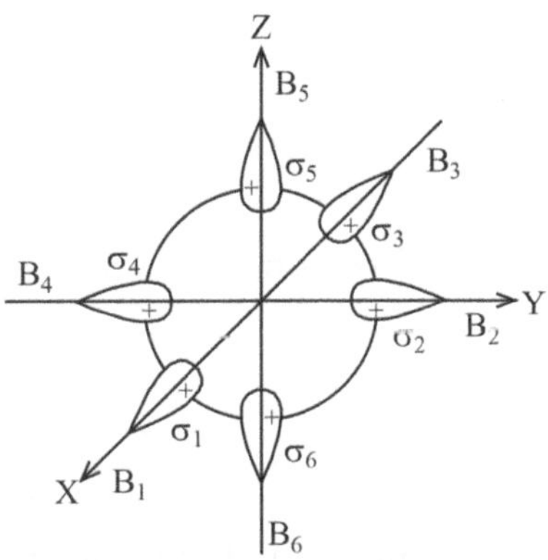

$$\sqrt{\frac{1}{6}}\,(\sigma_1+\sigma_2+\sigma_3+\sigma_4+\sigma_5+\sigma_6)$$

For T_{1u}, SALCs, each one must match to one of the p orbital of central atom.

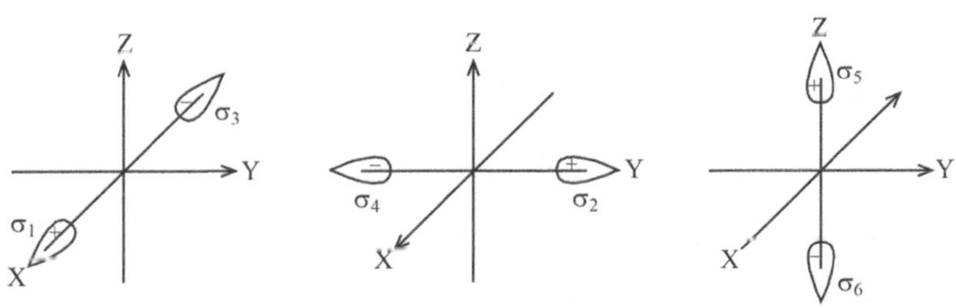

$$p_x=\sqrt{\frac{1}{2}}\,(\sigma_1-\sigma_3)\quad p_y=\sqrt{\frac{1}{2}}\,(\tilde{A}_2-\tilde{A}_4)\quad p_z=\sqrt{\frac{1}{2}}\,(\tilde{A}_5-\tilde{A}_6)$$

For E_g, SALCs, we required combination that match d_{z^2} and $d_{x^2-y^2}$.

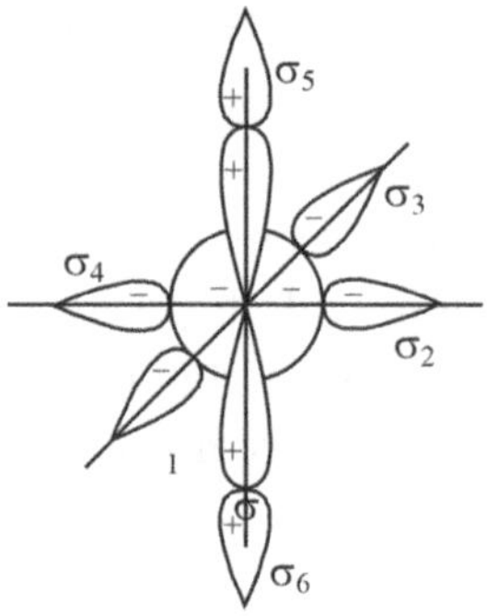

$$d_{z^2} = \sqrt{\frac{1}{12}}\,(2\sigma_5 + 2\sigma_6 - \sigma_1 - \sigma_2 - \sigma_3 - \sigma_4)$$

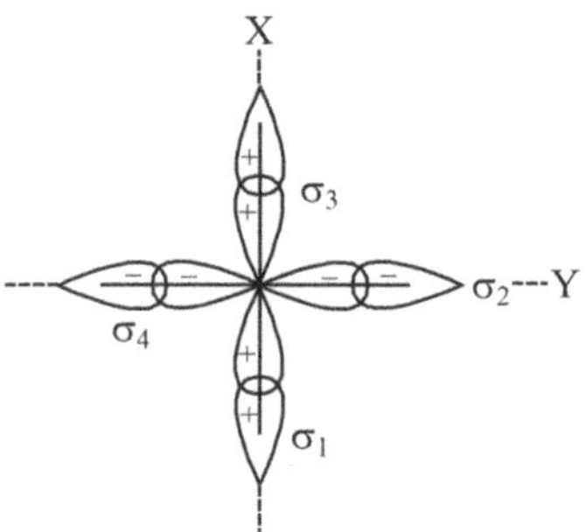

$$d_{x^2-y^2} = \frac{1}{2}(\sigma_1 - \sigma_2 + \sigma_3 - \sigma_4)$$

The MO energy can be obtained by combining AO of metal atom with ligand group orbitals (LGO) of ligand atoms having same symmetry.

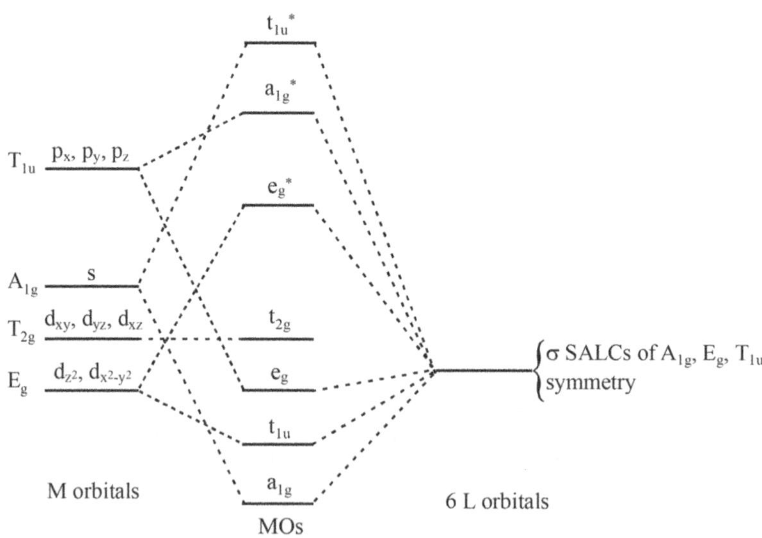

MO diagram for octahedral AB_6 complex, in which only σ bonds are there. Asterisk (*) denotes an antibonding orbital.

9.10 MOLECULAR ORBITALS FOR π-BONDING IN ML_n COMPLEXES

For many AB_n molecule, π-bonding as well as σ-bonding is important, specially, compounds with π-acceptor ligands such as metal carbonyls and oxometallates.

Example of this type are MnO_4^-, $Ni(CO)_4$, $Cr(CO)_6$, etc.

9.10.1 TETRAHEDRAL COMPLEXES (ML_4)

In tetrahedral complex, each ligand L contains two p orbitals, apart from a p σ orbital, which are capable of forming π bonds. Vector p_x and p_y are perpendicular to each other on each atom L and to M-L σ bond.

With this vector set as a basis, one obtains following results:

T_d	E	$8C_3$	$3C_2$	$6S_4$	$6\sigma_d$
Γ_π	8	−1	0	0	0

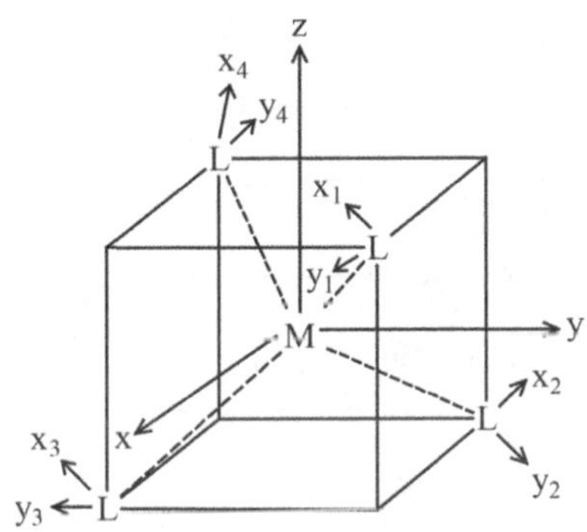

A set of vector representations; π-type p orbitals on the four L atoms of ML_4 complex (ML_4 type).

The reducible representation can now be reduced in terms of irreducible representation as:

$$\Gamma_\pi = E + T_1 + T_2$$

SALCs of E, T_1, and T_2 symmetries can be obtained from eight π orbitals from 4 L (ligands) atoms. But, π MOs form only, when appropriate symmetry AOs are present on atom M.

T_d character table shows that the following AOs are available:

Irreducible representation	**Orbitals**
E	d_{z^2}, $d_{x^2-y^2}$
T_1	None
T_2	p_x, p_y, p_z or d_{xy}, d_{yz}, d_{xz}

This results in two consequences:

- There is no orbital on atom M having T_1 symmetry.
- Atom M with orbitals having T_2 symmetry are better suited for σ-bonding. Here, two sets of T_2 type AOs are present; thus, it is possible to have σ and π MOs of T_2 symmetry.

Thus, MO energy level diagram representing σ- and π-bonding is:

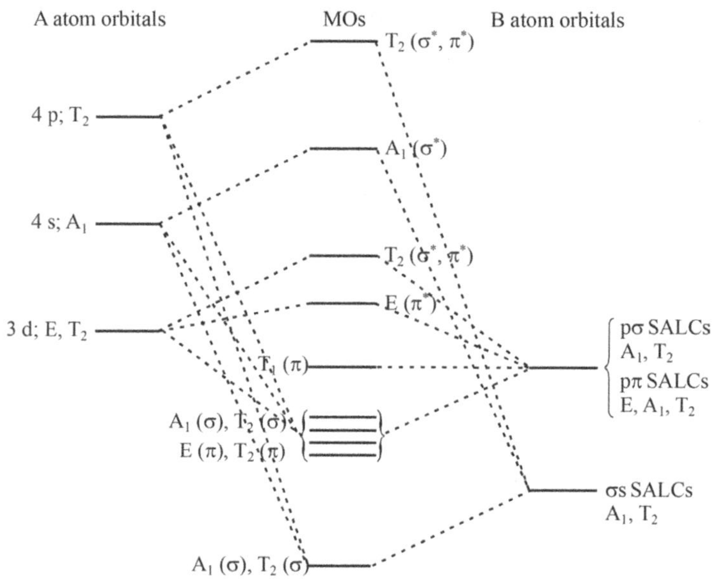

Approximate MO diagram for T_d (ML_4) complex

9.10.2 OCTAHEDRAL COMPLEXES (ML_6)

Each atom L has two p orbitals, which are mutually perpendicular and also perpendicular to M – L σ bond; thus, forming complete 12 sets of M – L π bond. The following results are obtained:

O_h	E	$8C_3$	$6C_2$	$6C_4$	$3C_2$	i	$6S_4$	$8S_6$	$3\sigma_h$	$6\sigma_d$
Γ_π	12	0	0	0	–4	0	0	0	0	0

Reducible representation Γ_π can be reduced to irreducible representations as:

$$\Gamma_\pi = T_{1g} + T_{2g} + T_{1u} + T_{2u}$$

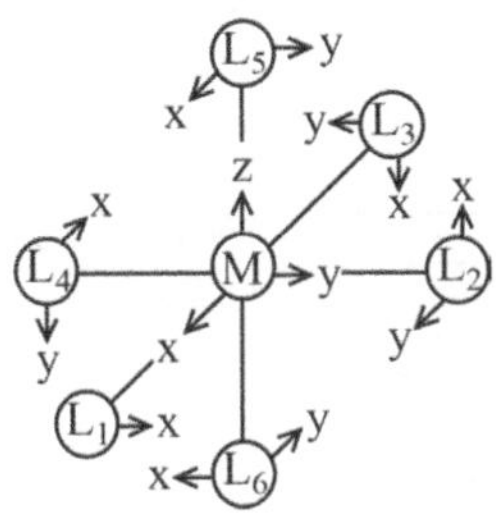

Coordinate system for an octahedral AB_6 complex

There are no M atom orbitals with T_{2u}, and T_{1g} symmetry. On the contrary, M atom orbitals with T_{1u} symmetry are better suited for σ overlap. Thus, we are left with only T_{2g} for M – L π bonding.

On inspection of character table of O_h group, the following AOs are available in it.

Irreducible representation	**Orbitals**
T_{2g}	d_{xy}, d_{xz}, d_{yz}
T_{1u}, T_{2u}, T_{1g}	None

$$\psi_{T_{2g}} = \left\{ \begin{array}{l} \frac{1}{2}(p_y^{\,1} + p_x^{\,5} + p_x^{\,3} + p_y^{\,6}) \\ \frac{1}{2}(p_x^{\,1} + p_y^{\,5} + p_y^{\,4} + p_x^{\,6}) \\ \frac{1}{2}(p_x^{\,1} + p_y^{\,2} + p_y^{\,3} + p_x^{\,4}) \end{array} \right\} \text{ Matching } \left\{ \begin{array}{l} d_{xz} \\ d_{yz} \\ d_{xy} \end{array} \right\}$$

Thus, there remains SALCs, T_{1u}, T_{2u}, and T_{2g} symmetry that are non-bonding in character.

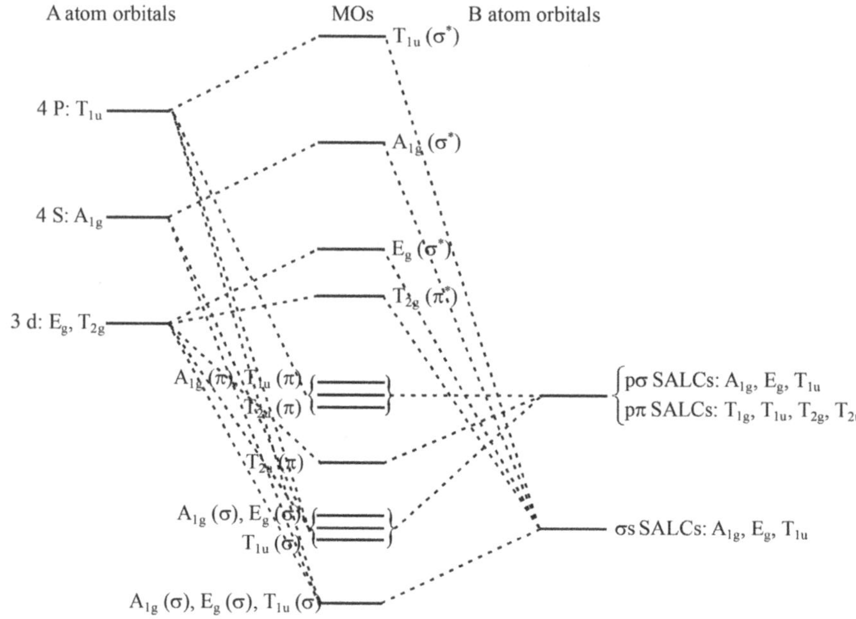

MO diagram for an octahedral ML_6 complex

9.11 METHOD OF DESCENDING SYMMETRY

The complete correlation diagram can be constructed but in this process, some problems are faced. A straightforward and general approach, called the method of descending symmetry, was developed that helps us to construct correlation diagram or any configuration.

In case of d^2, if ML_6 molecule of O_h symmetry is taken and as one descends (lower) the symmetry, the new point group as subgroup (D_{4h}, C_{4v}, C_{2v}, C_{3v}) of old one form. Here, we consider O_h symmetry, by descent of symmetry from D_{4h} (trans pair of ligand in an O_h ML_6 complex/molecule move out to a greater distance than the other four.

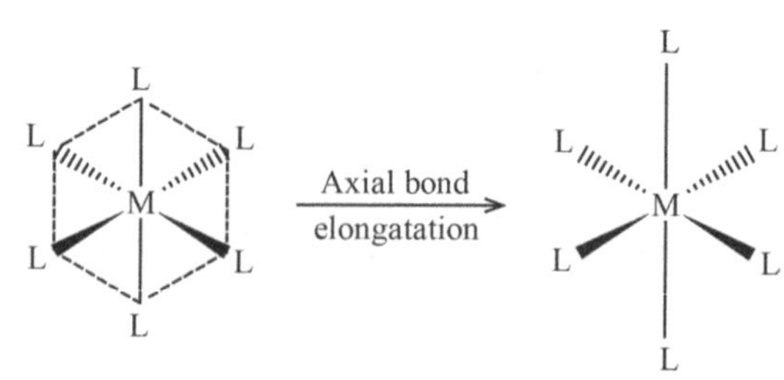

Direct product of e_g ($e_g \times e_g$) give A_{1g}, A_{2g} and E_g states as a result of electron interaction in O_h symmetry.

$$e_g \times e_g = e_g^{\,2} = A_{1g} + A_{2g} + E_g \ (O_h)$$

O_h	E	C_2	i	S_4
e_g	2	0	2	0
e_g	2	0	2	0
Γ_P	4	0	4	0

The direct product of degenerate representation is a reducible representation. No irreducible representation has order greater than 3, and therefore, the product Γ_P must be reducible representation. Now Γ_P decomposes as A_{1g}, A_{2g}, and E_g.

Therefore, on descending symmetry O_h to D_{4h}, e_g orbital split into a_{1g} and b_{1g}.

$$A_{1g} + B_{1g} + A_{1g} + B_{1g} (D_{4h})$$

Above state on lower symmetry show below correlation.

O_h	D_{4h}
A_{1g}	A_{1g}
A_{2g}	B_{1g}
E_g	$\begin{cases} A_{1g} \\ B_{1g} \end{cases}$

Again above decomposition can be proved in some way as previously deduced. On lowering symmetry cannot change spin degeneracies.

O_h	D_{4h}
$^1A_{1g}$	$^1A_{1g}$
$^3A_{2g}$	$^3B_{1g}$
^1Eg	$\begin{cases} ^1A_{1g} \\ ^1B_{1g} \end{cases}$

e_g^2 configuration with two electron can be placed in a_{1g} and b_{1g} level in number of ways.

The one electron in e_g orbital (O_h symmetry) goes over the level a_{1g} and b_{1g}, where as symmetry is lowered to D_{4h}.

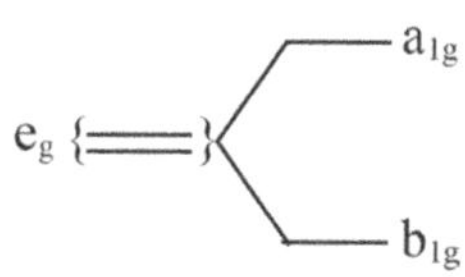

O_h D_{4h}

O_h	E	C_2	C_3	i	S_4
E_g	2	0	0	2	0

D_{4h}	E	C_2	C_3	i	S_4
A_{1g}	1	1	1	1	1
B_{1g}	1	−1	−1	−1	−1
	2	0	0	2	0

These tables show correlation of symmetry operation of O_h with sum of character of symmetry operation of A_{1g} and B_{1g}. Both are equal.

The number of ways in which we can place 2 electrons in a_{1g} and b_{1g}.

Case I: Both electrons are in a_{1g}

D_{4h}	E	C_2	C_4
a_{1g}	1	1	1
a_{1g}	1	1	1
A_{1g}	1	1	1

In D_{4h}, direct product of a_{1g}, a_{1g}, formed by multiplying the character of the two representations is the irreducible representation of A_{1g}.

If all the combined irreducible are non-degenerate, then the product will also be non-degenerate representation, i.e., the product of non-degenerate representation is non-degenerated.

Both electrons are in same levels. Therefore, A_{1g} must be singlet.

Case II: One electron in a_{1g} while the other electron is in b_{1g}

D_{4h}	E	C_2	C_4
a_{1g}	1	1	1
b_{1g}	1	1	−1
B_{1g}	1	1	−1

Direct product of a_{1g} and b_{1g} formed by multiplying the character of the two representations is the irreducible representation of B_{1g}.

In a_{1g}. b_{1g} configuration, two electrons have different orbital state, hence B_{1g} state resulting from $a_{1g}.b_{1g}$ configuration can be singlet $^1B_{1g}$ or triplet $^3B_{1g}$.

Case III: Both the electrons are in b_{1g}, case similar to the case a_{1g}. As per exclusion principle, now, b_{1g}^2 multiplicity must be singlet as both electrons are in same orbital.

Let us now proceed to the state arising from t_{2g}^2 configuration.

O_h	E	C_2	C_4
t_{2g}	3	1	−1
t_{2g}	3	1	−1
Γ_P	9	1	1

Since product representation is greater than 3, Γ_P is reducible representation and it decomposes as:

$$\Gamma_P = A_{1g} + E_g + T_{1g} + T_{2g}$$

This decomposition is that sum of the character for the irreducible representation is equal to the character of the reducible representation Γ_P.

Lowering of O_h results into subgroup C_{2h} and C_{2v}. Let us now proceed with C_{2h} representation. In case of C_{2h}, states E_g, T_{1g}, T_{2g} of O_h splits into different sum of one-dimensional representations.

O_h	C_{2h}
A_{1g}	A_{1g}
Eg	$\begin{cases} A_g \\ B_g \end{cases}$
T_{1g}	$\begin{cases} A_g \\ B_g \\ B_g \end{cases}$
T_{2g}	$\begin{cases} A_g\ (1) \\ A_g\ (2) \\ B_g \end{cases}$

Two electron in t_{2g}^2 goes into a_g (1), a_g (2) and b_g in C_{2h} in six different ways.

These are as follows:

Both the electrons are in first a_g.

$$a_g^2\ (1) = A_g$$

One electron in first a_g and another electron in second a_g.

$$a_g^1\ (1) \times a_g^{\ 1}\ (2) = A_g$$

One electron in first a_g and second electron in b_g.

$$a_g^1\ (1) \times b_g^1 = B_g$$

Both the electrons are in second a_g.

$$a_g^2\ (2) = A_g$$

One electron in second a_g and second electron in b_g.

$$a_g^1\ (2) \times b_g^1 = B_g$$

Both the electrons are in b_g.

$$b_g^2 = A_g$$

$$a_g^1\,(1) \times a_g^1\,(1) = A_g \qquad \text{(Singlet)}$$

$$a_g^1\,(1) \times a_g^1\,(2) = A_g \qquad \text{(Triplet)}$$

$$a_g^1\,(1) \times b_g^1\,(1) = B_g \qquad \text{(Triplet)}$$

$$a_g^1\,(2) \times a_g^1\,(2) = A_g \qquad \text{(Singlet)}$$

$$a_g^1\,(2) \times b_g^1 = B_g \qquad \text{(Triplet)}$$

$$b_g^1 \times b_g^1 = A_g \qquad \text{(Singlet)}$$

If both the electrons are in one level, they can exist only in paired form and therefore, a singlet state is obtained. It is true for a_g (1) × a_g (1), a_g (2) × a_g (2) and $b_g \times b_g$. On the other hand, it will give a triplet state, if the electrons are pressed in separate levels, i.e., a_g (1) × a_g (2), a_g (1) × b_g, and a_g (2) × b_g.

9.12 CAGE AND CLUSTER COMPOUNDS

Cluster compounds are ensemble of bound atoms, which form a polygonal or polyhedral array to which ligands are attached by direct or substantial bonding. In most of the cases, nothing is present in center, although in some cases, a small central atom such as H, Be, B, C, N, or Si is present.

9.12.1 *POLYHEDRAL BORANES ($B_6H_6^{2-}$)*

In boranes, each boron atom has four valence shell orbitals, s, p_x, p_y, and p_z. In this case, same coordination system is selected as was used in an octahedral ML_6 case.

Irreducible representations are obtained in the same manner as for working out SALCs for σ- and π-bonding in the case of octahedral ML_6.

Radial		**Tangential**
s orbitals	p_z orbitals	p_x, p_y orbitals
$A_{1g} + E_g + T_{1u}$	$A_{1g} + E_g + T_{1u}$	$T_{1g} + T_{2g} + T_{1u} + T_{2u}$

Radials are the orbitals that point directly in or out of the cluster while tangential are the one's mainly on the surface.

The set of B-H bonds is formed either with set of s orbital, the set of p_z orbital or some mix of two; thus, leaving only one set of radial orbital that is pointing in towards the center of the octahedron.

SALCs formed from these orbitals give following MOs:

A_{1g} Bonding MO
E_g and T_{1u} Distinctly antibonding

MOs formed by the 12 tangential (p_x and p_y) orbitals for T_{2g} type. The shape is already known from the π-bonding in an octahedral AB_6 MO and other orbitals are obtained by projection operation. These overlaps give following MOs.

T_{2g}, T_{1u} → Bonding
T_{1g}, T_{2u} → Antibonding

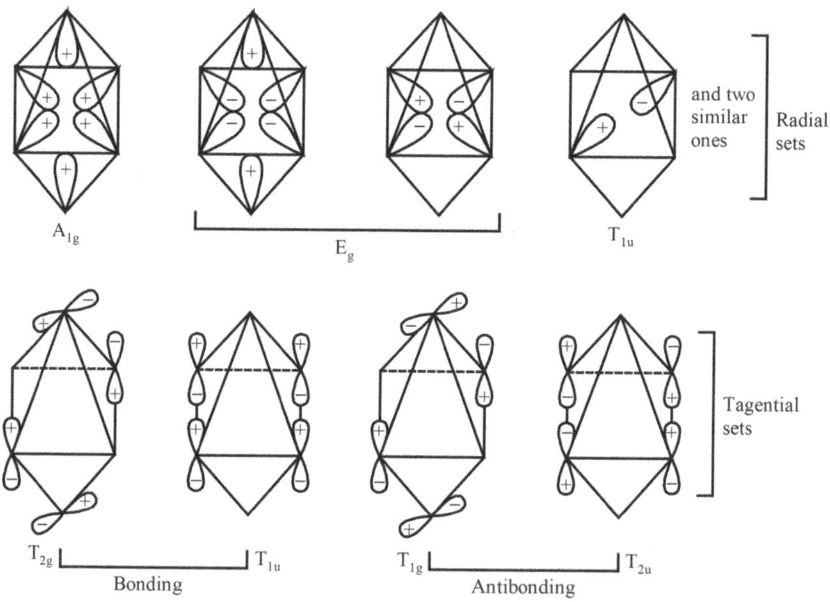

Schematic representation of MOs of $B_6H_6^{2-}$

It may be concluded that the B_6 cluster itself has the following bonding MOs, A_{1g}, T_{2g} and T_{1u}. These can hold 14 electrons (2, 6, 6, respectively) in bonding orbital. As each boron atom has 3 valence shell electrons, one of which is being used in B-H bond formation. Therefore, the 6 boron atoms can provide 12 electrons to form bonding MOs in B_6 cluster. Complete filling of bonding MOs requires 2 more electrons; thus, accounting for the stability of B_6H_6 unit as dianion.

Thus, B_nH_n cluster with n = 5–12, are closed polyhedral, i.e., closo boron clusters.

Closo borons have completely closed polyhedral, where all the vertices are occupied by B atoms. There are n + 1 bonding electron pairs within the cluster. Nido and arachno-boranes can be derived from closo borane by the removal of one and two vertex of polyhedral, respectively. Nido- and arachno- boranes have n + 2 and n + 3 bonding electron pair, respectively.

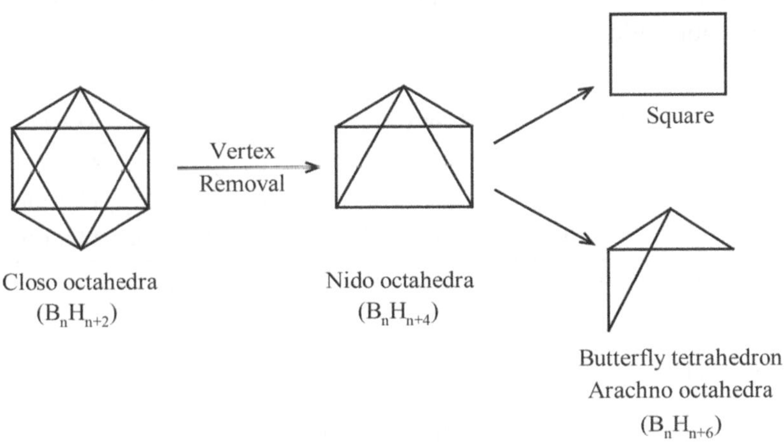

In term of bonding, removal of one vertex results simply in the elimination of one of the AOs, that lead to slight change in bonding and antibonding orbitals, i.e., the bonding orbitals become slightly less bonding and antibonding orbitals become less antibonding and infact, some of them disappear.

9.12.2 TOTAL ELECTRON COUNT

Total electron in cage and cluster is equal to sum of number of valence electrons in the molecule or ions. The main group elements have only four valence orbitals, one is out of them and other three are p's out of these four. Three orbitals from each atom combine to form the n + 1 bonding orbitals, which contain 2 n + 2 electrons, when they are filled. Therefore, one remaining orbital on each atom will have either a bond pair or lone pair (i.e., 2 electrons). Hence, 2 n electrons will be required to fill these n orbitals. Thus, the total electron count for closo cluster containing n vertices is 4 n + 2, for capped cluster 4 n, for nido cluster 4 n + 4, and for arachno cluster 4 n + 6.

A transition element has 5 *d* orbitals and therefore, 10 additional electrons are required per atom to fill the valence shell of each metal atom. A closo cluster contains 14 n + 2 valence electrons, capped cluster have 14 n, a nido cluster 14 n + 4, and an arachno cluster 14 n + 6.

The combination of main group atoms and transition atoms in any polyhedral lead to the combined formula $A_{n-m}\ M_m$ where n is atoms that contain m transition metal atoms and n – m main group atoms. Summarized form of electron count rule is:

	Capped	**Closo**	**Nido**	**Arachno**
A_n	4 n	4 n + 2	4 n + 4	4 n + 6
M_n	14 n	14 n + 2	14 n + 4	14 n + 6
$A_{n-m}\ M_m$	4 n + 10 m	4 n + 2 + 10 m	14 n + 4 + 10 m	14 n + 6 + 10 m

where A = main group element; and M = transition metal.

$[Rh_7\ (CO)_{16}]^{3-}$

$$7\ Rh \times 9\ e^-/Rh = 63\ e^-$$

$$16\ CO \times 2\ e^-/CO = 32\ e^-$$

$$\text{Charge} = 3\ e^-$$

$$\text{Total electron count} = 98\ e^-$$

Here n = 7; and therefore 14 n = 98 e^- is justified and it predicts a capped octahedron.

$Rh_6\ (CO)_{16}$

$$6\ Rh \times 9\ e^-/Rh = 54\ e^-$$

$$16\ CO \times 2\ e^-/CO = 32\ e^-$$

$$\text{Total electron count} = 86\ e^-$$

Here n = 6; and therefore 14 n + 2 = 86 e^- is justified and it predicts a closo octahedron.

$Os_5C(CO)_{15}$

$$5\ Os \times 8\ e^-/Os = 40\ e^-$$
$$15\ CO \times 2\ e^-/CO = 30\ e^-$$
$$1\ C \times 4\ e^-/C = 2\ e^-$$
$$\text{Total electron count} = 74\ e^-$$

Here n = 5; and therefore 14 n + 4 = 74 e^- is justified and it predicts a nido octahedron.

$[Os_4N(CO)_{12}]^-$

$$4\ Os \times 8\ e^-/Os = 32\ e^-$$
$$12\ CO \times 2\ e^-/CO = 24\ e^-$$
$$1\ N \times 5\ e^-/N = 5\ e^-$$
$$\text{Charge} = 1\ e^-$$
$$\text{Total electron count} = 62\ e^-$$

Here n = 4; and therefore 14 n + 6 = 62 e^- is justified and it predicts a arachno octahedron.

$Fe_3(CO)_9(S)_2$

$$3\ Fe \times 8\ e^-/Fe = 24\ e^-$$
$$9\ CO \times 2\ e^-/CO = 18\ e^-$$
$$2\ S \times 6\ e^-/S = 12\ e^-$$
$$\text{Total electron count} = 54\ e^-$$

Here n = 5; and therefore 4 n + 4 + 30 = 54 e^- is justified. It predicts a nido octahedron.

9.13 METAL SANDWICH COMPOUNDS

Metal sandwich (metallocene) compounds are made up of haptic covalent bonding between metal and arene ligands. These compounds are denoted by $(C_nH_n)_2$ M, where C_nH_n is an arene ligand and M is metal.

In this case, all the C-C bonds are of the same length and the rings are parallel. All compounds, in which at least one carbocyclic ring, C_nH_n, such as C_4H_4, C_5H_5, etc., is bound to a metal atom in such a way that the M atom lies along n-fold symmetry axis of the ring and is thus equivalently bonded to all the carbon atoms in the ring. However, there are also compounds, where rings are tilted with respect to one another, although it is believed that the metal-ring bonding is still symmetrical about axis of symmetry. Examples are Cp_2ReH, Cp_2TiCl_2, Cp_2 TaH_2, etc.

Using ferrocene, Cp_2Fe, as an example, one can demonstrate basic idea in the MO treatment of molecule.

9.13.1 FERROCENE

In formation of ferrocene, π MOs has to be considered. The set of ten pπ orbitals over two C_5H_5 ring combine to form ligand group orbitals (LGOs). The number and the symmetries of the ligand group orbitals can be worked out by performing the operation of D_{5d} point group on the 10 pπ orbitals. This gives following reducible representation of D_{5d} as:

D_{5d}	E	$2C_5$	$2C_5^2$	$5C_2$	i	$2S_{10}$	$2S_{10}^3$	$5\sigma_d$
Γ_π	10	0	0	0	0	0	0	2

This reducible representation can be reduced to following irreducible representation.

$$\Gamma_\pi = A_{1g} + A_{2u} + E_{1g} + E_{1u} + E_{2g} + E_{2u}$$

Normalised combination of A, E_1, and E_2 orbitals of individual rings are as follows:

$$\psi_{(A_{1g})} = \frac{1}{\sqrt{2}}\left[\psi_1(A) + \psi_2(A)\right]$$

$$\psi_{(A_{2u})} = \frac{1}{\sqrt{2}}\left[\psi_1(A) - \psi_2(A)\right]$$

$$\begin{cases} \psi_{(E_{1g}\,a)} = \dfrac{1}{\sqrt{2}}\left[\psi_1(E_1\,a) + \psi_2(E_1\,a)\right] \\ \psi_{(E_{1g}\,b)} = \dfrac{1}{\sqrt{2}}\left[\psi_1(E_1\,b) + \psi_2(E_1\,b)\right] \end{cases}$$

$$\begin{cases} \psi_{(E_{1u}\,a)} = \dfrac{1}{\sqrt{2}}\left[\psi_1(E_1\,a) - \psi_2(E_1\,a)\right] \\ \psi_{(E_{1u}\,b)} = \dfrac{1}{\sqrt{2}}\left[\psi_1(E_1\,b) - \psi_2(E_1\,b)\right] \end{cases}$$

$$\begin{cases} \psi_{(E_{2g}\,a)} = \dfrac{1}{\sqrt{2}}\left[\psi_1(E_2\,a) + \psi_2(E_2\,a)\right] \\ \psi_{(E_{2g}\,b)} = \dfrac{1}{\sqrt{2}}\left[\psi_1(E_2\,b) + \psi_2(E_2\,b)\right] \end{cases}$$

$$\underset{\text{I}}{\begin{bmatrix} e^{li\varphi} \\ e^{(l-1)i\varphi} \\ \vdots \\ e^{(1-l)i\varphi} \\ e^{-li\varphi} \end{bmatrix}} \xrightarrow[\text{angle } \alpha]{\text{Rotated by}} \underset{\text{II}}{\begin{bmatrix} e^{li(\varphi+\alpha)} \\ e^{(l-1)i(\varphi+\alpha)} \\ \vdots \\ e^{(1-l)i(\varphi+\alpha)} \\ e^{-li(\varphi+\alpha)} \end{bmatrix}}$$

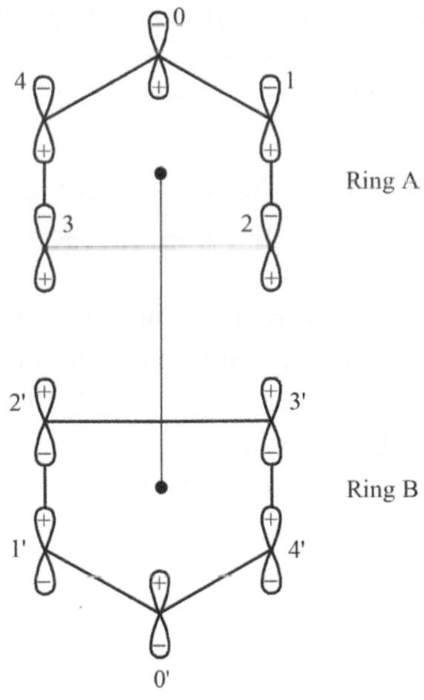

Skeleton showing the pπ orbitals on the two rings used to construct MOs for $(C_5H_5)_2M$ molecule.

The 10 pπ orbitals combine to form two non-degenerate and four pairs of doubly degenerate group orbitals. They combine with the orbital Fe atom of with same symmetry and form bonding and antibonding orbitals.

Irreducible representation	**Orbitals**
A_{1g}	4 s, 3 d_{z^2}
A_{2u}	4 p_z
E_{1g}	3 d_{xz}, 3 d_{yz}
E_{1u}	4 p_x, 4 p_y
E_{2g}	3 d_{xy}, 3 $d_{x^2-y^2}$
E_{2u}	None

In all, it is not necessary to solve a 19 × 19 secular determinant for 19 orbitals because of their symmetric properties and considering the degenerecies. In place of it, some small determinants of lower dimensions are to be solved. These are:

A_{1g} Molecule orbitals = 1 (3 × 3)
A_{2u} Molecule orbitals = 1 (2 × 2)
E_{1g} Molecule orbitals = 2 (2 × 2)
E_{1u} Molecule orbitals = 2 (2 × 2)
E_{2g} Molecule orbitals = 2 (2 × 2)

E_{2u} molecular orbital on ring has no E_{2u} MOs on metal to interact with. Therefore, E_{2u} MOs on the rings are in themselves E_{2u} MOs for the complete molecule.

LGO, A_{1g} match in energy with 4 s and 3 d_{z^2} to form 3 A_{1g} MOs. A_{2u} doesn't match in energy with p_z and hence, bonding MO A_{2u} is same as LGO A_{2u}^* and A_{2u} MOs is same as p_z orbital. LGOs E_{1g} combined with 3 d_{xz}, 3 d_{yz} and E_{1u} combined with 4 p_x, and 4 p_y orbitals form two pairs of bonding and antibonding MOs. LGOs E_{2g} combines with 3 d_{xy}, and 3 $d_{x^2-y^2}$ orbitals forming two bonding and antibonding MOs, respectively. There is no central AO corresponding to E_{2u} and hence, it remained non-bonding.

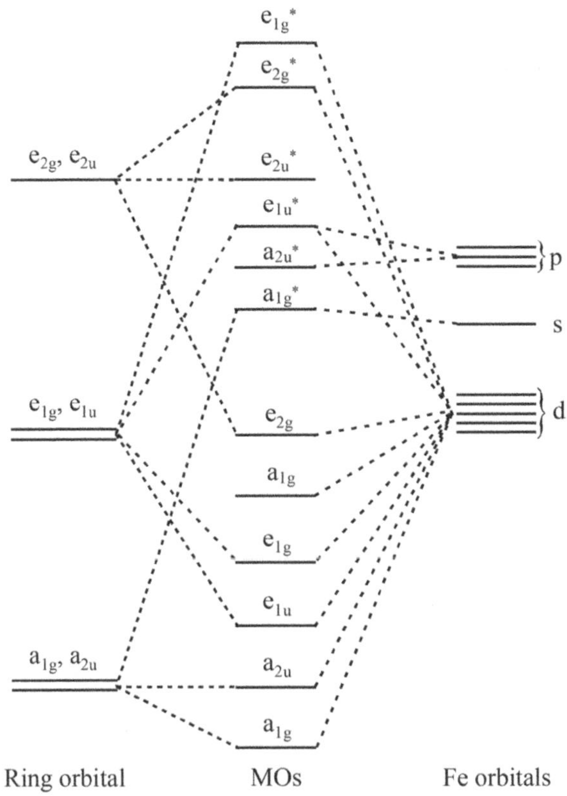

Energy level diagram for ferrocene

Group theory deals with not only symmetry and geometry of molecules, but it also explains many interesting facts like hybridization of molecules, molecular vibrations, spectroscopy, M.O. theory for carbocyclic systems, bonding in complexes and organometallics, etc. It provides a strong mathematical background to deal with chemical problems related to all these aspects.

KEYWORDS

- **Cage**
- **Cluster**
- **Descending symmetry**
- **Laporte rule**

- **Orgel diagram**
- **Sandwich**
- **Splitting**
- **Tanabe–Sugano diagram**

CHARACTER TABLES FOR CHEMICALLY IMPORTANT SYMMETRY GROUPS

1. The Nonaxial Groups

C_1	E
A	1

C_s	E	σ_h		
A'	1	1	x, y, R_z	x^2, y^2, z^2, xy
A"	1	−1	z, R_x, R_y	yz, xz

C_i	E	i		
A_g	1	1	R_x, R_y, R_z	$x^2, y^2, z^2, xy, yz, xz$
A_u	1	−1	x, y, z	

2. The C_n Groups

C_2	E	C_2		
A	1	1	z, R_z	x^2, y^2, z^2, xy
B	1	−1	x, y, R_x, R_y	yz, xz

C_3	E	C_3	C_3^2		$\varepsilon = \exp(2\pi i/3)$
A	1	1	1	z, R_z	$x^2 + y^2, z^2$
E	$\left\{\begin{matrix} 1 \\ 1 \end{matrix}\right.$	$\begin{matrix} \varepsilon \\ \varepsilon^* \end{matrix}$	$\left.\begin{matrix} \varepsilon^* \\ \varepsilon \end{matrix}\right\}$	$(x, y)\ (R_x, R_y)$	$(x^2 - y^2, xy)\ (yz, xz)$

C_4	E	C_4	C_2	C_4^3		
A	1	1	1	1	z, R_z	$x^2 + y^2, z^2$
B	1	−1	1	−1		$x^2 - y^2, xy$
E	$\left\{\begin{matrix} 1 \\ 1 \end{matrix}\right.$	$\begin{matrix} i \\ -i \end{matrix}$	$\begin{matrix} -1 \\ -1 \end{matrix}$	$\left.\begin{matrix} -i \\ i \end{matrix}\right\}$	$(x, y)\ (R_x, R_y)$	(yz, xz)

C_5	E	C_5	C_5^2	C_5^3	C_5^4		$\varepsilon = \exp(2\pi i/5)$
A	1	1	1	1	1	z, R_z	x^2+y^2, z^2
E_1	1	ε	ε^2	$\varepsilon^2{*}$	$\varepsilon{*}$	$(x, y)\ (R_x, R_y)$	(yz, xz)
	1	$\varepsilon{*}$	$\varepsilon^2{*}$	ε^2	ε		
E_2	1	ε^2	$\varepsilon{*}$	ε	$\varepsilon^2{*}$		x^2-y^2, xy
	1	$\varepsilon^2{*}$	ε	$\varepsilon{*}$	ε^2		

C_6	E	C_6	C_3	C_2	C_3^2	C_6^5		$\varepsilon = \exp(2\pi i/6)$
A	1	1	1	1	1	1	z, R_z	x^2+y^2, z^2
B	1	−1	1	−1	1	−1		
E_1	1	ε	$-\varepsilon{*}$	−1	$-\varepsilon$	$\varepsilon{*}$	$(x, y)\ (R_x, R_y)$	(yz, xz)
	1	$\varepsilon{*}$	$-\varepsilon$	−1	$-\varepsilon{*}$	ε		
E_2	1	$-\varepsilon{*}$	$-\varepsilon$	1	$-\varepsilon{*}$	$-\varepsilon$		(x^2-y^2, xy)
	1	$-\varepsilon$	$-\varepsilon{*}$	1	$-\varepsilon$	$-\varepsilon{*}$		

C_7	E	C_7	C_7^2	C_7^2	C_7^4	C_7^5	C_7^6		$\varepsilon = \exp(2\pi i/7)$
A	1	1	1	1	1	1	1	z, R_z	x^2+y^2, z^2
E_1	1	ε	ε^2	ε^3	$\varepsilon^3{*}$	$\varepsilon^2{*}$	$\varepsilon{*}$	(x, y)	(yz, xz)
	1	$\varepsilon{*}$	$\varepsilon^2{*}$	$\varepsilon^3{*}$	ε^3	ε^2	ε	(R_x, R_y)	
E_2	1	ε^2	$\varepsilon^3{*}$	$\varepsilon{*}$	ε	ε^3	$\varepsilon^2{*}$		(x^2-y^2, xy)
	1	$\varepsilon^2{*}$	ε^3	ε	$\varepsilon{*}$	$\varepsilon^3{*}$	ε^2		
E_3	1	ε^3	$\varepsilon{*}$	ε^2	$\varepsilon^2{*}$	ε	$\varepsilon^3{*}$		
	1	$\varepsilon^3{*}$	ε	$\varepsilon^2{*}$	ε^2	$\varepsilon{*}$	ε^3		

C_8	E	C_8	C_4	C_2	C_4^3	C_8^3	C_8^5	C_8^7		$\varepsilon = \exp(2\pi i/8)$
A	1	1	1	1	1	1	1	1	z, R_z	x^2+y^2, z^2
B	1	−1	1	1	1	−1	−1	−1		
E_1	$\begin{cases}1\\1\end{cases}$	ε ε^*	i $-i$	−1 −1	$-i$ i	$-\varepsilon^*$ $-\varepsilon$	$-\varepsilon$ $-\varepsilon^*$	ε^* ε	(x, y) (R_x, R_y)	(xz, yz)
E_2	$\begin{cases}1\\1\end{cases}$	i $-i$	−1 −1	1 1	−1 −1	$-i$ i	i $-i$	$-i$ i		$(x^2-y^2$, $xy)$
E_3	$\begin{cases}1\\1\end{cases}$	$-\varepsilon$ $-\varepsilon^*$	i $-i$	−1 −1	$-i$ i	ε^* ε	ε ε^*	$-\varepsilon^*$ $-\varepsilon$		

3. The D_n Groups

D_2	E	$C_2(z)$	$C_2(y)$	$C_2(x)$		
A	1	1	1	1		x^2, y^2, z^2
B_1	1	1	−1	−1	z, R_z	xy
B_2	1	−1	1	−1	y, R_y	xz
B_3	1	−1	−1	1	x, R_x	yz

D_3	E	$2C_3$	$3C_2$		
A_1	1	1	1		x^2+y^2, z^2
A_2	1	1	−1	z, R_z	
E	2	−1	0	$(x, y)(R_x, R_y)$	$(x^2-y^2, xy)(xz, yz)$

D_4	E	$2C_4$	$C_2(=C_4^2)$	$2C_2'$	$2C_2''$		
A_1	1	1	1	1	1		x^2+y^2, z^2
A_2	1	1	1	−1	−1	z, R_z	
B_1	1	−1	1	1	−1		x^2-y^2
B_2	1	−1	1	−1	1		xy
E	2	0	−2	0	0	$(x, y)(R_x, R_y)$	(xz, yz)

D_5	E	$2C_5$	$2C_5^2$	$5C_2$		
A_1	1	1	1	1		x^2+y^2, z^2
A_2	1	1	1	−1	z, R_z	
E_1	2	2cos72°	2cos144°	0	$(x, y)(R_x, R_y)$	(xz, yz)
E_2	2	2cos144°	2cos72°	0		(x^2-y^2, xy)

D_6	E	$2C_6$	$2C_3$	C_2	$3C_2'$	$3C_2''$		
A_1	1	1	1	1	1	1		x^2+y^2, z^2
A_2	1	1	1	1	−1	−1	z, R_z	
B_1	1	−1	1	−1	1	−1		
B_2	1	−1	1	−1	−1	1		
E_1	2	1	−1	−2	0	0	$(x, y)(R_x, R_y)$	(xz, yz)
E_2	2	−1	−1	2	0	0		(x^2-y^2, xy)

4. The C_{nv} Groups

C_{2V}	E	C_2	$\sigma_V(xz)$	$\sigma_V'(yz)$		
A_1	1	1	1	1	z	x^2, y^2, z^2
A_2	1	1	−1	−1	R_z	xy
B_1	1	−1	1	−1	x, R_y	xz
B_2	1	−1	−1	1	y, R_x	yz

C_{3V}	E	$2C_3$	$3\sigma_V$		
A_1	1	1	1	z	x^2+y^2, z^2
A_2	1	1	−1	R_z	
E	2	−1	0	$(x, y)(R_x, R_y)$	$(x^2-y^2, xy)(xz, yz)$

C_{4V}	E	$2C_4$	C_2	$2\sigma_V$	$2\sigma_d$		
A_1	1	1	1	1	1	z	x^2+y^2, z^2
A_2	1	1	1	−1	−1	R_z	
B_1	1	−1	1	1	−1		x^2-y^2
B_2	1	−1	1	−1	1		xy
E	2	0	−2	0	0	$(x, y)(R_x, R_y)$	(xz, yz)

C_{5V}	E	$2C_5$	$2C_5^2$	$5\sigma_V$		
A_1	1	1	1	1	z	x^2+y^2, z^2
A_2	1	1	1	−1	R_z	
E_1	2	2cos 72°	2cos 144°	0	$(x, y)\ (R_x, R_y)$	(xz, yz)
E_2	2	2cos 144°	2cos 72°	0		(x^2-y^2, xy)

C_{6V}	E	$2C_6$	$2C_3$	C_2	$3\sigma_V$	$3\sigma_d$		
A_1	1	1	1	1	1	1	z	x^2+y^2, z^2
A_2	1	1	1	1	−1	−1	R_z	
B_1	1	−1	1	−1	1	−1		
B_2	1	−1	1	−1	−1	1		
E_1	2	1	−1	−2	0	0	$(x, y)(R_x, R_y)$	(xz, yz)
E_2	2	−1	−1	2	0	0		(x^2-y^2, xy)

5. The C_{nh} Groups

C_{2h}	E	C_2	i	σ_h		
A_g	1	1	1	1	R_z	x^2, y^2, z^2, xy
B_g	1	−1	1	−1	R_x, R_y	xz, yz
A_u	1	1	−1	−1	z	
B_u	1	−1	−1	1	x, y	

C_{3h}	E	C_3	C_3^2	σ_h	S_3	S_3^5		$e=\exp(2\pi i/3)$
A'	1	1	1	1	1	1	R_z	x^2+y^2, z^2
E'	$\begin{cases}1\\1\end{cases}$	ε ε^*	ε^* ε	1 1	ε ε^*	ε^* ε	(x, y)	(x^2-y^2, xy)
A''	1	1	1	−1	−1	−1	z	
E''	$\begin{cases}1\\1\end{cases}$	ε ε^*	ε^* ε	−1 −1	$-\varepsilon$ $-\varepsilon^*$	$-\varepsilon^*$ $-\varepsilon$	(R_x, R_y)	(xz, yz)

C_{4h}	E	C_4	C_2	C_4^3	i	S_4^3	σ_h	S_4		
A_g	1	1	1	1	1	1	1	1	R_z	x^2+y^2, z^2
B_g	1	−1	1	−1	1	−1	1	−1		x^2-y^2, xy
E_g	1	i	−1	$-i$	1	1	−1	$-i$	(R_x, R_y)	(xz, yz)
	1	$-i$	−1	i	1	$-i$	−1	i		
A_u	1	1	1	1	−1	−1	−1	−1	z	
B_u	1	−1	1	−1	−1	1	−1	1		
E_u	1	i	−1	$-i$	−1	$-i$	1	i	(x, y)	
	1	$-i$	−1	i	−1	i	1	$-i$		

C_{5h}	E	C_5	C_5^2	C_5^3	C_5^4	σ_h	S_5	S_5^7	S_5^3	S_5^9		$\mu = \exp(2\pi i/5)$
A'	1	1	1	1	1	1	1	1	1	1	R_z	x^2+y^2, z^2
E_1'	1	ε	ε^2	ε^{2*}	ε^*	1	ε	ε^2	ε^{2*}	ε^*	(x, y)	
	1	ε^*	ε^{2*}	ε^2	ε	1	ε^*	ε^{2*}	ε^2	ε		
E_2'	1	ε^2	ε^*	ε	ε^{2*}	1	ε^2	ε^*	ε	ε^{2*}		$(x^2-y^2,$
	1	ε^{2*}	ε	ε^*	ε^2	1	ε^{2*}	ε	ε^*	ε^2		$xy)$
A''	1	1	1	1	1	−1	−1	−1	−1	−1	z	
E_1''	1	ε	ε^2	ε^{2*}	ε^*	−1	$-\varepsilon$	$-\varepsilon^2$	$-\varepsilon^{2*}$	$-\varepsilon^*$	(R_x, R_y)	(xz, yz)
	1	ε^*	ε^{2*}	ε^2	ε	−1	$-\varepsilon^*$	$-\varepsilon^{2*}$	$-\varepsilon^2$	$-\varepsilon$		
E_2''	1	ε^2	ε^*	ε	ε^{2*}	−1	$-\varepsilon^2$	$-\varepsilon^*$	$-\varepsilon$	$-\varepsilon^{2*}$		
	1	ε^{2*}	ε	ε^*	ε^2	−1	$-\varepsilon^{2*}$	$-\varepsilon$	$-\varepsilon^*$	$-\varepsilon^2$		

C_{6h}	E	C_6	C_3	C_2	C_3^2	C_6^5	i	S_3^5	S_6^5	σ_h	S_6	S_3		$\varepsilon = \exp(2\pi i/6)$
A_g	1	1	1	1	1	1	1	1	1	1	1	1	R_z	x^2+y^2, z^2
B_g	1	−1	1	−1	1	−1	1	−1	1	−1	−1	−1		
E_{1g}	1	ε	$-\varepsilon^*$	−1	$-\varepsilon$	ε^*	1	ε	$-\varepsilon^*$	−1	$-\varepsilon$	ε^*	(R_x, R_y)	(xz, yz)
	1	ε^*	$-\varepsilon$	−1	$-\varepsilon^*$	ε	1	ε^*	$-\varepsilon$	−1	$-\varepsilon^*$	ε		
E_{2g}	1	$-\varepsilon^*$	$-\varepsilon$	1	$-\varepsilon^*$	$-\varepsilon$	1	$-\varepsilon^*$	$-\varepsilon$	1	$-\varepsilon^*$	$-\varepsilon$		(x^2-y^2, xy)
	1	$-\varepsilon$	$-\varepsilon^*$	1	$-\varepsilon$	$-\varepsilon^*$	1	$-\varepsilon$	$-\varepsilon^*$	1	$-\varepsilon$	$-\varepsilon^*$		
A_u	1	1	1	1	1	1	−1	−1	−1	−1	−1	−1	z	
B_u	1	−1	1	−1	1	−1	−1	1	−1	1	−1	1		
E_{1u}	1	ε	$-\varepsilon^*$	−1	$-\varepsilon$	ε^*	−1	$-\varepsilon$	ε^*	1	ε	$-\varepsilon^*$	(x, y)	
	1	ε^*	$-\varepsilon$	−1	$-\varepsilon^*$	ε	−1	$-\varepsilon^*$	ε	1	ε^*	$-\varepsilon$		
E_{2u}	1	$-\varepsilon^*$	$-\varepsilon$	1	$-\varepsilon^*$	$-\varepsilon$	−1	ε^*	ε	−1	ε^*	ε		
	1	$-\varepsilon$	$-\varepsilon^*$	1	$-\varepsilon$	$-\varepsilon^*$	−1	ε	ε^*	−1	ε	ε^*		

6. The D_{nh} Groups

D_{2h}	E	$C_2(z)$	$C_2(y)$	$C_2(x)$	i	$\sigma(xy)$	$\sigma(xz)$	$\sigma(yz)$		
A_g	1	1	1	1	1	1	1	1		x^2, y^2, z^2
B_{1g}	1	1	−1	−1	1	1	−1	−1	R_z	xy
B_{2g}	1	−1	1	−1	1	−1	1	−1	R_y	xz
B_{3g}	1	−1	−1	1	1	−1	−1	1	R_x	yz
A_u	1	1	1	1	−1	−1	−1	−1		
B_{1u}	1	1	−1	−1	−1	−1	1	1	z	
B_{2u}	1	−1	1	−1	−1	1	−1	1	y	
B_{3u}	1	−1	−1	1	−1	1	1	−1	x	

D_{3h}	E	$2C_3$	$2C_2$	σ_h	$2S_3$	$3\sigma_v$		
A_1'	1	1	1	1	1	1		$x^2 + y^2, z^2$
A_2'	1	1	−1	1	1	−1	R_z	
E'	2	−1	0	2	−1	0	(x, y)	$(x^2 - y^2, xy)$
A_1''	1	1	1	−1	−1	−1		
A_2''	1	1	−1	−1	−1	1	z	
E''	2	−1	0	−2	1	0	(R_x, R_y)	(xz, yz)

D_{4h}	E	$2C_4$	C_2	$2C_2'$	$2C_2''$	i	$2S_4$	σ_h	$2\sigma_v$	$2\sigma_d$		
A_{1g}	1	1	1	1	1	1	1	1	1	1		x^2, y^2, z^2
A_{2g}	1	1	1	−1	−1	1	1	1	−1	−1	R_z	
B_{1g}	1	−1	1	1	−1	1	1	1	1	−1		$x^2 - y^2$
B_{2g}	1	−1	1	−1	1	1	−1	1	−1	1		xy
E_g	2	0	−2	0	0	2	0	−2	0	0	(R_x, R_y)	(xz, yz)
A_{1u}	1	1	1	1	1	−1	−1	−1	−1	−1		
A_{2u}	1	1	1	−1	−1	−1	−1	−1	1	1	z	
B_{1u}	1	−1	1	1	−1	−1	1	−1	−1	1		
B_{2u}	1	−1	1	−1	1	−1	1	−1	1	−1		
E_u	2	0	−2	0	0	−2	0	2	0	0	(x, y)	

D_{5h}	E	$2C_5$	$2C_5^2$	$5C_2$	σ_h	$2S_5$	$2S_5^3$	$5\sigma_v$		
A_1'	1	1	1	1	1	1	1	1		x^2+y^2, z^2
A_2'	1	1	1	−1	1	1	1	−1	R_z	
E_1'	2	2 cos 72°	2 cos 144°	0	2	2 cos 72°	2 cos 144°	0	(x, y)	
E_2'	2	2 cos 144°	2 cos 72°	0	2	2 cos 144°	2 cos 72°	0		(x^2-y^2, xy)
A_1''	1	1	1	1	−1	−1	−1	−1		
A_2''	1	1	1	−1	−1	−1	−1	1	z	
E_1''	2	2 cos 72°	2 cos 144°	0	−2	−2 cos 72°	−2 cos 144°	0	(R_x, R_y)	(xz, yz)
E_2''	2	2 cos 144°	2 cos 72°	0	−2	−2 cos 144°	−2 cos 72°	0		

D_{6h}	E	$2C_6$	$2C_3$	C_2	$3C_2'$	$3C_2''$	i	$2S_3$	$2S_6$	σ_h	$3\sigma_d$	$3\sigma_v$		
A_{1g}	1	1	1	1	1	1	1	1	1	1	1	1		x^2+y^2, z^2
A_{2g}	1	1	1	1	−1	−1	1	1	1	1	−1	−1	R_z	
B_{1g}	1	−1	1	−1	1	−1	1	−1	1	−1	1	−1		
B_{2g}	1	−1	1	−1	−1	1	1	−1	1	−1	−1	1		
E_{1g}	2	1	−1	−2	0	0	2	1	−1	−2	0	0	(R_x, R_y)	(xz, yz)
E_{2g}	2	−1	−1	2	0	0	2	−1	−1	2	0	0		(x^2-y^2, xy)
A_{1u}	1	1	1	1	1	1	−1	−1	−1	−1	−1	−1		
A_{2u}	1	1	1	1	−1	−1	−1	−1	−1	−1	1	1	z	
B_{1u}	1	−1	1	−1	1	−1	−1	1	−1	1	−1	1		
B_{2u}	1	−1	1	−1	−1	1	−1	1	−1	1	1	−1		
E_{1u}	2	1	−1	−2	0	0	−2	−1	1	2	0	0	(x, y)	
E_{2u}	2	−1	−1	2	0	0	−2	1	1	−2	0	0		

D_{8h}	E	$2C_8$	$2C_8^3$	$2C_4$	C_2	$4C_2'$	$4C_2''$	i	$2S_8$	$2S_8^3$	$2S_4$	σ_h	$4\sigma_d$	$4\sigma_v$		
A_{1g}	1	1	1	1	1	1	1	1	1	1	1	1	1	1		x^2+y^2, z^2
A_{2g}	1	1	1	1	1	−1	−1	1	1	1	1	1	−1	−1	R_z	
B_{1g}	1	−1	−1	1	1	1	−1	1	−1	−1	1	1	1	−1		
B_{2g}	1	−1	−1	1	1	−1	1	1	−1	−1	1	1	−1	1		
E_{1g}	2	$\sqrt{2}$	$-\sqrt{2}$	0	−2	0	0	2	$\sqrt{2}$	$-\sqrt{2}$	0	−2	0	0	(R_x, R_y)	(xz, yz)
E_{2g}	2	0	0	−2	1	0	0	2	0	0	−2	2	0	0		(x^2-y^2, xy)
E_{3g}	2	$-\sqrt{2}$	$\sqrt{2}$	0	−2	0	0	2	$-\sqrt{2}$	$\sqrt{2}$	0	−2	0	0		
A_{1u}	1	1	1	1	1	1	1	−1	−1	−1	−1	−1	−1	−1		
A_{2u}	1	1	1	1	1	−1	−1	−1	−1	−1	−1	−1	1	1	z	
B_{1u}	1	−1	−1	1	1	1	−1	−1	1	1	−1	−1	−1	1		
B_{2u}	1	−1	−1	1	1	−1	1	−1	1	1	−1	−1	1	−1		
E_{1u}	2	$\sqrt{2}$	$-\sqrt{2}$	0	−2	0	0	−2	$-\sqrt{2}$	$\sqrt{2}$	0	2	0	0	(x, y)	
E_{2u}	2	0	0	−2	2	0	0	−2	0	0	2	−2	0	0		
E_{3u}	2	$-\sqrt{2}$	$\sqrt{2}$	0	−2	0	0	−2	$\sqrt{2}$	$-\sqrt{2}$	0	2	0	0		

7. The D_{nd} Groups

D_{2d}	E	$2S_4$	C_2	$2C_2'$	$2\sigma_d$		
A_1	1	1	1	1	1		x^2+y^2, z^2
A_2	1	1	1	−1	−1	R_z	
B_1	1	−1	1	1	−1		x^2-y^2
B_2	1	−1	1	−1	1	z	xy
E	2	0	−2	0	0	$(x, y)\ (R_x, R_y)$	(xz, yz)

D_{3d}	E	$2C_3$	$3C_2$	i	$2S_6$	$3\sigma_d$		
A_{1g}	1	1	1	1	1	1		x^2+y^2, z^2
A_{2g}	1	1	−1	1	1	−1	R_z	
E_g	2	−1	0	2	−1	0	(R_x, R_y)	$(x^2-y^2, xy); (xz, yz)$
A_{1u}	1	1	1	−1	−1	−1		
A_{2u}	1	1	−1	−1	−1	1	z	
E_u	2	−1	0	−2	1	0	(x, y)	

D_{4d}	E	$2C_8$	$2C_4$	$2S_8^3$	C_2	$4C_2'$	$4\sigma_d$		
A_1	1	1	1	1	1	1	1		x^2+y^2, z^2
A_2	1	1	1	1	1	−1	−1	R_z	
B_1	1	−1	1	−1	1	1	−1		
B_2	1	−1	1	−1	1	−1	1	z	
E_1	2	$\sqrt{2}$	0	$-\sqrt{2}$	−2	0	0	(x, y)	
E_2	2	0	−2	0	2	0	0		(x^2-y^2, xy)
E_3	2	$-\sqrt{2}$	0	$\sqrt{2}$	−2	0	0	(R_x, R_y)	(xz, yz)

D_{5d}	E	$2C_5$	$2C_5^2$	$5C_2$	i	$2S_{10}^3$	$2S_{10}$	$5\sigma_d$		
A_{1g}	1	1	1	1	1	1	1	1		x^2+y^2, z^2
A_{2g}	1	1	1	−1	1	1	1	−1	R_z	
E_{1g}	2	2 cos 72°	2 cos 144°	0	2	2 cos 72°	2 cos 144°	0	(R_x, R_y)	(xz, yz)
E_{2g}	2	2 cos 144°	2 cos 72°	0	2	2 cos 144°	2 cos 72°	0		(x^2-y^2, xy)
A_{1u}	1	1	1	1	−1	−1	−1	−1		
A_{2u}	1	1	1	−1	−1	−1	−1	1	z	
E_{1u}	2	2 cos 72°	2 cos 144°	0	−2	−2 cos 72°	−2 cos 144°	0	(x, y)	
E_{2u}	2	2 cos 144°	2 cos 72°	0	−2	−2 cos 144°	−2 cos 72°	0		

D_{6d}	E	$2S_{12}$	$2C_6$	$2S_4$	$2C_3$	$2S_{12}^5$	C_2	$6C_2'$	$6\sigma_d$		
A_1	1	1	1	1	1	1	1	1	1		x^2+y^2, z^2
A_2	1	1	1	1	1	1	1	−1	−1	R_z	
B_1	1	−1	1	−1	1	−1	1	1	−1		
B_2	1	−1	1	−1	1	−1	1	−1	1	z	
E_1	2	$\sqrt{3}$	1	0	−1	$-\sqrt{3}$	−2	0	0	(x, y)	
E_2	2	1	−1	−2	−1	1	2	0	0		(x^2-y^2, xy)
E_3	2	0	−2	0	2	0	−2	0	0		
E_4	2	−1	−1	2	−1	−1	2	0	0		
E_5	2	$-\sqrt{3}$	1	0	−1	$\sqrt{3}$	−2	0	0	(R_x, R_y)	(xz, yz)

8. The S_n Groups

S_4	E	S_4	C_2	S_4^3		
A	1	1	1	1	R_z	x^2+y^2, z^2
B	1	−1	1	−1	z	x^2-y^2, xy
E	1	i	−1	$-i$	$(x, y); (R_x, R_y)$	(xz, yz)
	1	$-i$	−1	i		

S_6	E	C_3	C_3^2	i	S_6^5	S_6		$\varepsilon = \exp(2\pi i/3)$
A_g	1	1	1	1	1	1	R_z	x^2+y^2, z^2
E_g	1	ε	ε^*	1	ε	ε^*	(R_x, R_y)	$(x^2-y^2, xy); (xz, yz)$
	1	ε^*	ε	1	ε^*	ε		
A_u	1	1	1	−1	−1	−1	z	
E_u	1	ε	ε^*	−1	$-\varepsilon$	$-\varepsilon^*$	(x, y)	
	1	ε^*	ε	−1	$-\varepsilon^*$	$-\varepsilon$		

S_8	E	S_8	C_4	C_8^3	C_2	S_8^5	C_4^3	S_8^7		$\varepsilon = \exp(2\pi i/8)$
A	1	1	1	1	1	1	1	1	R_z	x^2+y^2, z^2
B	1	−1	1	−1	1	−1	1	−1	z	
E_1	1	ε	i	$-\varepsilon^*$	−1	$-\varepsilon$	$-i$	ε^*	$(x, y); (R_x, R_y)$	
	1	ε^*	$-i$	$-\varepsilon$	−1	$-\varepsilon^*$	i	ε		
E_2	1	i	−1	$-i$	1	i	−1	$-i$		(x^2-y^2, xy)
	1	$-i$	−1	i	1	$-i$	−1	i		
E_3	1	$-\varepsilon^*$	$-i$	ε	−1	ε^*	i	$-\varepsilon$		(xz, yz)
	1	$-\varepsilon$	i	ε^*	−1	ε	$-i$	$-\varepsilon^*$		

9. The Cubic Groups

T	E	$4C_3$	$4C_3^2$	$3C_2$		$e = \exp(2\pi i/3)$
A	1	1	1	1		$x^2+y^2+z^2$
E	$\begin{cases}1\\1\end{cases}$	$\begin{matrix}\varepsilon\\\varepsilon^*\end{matrix}$	$\begin{matrix}\varepsilon^*\\\varepsilon\end{matrix}$	$\begin{matrix}1\\1\end{matrix}$		$(2z^2-x^2-y^2, x^2-y^2)$
T	3	0	0	−1	$(R_x, R_y, R_z); (x, y, z)$	(xy, xz, yz)

T_h	E	$4C_3$	$4C_3^2$	$3C_2$	i	$4S_6$	$4S_6^5$	$3\sigma_h$		$\varepsilon = \exp(2\pi i/3)$
A_g	1	1	1	1	1	1	1	1		$x^2+y^2+z^2$
A_u	1	1	1	1	−1	−1	−1	−1		
E_g	$\begin{cases}1\\1\end{cases}$	$\begin{matrix}\varepsilon\\\varepsilon^*\end{matrix}$	$\begin{matrix}\varepsilon^*\\\varepsilon\end{matrix}$	$\begin{matrix}1\\1\end{matrix}$	$\begin{matrix}1\\1\end{matrix}$	$\begin{matrix}\varepsilon\\\varepsilon^*\end{matrix}$	$\begin{matrix}\varepsilon^*\\\varepsilon\end{matrix}$	$\begin{matrix}1\\1\end{matrix}$		$(2z^2-x^2-y^2, x^2-y^2)$
E_u	$\begin{cases}1\\1\end{cases}$	$\begin{matrix}\varepsilon\\\varepsilon^*\end{matrix}$	$\begin{matrix}\varepsilon^*\\\varepsilon\end{matrix}$	$\begin{matrix}1\\1\end{matrix}$	$\begin{matrix}-1\\-1\end{matrix}$	$\begin{matrix}-\varepsilon\\-\varepsilon^*\end{matrix}$	$\begin{matrix}-\varepsilon^*\\-\varepsilon\end{matrix}$	$\begin{matrix}-1\\-1\end{matrix}$		
T_g	3	0	0	−1	1	0	0	−1	(R_x, R_y, R_z)	(xy, xz, yz)
T_u	3	0	0	−1	−1	0	0	1	(x, y, z)	

T_d	E	$8C_3$	$3C_2$	$6S_4$	$6\sigma_d$		
A_1	1	1	1	1	1		$x^2+y^2+z^2$
A_2	1	1	1	−1	−1		
E	2	−1	2	0	0		$(2z^2-x^2-y^2, x^2-y^2)$
T_1	3	0	−1	1	−1	(R_x, R_y, R_z)	
T_2	3	0	−1	−1	1	(x, y, z)	(xy, xz, yz)

O	E	$6C_4$	$3C_2(=C_4^2)$	$8C_3$	$6C_2$		
A_1	1	1	1	1	1		$x^2+y^2+z^2$
A_2	1	−1	1	1	−1		
E	2	0	2	−1	0		$(2z^2-x^2-y^2, x^2-y^2)$
T_1	3	1	−1	0	−1	(R_x, R_y, R_z);	
T_2	3	−1	−1	0	1	(x, y, z)	(xy, xz, yz)

O_h	E	$8C_3$	$6C_2$	$6C_4$	$3C_2(=C_4^2)$	i	$6S_4$	$8S_6$	$3\sigma_h$	$6\sigma_d$		
A_{1g}	1	1	1	1	1	1	1	1	1	1		$x^2+y^2+z^2$
A_{2g}	1	1	−1	−1	1	1	−1	1	1	−1		
E_g	2	−1	0	0	2	2	0	−1	2	0		$(2z^2-x^2-y^2, x^2-y^2)$
T_{1g}	3	0	−1	1	−1	3	1	0	−1	−1	(R_x, R_y, R_z)	
T_{2g}	3	0	1	−1	−1	3	−1	0	−1	1		(xz, yz, xy)
A_{1u}	1	1	1	1	1	−1	−1	−1	−1	−1		
A_{2u}	1	1	−1	−1	1	−1	1	−1	−1	1		
E_u	2	−1	0	0	2	−2	0	1	−2	0		
T_{1u}	3	0	−1	1	−1	−3	−1	0	1	1	(x, y, z)	
T_{2u}	3	0	1	−1	−1	−3	1	0	1	−1		

10. The Groups C_v and D_h for Linear Molecules

$C_{\infty v}$	E	$2C_\infty^\Phi$	…	$\infty\sigma_v$		
$A_1 \equiv \Sigma^+$	1	1	…	1	z	x^2+y^2, z^2
$A_2 \equiv \Sigma^-$	1	1	…	−1	R_z	
$E_1 \equiv \Pi$	2	$2\cos\Phi$	…	0	$(x, y); (R_x, R_y)$	(xz, yz)
$E_2 \equiv \Delta$	2	$2\cos 2\Phi$	…	0		(x^2-y^2, xy)
$E_3 \equiv \Phi$	2	$2\cos 3\Phi$	…	0		
…	…	…	…	…		

$D_{\infty h}$	E	$2C_\infty^\Phi$	…	$\infty\sigma_v$	i	$2S_\infty^\Phi$	…	∞C_2		
$A_{1g} \equiv \Sigma_g^+$	1	1	…	1	1	1	…	1		x^2+y^2, z^2
$A_{2g} \equiv \Sigma_g^-$	1	1	…	−1	1	1	…	−1	R_z	
$E_{1g} \equiv \Pi_g$	2	$2\cos\Phi$	…	0	2	$-2\cos\Phi$	…	0	(R_x, R_y)	(xz, yz)
$E_{2g} \equiv \Delta_g$	2	$2\cos 2\Phi$	…	0	2	$2\cos 2\Phi$	…	0		(x^2-y^2, xy)
…	…	…	…	…	…	…	…	…		
$A_{1u} \equiv \Sigma_u^+$	1	1	…	1	−1	−1	…	−1	z	
$A_{2u} \equiv \Sigma_u^-$	1	1	…	−1	−1	−1	…	1		
$E_{1u} \equiv \Pi_u$	2	$2\cos\Phi$	…	0	−2	$2\cos\Phi$	…	0	(x, y)	
$E_{2u} \equiv \Delta_u$	2	$2\cos 2\Phi$	…	0	−2	$-2\cos 2\Phi$	…	0		
…	…	…	…	…	…	…	…	…		

11. The Icosahedral Group

I	E	$12C_5$	$12C_5^2$	$20C_3$	$15C_2$		
A	1	1	1	1	1		$x^2+y^2+z^2$
T_1	3	$\frac{1}{2}(1+\sqrt{5})$	$\frac{1}{2}(1-\sqrt{5})$	0	−1	(R_x, R_y, R_z); (x, y, z)	
T_2	3	$\frac{1}{2}(1-\sqrt{5})$	$\frac{1}{2}(1+\sqrt{5})$	0	−1		
G	4	−1	−1	1	0		
H	5	0	0	−1	1		$(2z^2-x^2-y^2, x^2-y^2, xy, xz, yz)$

I_h	E	$12C_5$	$12C_5^2$	$20C_3$	$15C_2$	i	$12S_{10}$	$12S_{10}^3$	$20S_6$	15σ		
A_g	1	1	1	1	1	1	1	1	1	1		$x^2+y^2+z^2$
T_{1g}	3	$\frac{1}{2}(1+\sqrt{5})$	$\frac{1}{2}(1-\sqrt{5})$	0	−1	3	$\frac{1}{2}(1-\sqrt{5})$	$\frac{1}{2}(1+\sqrt{5})$	0	−1	(R_x, R_y, R_z)	
T_{2g}	3	$\frac{1}{2}(1-\sqrt{5})$	$\frac{1}{2}(1+\sqrt{5})$	0	−1	3	$\frac{1}{2}(1+\sqrt{5})$	$\frac{1}{2}(1-\sqrt{5})$	0	−1		
G_g	4	−1	−1	1	0	4	−1	−1	1	0		
H_g	5	0	0	−1	1	5	0	0	−1	1		$(2z^2-x^2-y^2,$ $x^2-y^2, xy, xz, yz)$
A_u	1	1	1	1	1	−1	−1	−1	−1	−1		
T_{1u}	3	$\frac{1}{2}(1+\sqrt{5})$	$\frac{1}{2}(1-\sqrt{5})$	0	−1	−3	$\frac{1}{2}(1-\sqrt{5})$	$-\frac{1}{2}(1+\sqrt{5})$	0	1	(x, y, z)	
T_{2u}	3	$\frac{1}{2}(1-\sqrt{5})$	$\frac{1}{2}(1+\sqrt{5})$	0	−1	−3	$\frac{1}{2}(1+\sqrt{5})$	$-\frac{1}{2}(1-\sqrt{5})$	0	1		
G_u	4	−1	−1	1	0	−4	1	1	−1	0		
H_u	5	0	0	−1	1	−5	0	0	1	−1		

SUGGESTED FURTHER READING

Aldersey-Williams, H. (1994). The Beautiful Molecule—An Adventure in Chemistry, London: Aurum Press.

Baggott, J. (1994). Perfect Symmetry, Oxford: Oxford University Press.

Bishop, D. M. (1993). Group Theory and Chemistry, Oxford: Clarendon Press.

Carter, R. L. (1997). Molecular Symmetry and Group Theory, New York: John Wiley.

Cotton, F. A. (2008). Chemical Applications of Group Theory, New York: John Wiley.

Ferraro, J. R. & Ziomek, J. S. (2012). Introductory Group Theory and its Application to Molecular Structure, New York: Plenum Press.

Hargittai, I. & Hrgittai, M. (2010). Symmetry through the Eyes of a Chemist, Berlin: Springer-Verlag.

Lad, M. (1998). Symmetry and Group Theory in Chemistry, Hemstead, UK: Horwood.

Ledermann, W. & Weir, A. J. (1996). Introduction to Group Theory, Harlow: Pearson Education Ltd.

Lesk, A. M. (2004). Introduction to Symmetry and Group Theory for Chemist, New York: Kluwer Academic Publishers.

McWeeny, R. (2002). Symmetry: An Introduction to Group Theory and its Applications, Oxford: Pergamon Press.

Molloy, K. C. (2004). Group Theory for Chemists: Fundamental Theory and Applications, Hemstead, UK: Horwood.

Molloy, K. C. (2010). Group Theory for Chemists: Fundamental Theory and Applications, Cambridge: Woodhead Publisher.

Roman, S. (2011). Fundamentals of Group Theory: An Advanced Approach, Switzerland: Birkhäuser.

Schwerdtfeger, H. (1976). Introduction to Group Theory, Leyden: Noordhoff International.

Tsukerblat, B. S. (2006). Group Theory in Chemistry and Spectroscopy: A Simple Guide to Advanced Usage, New York: Dover Publication.

Vincent, A. (2000). Molecular Symmetry and Group Theory: A Programmed Introduction to Chemical Applications, Chichester: Wiley-Blackwell.

INDEX